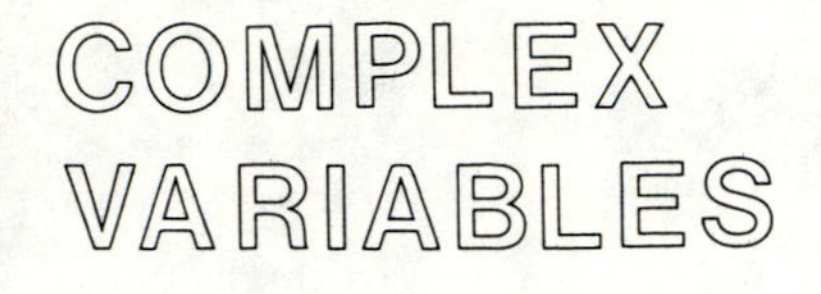
COMPLEX
VARIABLES

COMPLEX VARIABLES for Mathematics and Engineering

Second Edition

JOHN H. MATHEWS
California State University
Fullerton

wcb
WM. C. BROWN PUBLISHERS
Dubuque, Iowa

To my wife Frances and my
three sons Robert, Daniel, and James

A previous version of *Complex Variables for Mathematics and Engineering* was class-tested by the author, John H. Mathews, © 1977.

Library of Congress Catalog Card Number 87-26997
ISBN 0-697-06764-5

Printed in the United States of America.

10 9 8 7 6 5 4 3 2

Contents

Preface ix

Chapter 1 Complex Numbers 1

1.1 Preliminaries 1
1.2 The Algebraic Properties of Complex Numbers 2
1.3 Cartesian Plane and Vector Representation 5
1.4 Polar Representation of Complex Numbers 10
1.5 Powers, Roots, and the Quadratic Formula 15
1.6 Curves, Regions, and Domains in the Plane 21

Chapter 2 Complex Functions 27

2.1 Functions of a Complex Variable 27
2.2 Transformations and Linear Mappings 30
2.3 The Mappings $w = z^n$ and $w = z^{1/n}$ 36
2.4 Limits and Continuity 42
2.5 The Reciprocal Transformation $w = 1/z$ 49

Chapter 3 Analytic and Harmonic Functions 56

3.1 Differentiable Functions 56
3.2 The Cauchy-Riemann Equations 61
3.3 Analytic Functions and Harmonic Functions 67

Chapter 4 Elementary Functions 77

4.1 Branches of Functions 77
4.2 The Exponential Function 83
4.3 Trigonometric and Hyperbolic Functions 88
4.4 Branches of the Logarithm Function 96

4.5 Complex Exponents 101
4.6 Inverse Trigonometric and Hyperbolic Functions 104

Chapter 5 Complex Integration 108

5.1 Complex Integrals 108
5.2 Contours and Contour Integrals 111
5.3 The Cauchy-Goursat Theorem 123
5.4 The Fundamental Theorems of Integration 135
5.5 Integral Representations for Analytic Functions 140
5.6 The Theorems of Morera and Liouville and Some Applications 146
5.7 Harmonic Functions and the Dirichlet Problem 153
Appendix A Goursat's Proof of the Cauchy-Goursat Theorem 158

Chapter 6 Series Representations 162

6.1 Convergence of Sequences and Series 162
6.2 Taylor Series Representations 173
6.3 Laurent Series Representations 180
6.4 Power Series 186
6.5 Singularities, Zeros, and Poles 199
6.6 Some Theoretical Results 205

Chapter 7 Residue Theory 208

7.1 The Residue Theorem 208
7.2 Calculation of Residues 210
7.3 Trigonometric Integrals 216
7.4 Improper Integrals of Rational Functions 219
7.5 Improper Integrals Involving Trigonometric Functions 224
7.6 Indented Contour Integrals 227
7.7 Integrands with Branch Points 232
7.8 The Argument Principle and Rouché's Theorem 236

Chapter 8 Conformal Mapping 243

8.1 Basic Properties of Conformal Mappings 243
8.2 Bilinear Transformations 249
8.3 Mappings Involving Elementary Functions 255
8.4 Mapping by Trigonometric Functions 260

Chapter 9 Applications of Harmonic Functions 266

9.1 Preliminaries 266
9.2 Invariance of Laplace's Equation and the Dirichlet Problem 268
9.3 Poisson's Integral Formula for the Upper Half-Plane 277
9.4 Two-Dimensional Mathematical Models 280
9.5 Steady State Temperatures 283
9.6 Two-Dimensional Electrostatics 295
9.7 Two-Dimensional Fluid Flow 302
9.8 The Schwarz-Christoffel Transformation 314
9.9 Image of a Fluid Flow 324
9.10 Sources and Sinks 328

Answers to Selected Problems 340

Bibliography of Articles 350

Bibliography of Books 351

Index 353

Preface

This text is intended for undergraduate students in mathematics, physics, and engineering. It strikes a balance between the pure and applied aspects of mathematics. Enough mathematical structure is developed so that theorems can be honestly presented, yet the proofs are kept elementary and understandable for students. Sufficient applications are included to show how complex variables are used in science and engineering. A wealth of exercises serve to develop the student's computational skills, theoretical understanding, and practical knowledge of complex variables.

The purpose of the first five chapters is to lay the foundations for the study of complex variables and develop the topics of analytic and harmonic functions, the elementary functions, and contour integration. Proofs are presented in a manner that is easy for students to read. In Section 3.2 the Cauchy-Riemann equations are carefully developed. In Section 5.3 an intuitive proof of the Cauchy-Goursat Theorem is given using Green's Theorem. For advanced readers the proof by Goursat is included in the appendix at the end of Chapter 5. In Section 5.7 the solution of the Dirichlet problem illustrates how complex variables can be applied in a practical situation.

Chapters 6 and 7 can then be used to cover the topics of series and the residue calculus. A brief introduction to uniform convergence supplies the prerequisites for understanding the proofs of the theorems of Taylor and Laurent. Rouché's Theorem and an application concerning the Nyquist criteria are included.

If conformal mappings and applications to harmonic functions are to be studied, then Chapters 8 and 9 can be used after Chapter 5. Boundary value problems for harmonic functions are solved in the upper half-plane, and conformal mappings are used to find solutions in other domains. The Schwarz-Christoffel Formula is outlined, and the two-dimensional models of ideal fluid flow, steady state temperatures, and electrostatics are given.

I would like to express my gratitude to all the people whose efforts contributed to the first and second editions of this book. I want to thank my students, who have made numerous suggestions for improvements, and my colleagues Vuryl Klassen, Gerald Marley, and Harris Shultz here at California

State University, Fullerton. I wish to thank the following individuals for their help and suggestions for the second edition: Arlo Davis, Indiana University of Pennsylvania; R. E. Williamson, Dartmouth College; Calvin Wilcox, University of Utah; Robert D. Brown, University of Kansas; Geoffrey Price, U.S. Naval Academy; Elgin H. Johnston, Iowa State University.

I would appreciate receiving correspondence regarding the book. Suggestions for improvement are always welcome.

John H. Mathews

1

Complex Numbers

1.1 Preliminaries

The real number system was developed as an extension of the set of rational numbers so that polynomial equations such as $x^2 - 2 = 0$ would have solutions. Historically, in the sixteenth century the search for solutions of quadratic and cubic equations led Girolamo Cardano to invent a symbol for the square roots of negative numbers. He noticed that if $\sqrt{-1}$ was treated as an ordinary number with the added rule that $\sqrt{-1}\sqrt{-1} = -1$, then the quadratic equation $x^2 + 1 = 0$ could be solved. Cardano's "imaginary" quantities were ignored by mathematicians for almost 300 years. The notation i for $\sqrt{-1}$ was introduced in the eighteenth century by Leonhard Euler and is widely used in mathematics. (Sometimes electrical engineers prefer to use $j = \sqrt{-1}$ so that the symbol i can be used for current.) In the nineteenth century, Carl Friedrich Gauss showed that a system of complex numbers could be developed that is an extension of the real numbers, and square roots of negative numbers lost the air of mystery as being "imaginary." Therefore it is justified to write

(1) $\quad i^2 = -1.$

We will use linear combinations of real numbers and the symbol i to construct a *complex number* z, and we shall write

(2) $\quad z = x + iy,$

where x and y are real numbers. The number x is called the *real part* of the complex number z, and the number y is called the *imaginary part* of z. We use the notational convention

(3) $\quad \operatorname{Re} z = x \quad \text{and} \quad \operatorname{Im} z = y$

to express the real and imaginary parts of z.

Gauss sought solutions of the polynomial equations

$$a_0 + a_1x + a_2x^2 + \cdots + a_nx^n = 0$$

and was surprised to find that all solutions are complex numbers. This result, known as the Fundamental Theorem of Algebra, is discussed in Chapter 5.

EXAMPLE 1.1 If $z = 4 - 6i$, then

$$\text{Re}(4 - 6i) = 4 \qquad \text{and} \qquad \text{Im}(4 - 6i) = -6.$$

If we treat i as an algebraic symbol and group similar terms, then we feel comfortable with the calculation

$$(2 - 7i) + (4 + 3i) = 2 + 4 + (-7 + 3)i = 6 - 4i.$$

EXAMPLE 1.2 If we use the distributive rules for multiplication and $i^2 = -1$, then the following computation seems correct:

$$(2 - 3i)(5 + 2i) = (2)(5) + (-3i)(5) + (2)(2i) + (-3i)(2i)$$
$$= 10 - 15i + 4i - 6i^2 = 16 - 11i.$$

EXAMPLE 1.3 A process similar to rationalizing the denominator will allow us to perform division. This is illustrated by the calculation

$$\frac{7 - 4i}{3 + 2i} = \frac{(7 - 4i)(3 - 2i)}{(3 + 2i)(3 - 2i)} = \frac{21 - 12i - 14i + 8i^2}{9 - 4i^2}$$
$$= \frac{13 - 26i}{13} = 1 - 2i.$$

The above calculations are accurate for practical purposes. We will now make these concepts more rigorous.

1.2 The Algebraic Properties of Complex Numbers

The complex number z can be considered as an ordered pair

(1) $\qquad z = (x, y)$

where x and y are real numbers. Complex numbers of the form $(x, 0)$ will behave like *real numbers*, and complex numbers of the form $(0, y)$ are called *pure imaginary numbers*. The real and imaginary parts of z in equation (1) are denoted by

(2) $\qquad \text{Re}\, z = x \qquad \text{and} \qquad \text{Im}\, z = y.$

We define two complex numbers $z_1 = (x_1, y_1)$ and $z_2 = (x_2, y_2)$ to be *equal* if and only if they have the same real part and imaginary part. That is,

(3) $\qquad (x_1, y_1) = (x_2, y_2) \qquad$ if and only if $x_1 = x_2$ and $y_1 = y_2$.

The operation of *addition of complex numbers* is performed on each coordinate of the ordered pair and is defined by the rule

(4) $\qquad z_1 + z_2 = (x_1, y_1) + (x_2, y_2) = (x_1 + x_2, y_1 + y_2).$

Since $(x_1, y_1) + (0, 0) = (x_1 + 0, y_1 + 0) = (x_1, y_1)$ and $(x_1, y_1) + (-x_1, -y_1) = (x_1 - x_1, y_1 - y_1) = (0, 0)$, we say that the *additive identity* is

$$0 = (0, 0) \tag{5}$$

and that the *additive inverse* of z_1 is

$$-z_1 = (-x_1, -y_1). \tag{6}$$

The operation of *subtraction* of complex numbers is defined to be

$$z_1 - z_2 = (x_1, y_1) - (x_2, y_2) = (x_1 - x_2, y_1 - y_2). \tag{7}$$

Addition of complex numbers obeys the commutative and associative laws given by the equations

$$z_1 + z_2 = z_2 + z_1 \qquad \text{(commutative law)}, \tag{8}$$

$$z_1 + (z_2 + z_3) = (z_1 + z_2) + z_3 \qquad \text{(associative law)}. \tag{9}$$

The operation of *multiplication* of complex numbers is defined by the rule

$$z_1 z_2 = (x_1, y_1)(x_2, y_2) = (x_1 x_2 - y_1 y_2, x_1 y_2 + x_2 y_1). \tag{10}$$

Since $(x, y)(1, 0) = (x - 0, y + 0) = (x, y)$, we say that the *multiplicative identity* is

$$1 = (1, 0). \tag{11}$$

Let $z = (x, y)$ be a nonzero complex number. Then there exists a complex number z^{-1} such that $zz^{-1} = 1$. The existence of z^{-1} is established by the following argument. Let $z^{-1} = (x_1, y_1)$. Then

$$(x, y)(x_1, y_1) = (xx_1 - yy_1, xy_1 + yx_1) = (1, 0).$$

Equating the real and imaginary parts results in the following system of linear equations:

$$xx_1 - yy_1 = 1, \qquad yx_1 + xy_1 = 0.$$

Cramer's rule can be used to solve for x_1 and y_1, and we obtain

$$x_1 = \frac{\begin{vmatrix} 1 & -y \\ 0 & x \end{vmatrix}}{\begin{vmatrix} x & -y \\ y & x \end{vmatrix}} = \frac{x}{x^2 + y^2} \qquad \text{and} \qquad y_1 = \frac{\begin{vmatrix} x & 1 \\ y & 0 \end{vmatrix}}{\begin{vmatrix} x & -y \\ y & x \end{vmatrix}} = \frac{-y}{x^2 + y^2}.$$

Therefore the *multiplicative inverse* is given by

$$z^{-1} = \left(\frac{x}{x^2 + y^2}, \frac{-y}{x^2 + y^2} \right) \qquad \text{where } z \neq 0. \tag{12}$$

Division by a nonzero complex number is now defined by the rule

$$\frac{z_1}{z_2} = z_1 (z_2)^{-1} \qquad \text{where } z_2 \neq 0. \tag{13}$$

Other familiar properties of multiplication and addition also hold for complex numbers. They include

(14) $z_1 z_2 = z_2 z_1$ (commutative law),

(15) $z_1(z_2 z_3) = (z_1 z_2) z_3$ (associative law),

(16) $z_1(z_2 + z_3) = z_1 z_2 + z_1 z_3$ (distributive law).

Now let us find out exactly how the real number system is embedded in the complex number system. For any real number x we make the correspondence

(17) $x = (x, 0)$.

Then the correspondences $x_1 + x_2 = (x_1 + x_2, 0)$ and $x_1 x_2 = (x_1 x_2, 0)$ are seen to be compatible with the addition and multiplication formulas (4) and (10); that is,

$$(x_1, 0) + (x_2, 0) = (x_1 + x_2, 0) \quad \text{and} \quad (x_1, 0)(x_2, 0) = (x_1 x_2, 0).$$

Consequently, the complex number system is an extension of the real number system. If we make the correspondence

(18) $i = (0, 1)$,

and use the correspondence $-1 = (-1, 0)$, then $i^2 = -1$ is verified by the calculation $(0, 1)(0, 1) = (0 - 1, 0) = (-1, 0)$. Hence the "pure imaginary" numbers are embedded in the complex number system. For completeness, observe that if $z = x + iy$, then we use the correspondence

(19) $x + iy = (x, y)$.

This development of the complex number system by means of ordered pairs is rigorous enough for our purposes. Hence we have justified the notation that was introduced in equation (2) of Section 1.1. As an application, we mention that computers store complex numbers as ordered pairs and are programmed to use formulas (4) and (10) to perform complex addition and multiplication, respectively.

EXERCISES FOR SECTION 1.2

1. Perform the required calculation and express the answer in the form $a + ib$.

   (a) $(3 - 2i) - i(4 + 5i)$ **(b)** $(7 - 2i)(3i + 5)$

   (c) $(1 + i)(2 + i)(3 + i)$ **(d)** $(3 + i)/(2 + i)$

   (e) $(i - 1)^3$ **(f)** i^5

   (g) $\dfrac{1 + 2i}{3 - 4i} - \dfrac{4 - 3i}{2 - i}$ **(h)** $(1 + i)^{-2}$

   (i) $\dfrac{(4 - i)(1 - 3i)}{-1 + 2i}$ **(j)** $(1 + i\sqrt{3})(i + \sqrt{3})$

2. Find the following quantities.

(a) $\text{Re}(1+i)(2+i)$ (b) $\text{Im}(2+i)(3+i)$

(c) $\text{Re}\,\dfrac{4-3i}{2-i}$ (d) $\text{Im}\,\dfrac{1+2i}{3-4i}$

(e) $\text{Re}(i-1)^3$ (f) $\text{Im}(1+i)^{-2}$

(g) $\text{Re}(x_1-iy_1)^2$ (h) $\text{Im}\,\dfrac{1}{x_1-iy_1}$

(i) $\text{Re}(x_1+iy_1)(x_1-iy_1)$ (j) $\text{Im}(x_1+iy_1)^3$

3. Prove that addition of complex numbers is associative.

4. Show that $(z-i)^3 = z^3 - 3iz^2 - 3z + i$.

5. Prove that

(a) $\text{Re}(iz) = -\text{Im}\,z$, and

(b) $1/z^{-1} = z$ provided that $z \neq 0$.

6. Find values of z that satisfy the equations

(a) $z^2 = -9$ (b) $z^4 = 16$

7. Show that $1+2i$ and $1-2i$ are solutions of the equation $z^2 - 2z + 5 = 0$.

8. If $i^n = i$ where n is an integer, show that $n = 4k+1$ where k is an integer.

9. Prove that multiplication of complex numbers is commutative and associative.

10. Let $z_1 z_2 = z_1 z_3$. If $z_1 \neq 0$, show that $z_2 = z_3$.

11. Prove that multiplication is distributive over addition.

1.3 Cartesian Plane and Vector Representation

It is natural to associate the complex number $z = x + iy$ with the point (x, y) in the xy plane as suggested by identity (18) of Section 1.2. Each point in the plane corresponds to a unique complex number, and conversely. For example, the point $(3, 1)$ corresponds to the complex number $3 + i$, as is indicated in Figure 1.1. The number $z = x + iy$ can also be thought of as a position vector in the xy plane with tail at the origin and head at the point (x, y). When the xy plane is used for displaying complex numbers, it is called the *complex plane* or, more simply, *the z plane*. The x axis is called the *real axis*, and the y axis is called the *imaginary axis*.

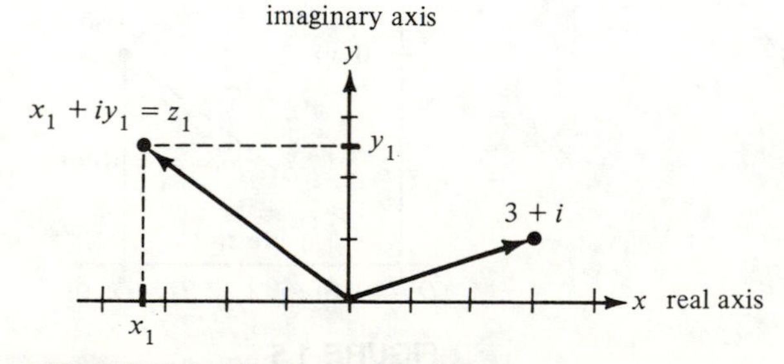

FIGURE 1.1 The complex plane.

Addition of complex numbers is analogous to addition of vectors in the plane. The sum of $z_1 = x_1 + iy_1$ and $z_2 = x_2 + iy_2$ corresponds to the point $(x_1 + x_2,\ y_1 + y_2)$. Hence $z_1 + z_2$ can be obtained vectorially by using the parallelogram law, which is illustrated in Figure 1.2. The difference $z_1 - z_2$ can be represented by the displacement vector from the point (x_2, y_2) to the point (x_1, y_1) and is shown in Figure 1.3.

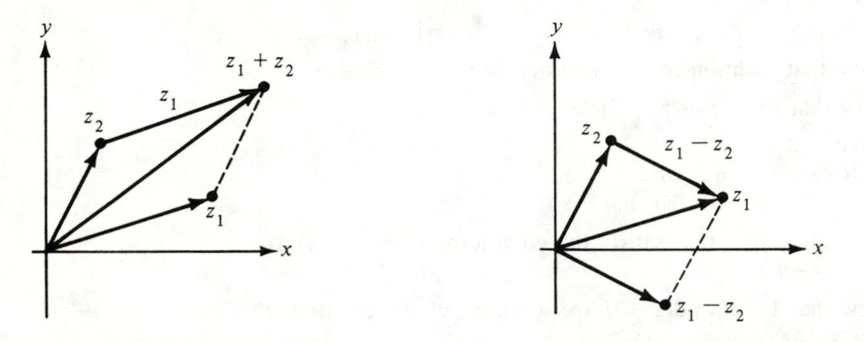

FIGURE 1.2 The sum $z_1 + z_2$. **FIGURE 1.3** The difference $z_1 - z_2$.

The *modulus*, or *absolute value*, of the complex number $z = x + iy$ is a nonnegative real number denoted by $|z|$ and is given by the equation

$$(1) \qquad |z| = \sqrt{x^2 + y^2}.$$

The number $|z|$ is the distance between the origin and the point (x, y). The only complex number with modulus zero is the number 0. The number $z = 4 + 3i$ has modulus 5 and is pictured in Figure 1.4. The numbers $|\text{Re } z|$, $|\text{Im } z|$, and $|z|$ are the lengths of the sides of the right triangle OPQ, which is shown in Figure 1.5. The inequality $|z_1| < |z_2|$ means that the point z_1 is closer to the origin than the point z_2, and it follows that

$$(2) \qquad |x| = |\text{Re } z| \leqq |z| \quad \text{and} \quad |y| = |\text{Im } z| \leqq |z|.$$

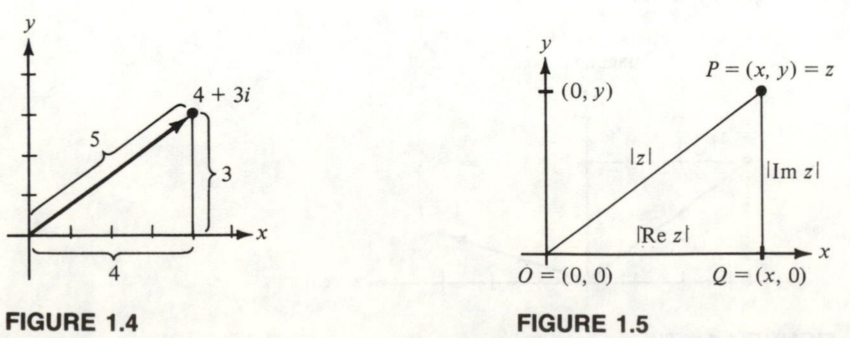

FIGURE 1.4 **FIGURE 1.5**

Since the difference $z_1 - z_2$ can represent the displacement vector from z_2 to z_1, it is evident that the *distance between* z_1 *and* z_2 is given by $|z_1 - z_2|$. This can be obtained by using equation (7) of Section 1.2 and definition (1) to obtain the familiar formula

(3) $$\text{dist}(z_1, z_2) = |z_1 - z_2| = \sqrt{(x_1 - x_2)^2 + (y_1 - y_2)^2}.$$

The *complex conjugate* of the complex number $x + iy$ is defined to be the complex number $x - iy$ and is denoted by $\bar{z}$; and we write

(4) $$\bar{z} = x - iy.$$

The number $\bar{z}$ corresponds to the point $(x, -y)$ and is the reflection of $z = (x, y)$ through the x axis. It is useful to observe that $z\bar{z} = (x + iy)(x - iy) = x^2 + y^2$, which leads to the important identity

(5) $$|z|^2 = z\bar{z}.$$

Other relationships involving the complex conjugate are the identities

(6) $$\text{Re}\, z = \frac{z + \bar{z}}{2} \quad \text{and} \quad \text{Im}\, z = \frac{z - \bar{z}}{2i}.$$

The following properties of the conjugate are easy to verify and are left as exercises:

(7) $$\overline{z_1 + z_2} = \overline{z_1} + \overline{z_2},$$

(8) $$\overline{z_1 z_2} = \overline{z_1}\,\overline{z_2},$$

(9) $$\overline{\left(\frac{z_1}{z_2}\right)} = \frac{\overline{z_1}}{\overline{z_2}} \quad \text{where } z_2 \neq 0.$$

From identities (5) and (8) it is easy to see that

$$|z_1 z_2|^2 = (z_1 z_2)(\overline{z_1 z_2}) = (z_1 \overline{z_1})(z_2 \overline{z_2}) = |z_1|^2 |z_2|^2.$$

Taking the square root of the terms on the left and right establishes another important identity:

(10) $$|z_1 z_2| = |z_1||z_2|.$$

In a similar fashion we can use (5) and (9) to show that

(11) $$\left|\frac{z_1}{z_2}\right| = \frac{|z_1|}{|z_2|} \quad \text{where } z_2 \neq 0.$$

If z_1, z_2, and $z_1 + z_2$ are the three sides of a triangle, then a well-known result from geometry states that the sum of the lengths of the two sides z_1 and z_2 is greater than or equal to the length of the third side $z_1 + z_2$, as is shown in Figure 1.6. This gives motivation for the following inequality, known as the *triangle inequality*:

(12) $$|z_1 + z_2| \leq |z_1| + |z_2|.$$

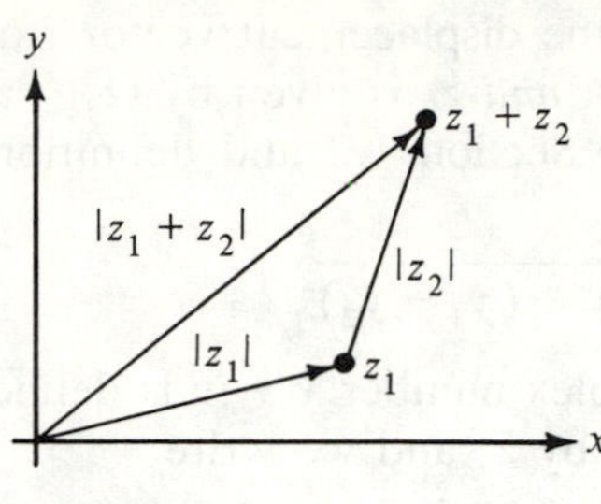

FIGURE 1.6 The triangle inequality.

The algebraic proof of inequality (12) is a beautiful application of elementary techniques and is given by the following:

$$\begin{aligned}|z_1 + z_2|^2 &= (z_1 + z_2)(\overline{z_1} + \overline{z_2}) \\ &= |z_1|^2 + (z_1\overline{z_2} + \overline{z_1\overline{z_2}}) + |z_2|^2 \\ &= |z_1|^2 + 2\,\mathrm{Re}(z_1\overline{z_2}) + |z_2|^2.\end{aligned}$$

Using (2) and (10), we obtain $2\,\mathrm{Re}(z_1\overline{z_2}) \leqq 2|z_1\overline{z_2}| = 2|z_1||z_2|$, which can be used with the above result to obtain $|z_1 + z_2|^2 \leqq |z_1|^2 + 2|z_1||z_2| + |z_2|^2 = (|z_1| + |z_2|)^2$. Taking square roots yields the desired inequality.

EXERCISES FOR SECTION 1.3

1. Locate the numbers z_1 and z_2 vectorially, and use vectors to find $z_1 + z_2$ and $z_1 - z_2$ when
   (a) $z_1 = 2 + 3i$ and $z_2 = 4 + i$
   (b) $z_1 = -1 + 2i$ and $z_2 = -2 + 3i$
   (c) $z_1 = 1 + i\sqrt{3}$ and $z_2 = -1 + i\sqrt{3}$
2. Find the following quantities.
   (a) $|(1 + i)(2 + i)|$ **(b)** $\overline{(1 + i)(2 + i)}$
   (c) $\left|\dfrac{4 - 3i}{2 - i}\right|$ **(d)** $\overline{\left(\dfrac{1 + 2i}{3 - 4i}\right)}$
   (e) $|(1 + i)^{50}|$ **(f)** $\overline{(1 + i)^3}$
   (g) $|z\bar{z}|$ **(h)** $|z - 1|^2$
3. Determine which of the following points lie inside the circle $|z - i| = 1$.
   (a) $\dfrac{1}{2} + i$ **(b)** $1 + \dfrac{i}{2}$
   (c) $\dfrac{1}{2} + \dfrac{i\sqrt{2}}{2}$ **(d)** $\dfrac{-1}{2} + i\sqrt{3}$
4. Show that the point $(z_1 + z_2)/2$ is the midpoint of the line segment joining z_1 to z_2.

5. Sketch the set of points determined by the following relations.
(a) $|z + 1 - 2i| = 2$ **(b)** $\text{Re}(z + 1) = 0$
(c) $|z + 2i| \leqq 1$ **(d)** $\text{Im}(z - 2i) > 6$

6. Show that the equation of the line through the points z_1 and z_2 can be expressed in the form $z = z_1 + t(z_2 - z_1)$ where t is a real number.

7. Show that the vector z_1 is perpendicular to the vector z_2 if and only if $\text{Re}(z_1\overline{z_2}) = 0$.

8. Show that the vector z_1 is parallel to the vector z_2 if and only if $\text{Im}(z_1\overline{z_2}) = 0$.

9. Show that the four points z, $\overline{z}$, $-z$, and $-\overline{z}$ are the vertices of a rectangle with its center at the origin.

10. Show that the four points z, iz, $-z$, and $-iz$ are the vertices of a square with its center at the origin.

11. Prove that
(a) $\overline{z_1 + z_2} = \overline{z_1} + \overline{z_2}$ **(b)** $\overline{z_1 z_2} = \overline{z_1}\,\overline{z_2}$ **(c)** $\overline{(z_1/z_2)} = \overline{z_1}/\overline{z_2}$
where $z_2 \neq 0$.

12. Prove that $\sqrt{2}|z| \geqq |\text{Re}\, z| + |\text{Im}\, z|$.

13. Show that $|z_1 - z_2| \leqq |z_1| + |z_2|$.

14. Show that $|z_1 z_2 z_3| = |z_1||z_2||z_3|$.

15. Show that $|z^n| = |z|^n$ where n is an integer.

16. Show that $||z_1| - |z_2|| \leqq |z_1 - z_2|$.

17. Prove that $|z| = 0$ if and only if $z = 0$.

18. Show that $z_1\overline{z_2} + \overline{z_1}z_2$ is a real number.

19. Prove that $|z_1 + z_2| = |z_1| + |z_2|$ if and only if z_1 and z_2 lie on the same ray through the origin; that is, $\arg z_1 = \arg z_2$.

20. Prove that $\overline{\overline{z}} = z$.

21. Prove that $|z_1 - z_2|^2 = |z_1|^2 - 2\,\text{Re}(z_1\overline{z_2}) + |z_2|^2$.

22. Use mathematical induction to prove that

$$\left|\sum_{k=1}^{n} z_k\right| \leqq \sum_{k=1}^{n} |z_k|.$$

23. Let z_1 and z_2 be two distinct points in the complex plane. Let K be a positive real constant that is greater than the distance between z_1 and z_2. Show that the set of points $\{z\colon |z - z_1| + |z - z_2| = K\}$ is an ellipse with foci z_1 and z_2.

24. Use Exercise 23 to find the equation of the ellipse with foci $\pm 2i$ that goes through the point $3 + 2i$.

25. Use Exercise 23 to find the equation of the ellipse with foci $\pm 3i$ that goes through the point $8 - 3i$.

26. Let z_1 and z_2 be two distinct points in the complex plane. Let K be a positive real constant that is less than the distance between z_1 and z_2. Show that the set of points $\{z\colon ||z - z_1| - |z - z_2|| = K\}$ is a hyperbola with foci z_1 and z_2.

27. Use Exercise 26 to find the equation of the hyperbola with foci ± 2 that goes through the point $2 + 3i$.

28. Use Exercise 26 to find the equation of the hyperbola with foci ± 25 that goes through the point $7 + 24i$.

1.4 Polar Representation of Complex Numbers

Let r and θ denote the polar coordinates of the point (x, y) in the xy plane. The identities relating r and θ to x and y are

(1) $\quad x = r\cos\theta \quad \text{and} \quad y = r\sin\theta.$

If $z = x + iy$ denotes a complex number, then z has the *polar representation*

(2) $\quad z = r\cos\theta + ir\sin\theta = r(\cos\theta + i\sin\theta).$

The coordinate r is the *modulus* of z; that is, $r = |z|$. The coordinate r is also the distance from the point z to the origin. The coordinate θ is undefined if $z = 0$; if $z \neq 0$, then θ is any value for which the identities $\cos\theta = x/r$ and $\sin\theta = y/r$ hold true. The number θ is called *an argument of* z and is denoted by $\theta = \arg z$. The directed angle θ is measured from the positive x axis to the segment OP joining the origin and the point (x, y) as indicated in Figure 1.7. Angles will be measured in radians and are chosen to be positive in the counterclockwise sense. Hence $\arg z$ can take on an infinite number of values and is determined up to multiples of 2π.

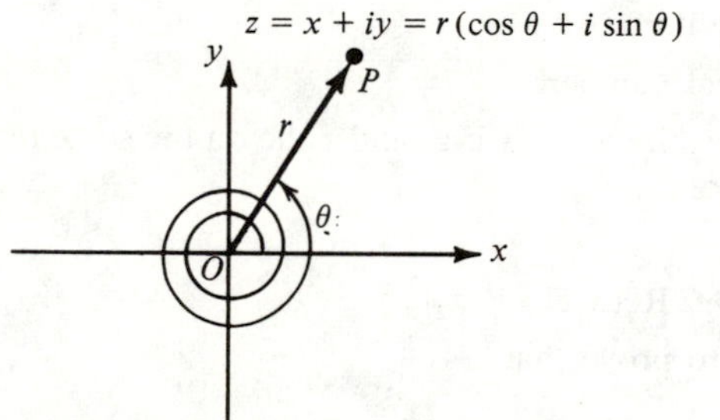

FIGURE 1.7 Polar representation of complex numbers.

As in trigonometry, we have

(3) $\quad \theta = \arg z = \arctan\dfrac{y}{x},$

but care must be taken to specify the choice of $\arctan(y/x)$ so that the point z corresponding to r and θ lies in the appropriate quadrant. The reason that we must be careful is because $\tan\theta$ has period π, while $\cos\theta$ and $\sin\theta$ have period 2π.

EXAMPLE 1.4 $\sqrt{3} + i = 2\cos\dfrac{\pi}{6} + i2\sin\dfrac{\pi}{6} = 2\cos\dfrac{13\pi}{6} + i2\sin\dfrac{13\pi}{6}.$

EXAMPLE 1.5 If $z = -\sqrt{3} - i$, then

$$r = |z| = |-\sqrt{3} - i| = 2 \qquad \text{and} \qquad \theta = \arctan \frac{y}{x} = \arctan \frac{-1}{-\sqrt{3}} = \frac{7\pi}{6},$$

so

$$-\sqrt{3} - i = 2\cos\frac{7\pi}{6} + i2\sin\frac{7\pi}{6}.$$

If θ_0 is a value of $\arg z$, then we can display all values of $\arg z$ as follows:

$$(4) \qquad \arg z = \theta_0 + 2\pi n \qquad \text{where } n \text{ is an integer.}$$

For a given nonzero complex number z, the value of $\arg z$ that lies in the range $-\pi < \theta \leqq \pi$ is called the *principal value of* $\arg z$ and is denoted by $\operatorname{Arg} z$. We write

$$(5) \qquad \operatorname{Arg} z = \theta \qquad \text{where } -\pi < \theta \leqq \pi.$$

Using (4) and (5), we can easily see a relation between $\arg z$ and $\operatorname{Arg} z$:

$$(6) \qquad \arg z = \operatorname{Arg} z + 2\pi n \qquad \text{where } n \text{ is an integer.}$$

We observe that $\operatorname{Arg} z$ is a discontinuous function of z and that it jumps by an amount 2π as we cross the negative real axis.

The polar form lends very useful geometric interpretation for multiplication of complex numbers. An argument of the product $z_1 z_2$ is the sum of an argument of z_1 and an argument of z_2. This is expressed by

$$(7) \qquad \arg(z_1 z_2) = \arg(z_1) + \arg(z_2).$$

To prove identity (7), we let $z_1 = r_1(\cos\theta_1 + i\sin\theta_1)$ and $z_2 = r_2(\cos\theta_2 + i\sin\theta_2)$. Direct multiplication of z_1 by z_2 and use of the trigonometric identities for $\cos(\theta_1 + \theta_2)$ and $\sin(\theta_1 + \theta_2)$ yield the following calculation:

$$\begin{aligned} z_1 z_2 &= r_1 r_2(\cos\theta_1 + i\sin\theta_1)(\cos\theta_2 + i\sin\theta_2) \\ &= r_1 r_2[\cos\theta_1\cos\theta_2 - \sin\theta_1\sin\theta_2 + i(\cos\theta_1\sin\theta_2 + \sin\theta_1\cos\theta_2)] \\ &= r_1 r_2[\cos(\theta_1 + \theta_2) + i\sin(\theta_1 + \theta_2)]. \end{aligned}$$

We have already seen that the modulus of the product is the product of the moduli; that is, $|z_1 z_2| = |z_1||z_2|$. The new result shows that an argument of z_1 plus an argument of z_2 is an argument of $z_1 z_2$. This allows us to obtain a picture of the situation as shown in Figure 1.8.

If $z = r(\cos\theta + i\sin\theta)$, then $\overline{z}$ has the form

$$(8) \qquad \overline{z} = r(\cos\theta - i\sin\theta) = r(\cos(-\theta) + i\sin(-\theta)).$$

Hence we see that the argument of the conjugate of a complex number is the negative of the argument of the number. Furthermore, the inverse z^{-1} is given by

$$(9) \qquad z^{-1} = \frac{1}{r(\cos\theta + i\sin\theta)} = \frac{1}{r}(\cos(-\theta) + i\sin(-\theta)).$$

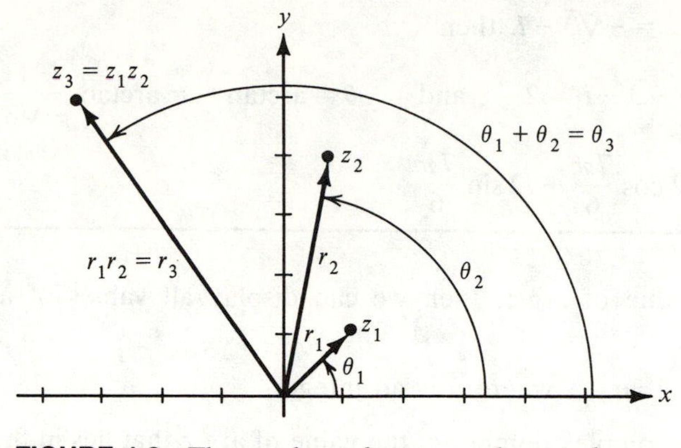

FIGURE 1.8 The product of two complex numbers $z_3 = z_1z_2$.

Hence the vectors z^{-1} and $\bar{z}$ are pointed in the same direction. From (9) we see that the modulus of the reciprocal $1/z$ is $1/r$ and is the reciprocal of the modulus of z. The numbers z, $\bar{z}$, and z^{-1} are shown in Figure 1.9.

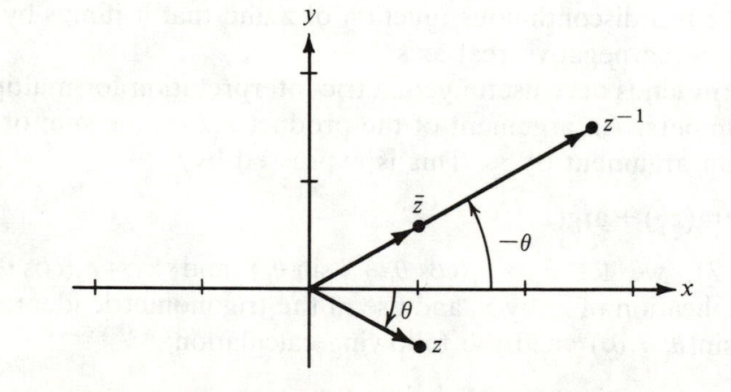

FIGURE 1.9 Graph of the numbers z, $\bar{z}$, and z^{-1}.

EXAMPLE 1.6 If $z = 5 + 12i$, then $r = 13$ and $z^{-1} = \frac{1}{13}(\frac{5}{13} - (12i/13))$ has modulus $\frac{1}{13}$. These results could also be inferred from

$$(10) \qquad z^{-1} = \frac{1}{z} = \frac{\bar{z}}{z\bar{z}} = \frac{1}{r^2}\,\bar{z}.$$

Since $z_1/z_2 = z_1z_2^{-1}$, we can use (9) to obtain

$$(11) \qquad \frac{z_1}{z_2} = \frac{r_1}{r_2}\,(\cos(\theta_1 - \theta_2) + i\sin(\theta_1 - \theta_2)).$$

EXAMPLE 1.7 If $z_1 = 8i$ and $z_2 = 1 + i\sqrt{3}$, then the polar forms are $z_1 = 8(\cos(\pi/2) + i\sin(\pi/2))$ and $z_2 = 2(\cos(\pi/3) + i\sin(\pi/3))$. So we have

$$\frac{z_1}{z_2} = \frac{8}{2}\left(\cos\left(\frac{\pi}{2} - \frac{\pi}{3}\right) + i\sin\left(\frac{\pi}{2} - \frac{\pi}{3}\right)\right) = 4\left(\cos\frac{\pi}{6} + i\sin\frac{\pi}{6}\right)$$
$$= 2\sqrt{3} + 2i.$$

In Chapter 4 we will show that the complex exponential function is defined as follows:

$$e^z = e^{x+iy} = e^x(\cos y + i\sin y).$$

When $z = 0 + iy$, we have $e^{iy} = \cos y + i\sin y$. For this reason it is convenient to use the identity

$$(12) \qquad e^{i\theta} = \cos\theta + i\sin\theta,$$

which is known as *Euler's Formula*. It can be obtained formally with series. Start with the two familiar expansions:

$$(13) \qquad \cos\theta = 1 - \frac{\theta^2}{2!} + \frac{\theta^4}{4!} - \frac{\theta^6}{6!} + - \cdots$$

and

$$(14) \qquad \sin\theta = \theta - \frac{\theta^3}{3!} + \frac{\theta^5}{5!} - \frac{\theta^7}{7!} + - \cdots.$$

In (13), use $-\theta^2 = (i\theta)^2$, $\theta^4 = (i\theta)^4$, $-\theta^6 = (i\theta)^6$, ... to get

$$(15) \qquad \cos\theta = 1 + \frac{(i\theta)^2}{2!} + \frac{(i\theta)^4}{4!} + \frac{(i\theta)^6}{6!} + \cdots.$$

Now multiply both sides of (14) by i and use $-i\theta^3 = (i\theta)^3$, $i\theta^5 = (i\theta)^5$, $-i\theta^7 = (i\theta)^7$, ... and obtain

$$(16) \qquad i\sin\theta = i\theta + \frac{(i\theta)^3}{3!} + \frac{(i\theta)^5}{5!} + \frac{(i\theta)^7}{7!} + \cdots.$$

Finally, add the two series (15) and (16); the result is

$$\cos\theta + i\sin\theta = 1 + i\theta + \frac{(i\theta)^2}{2!} + \frac{(i\theta)^3}{3!} + \frac{(i\theta)^4}{4!} + \frac{(i\theta)^5}{5!} + \cdots = e^{i\theta}.$$

The choice of notation $e^{i\theta}$ permits us to use the familiar rule of exponents:

$$(17) \qquad e^{i\theta_1 + i\theta_2} = (e^{i\theta_1})(e^{i\theta_2}),$$

which can also be verified directly by the calculation

$$\cos(\theta_1 + \theta_2) + i\sin(\theta_1 + \theta_2) = (\cos\theta_1 + i\sin\theta_1)(\cos\theta_2 + i\sin\theta_2).$$

This calculation is a special case of $z_1 z_2$ where $r_1 = r_2 = 1$, and it has already been established. With this exponential notation a nonzero complex number z has the form

(18) $z = re^{i\theta}$ where $r = |z|$ and $\theta = \arg z$.

By using this notation the arithmetic properties are easily summarized:

(19) $z_1 z_2 = r_1 r_2 e^{i(\theta_1 + \theta_2)}$,

(20) $\dfrac{z_1}{z_2} = \dfrac{r_1}{r_2} e^{i(\theta_1 - \theta_2)}$ where $z_2 \neq 0$.

The complex conjugate is expressed by

(21) $\overline{z} = re^{-i\theta}$.

EXERCISES FOR SECTION 1.4

1. Find Arg z for the following values of z.
   (a) $1 - i$ **(b)** $-\sqrt{3} + i$ **(c)** $(-1 - i\sqrt{3})^2$
   (d) $(1 - i)^3$ **(e)** $\dfrac{2}{1 + i\sqrt{3}}$ **(f)** $\dfrac{2}{i - 1}$
   (g) $\dfrac{1 + i\sqrt{3}}{(1 + i)^2}$ **(h)** $(1 + i\sqrt{3})(1 + i)$
2. Represent the following complex numbers in polar form.
   (a) -4 **(b)** $6 - 6i$ **(c)** $-7i$
   (d) $-2\sqrt{3} - 2i$ **(e)** $\dfrac{1}{(1 - i)^2}$ **(f)** $\dfrac{6}{i + \sqrt{3}}$
   (g) $(5 + 5i)^3$ **(h)** $3 + 4i$
3. Express the following in $a + ib$ form.
   (a) $e^{i\pi/2}$ **(b)** $4e^{-i\pi/2}$ **(c)** $8e^{i7\pi/3}$
   (d) $-2e^{i5\pi/6}$ **(e)** $2ie^{-i3\pi/4}$ **(f)** $6e^{i2\pi/3}e^{i\pi}$
   (g) $e^2 e^{i\pi}$ **(h)** $e^{i\pi/4}e^{-i\pi}$
4. Use the exponential notation to show that
   (a) $(\sqrt{3} - i)(1 + i\sqrt{3}) = 2\sqrt{3} + 2i$ **(b)** $(1 + i)^3 = -2 + 2i$
   (c) $2i(\sqrt{3} + i)(1 + i\sqrt{3}) = -8$ **(d)** $8/(1 + i) = 4 - 4i$
5. Show that $\arg(z_1 z_2 z_3) = \arg z_1 + \arg z_2 + \arg z_3$. *Hint*: Use property (7).
6. Let $z = \sqrt{3} + i$. Plot the points z, iz, $-z$, and $-iz$ and describe a relationship among their arguments.
7. Let $z_1 = -1 + i\sqrt{3}$ and $z_2 = -\sqrt{3} + i$. Show that the equation $\operatorname{Arg}(z_1 z_2) = \operatorname{Arg} z_1 + \operatorname{Arg} z_2$ *does not* hold for the specific choice of z_1 and z_2.
8. Show that the equation $\operatorname{Arg}(z_1 z_2) = \operatorname{Arg} z_1 + \operatorname{Arg} z_2$ is true if we require that $-\pi/2 < \operatorname{Arg} z_1 < \pi/2$ and $-\pi/2 < \operatorname{Arg} z_2 < \pi/2$.
9. Show that $\arg z_1 = \arg z_2$ if and only if $z_2 = cz_1$ where c is a positive real constant.
10. Establish identity (19).
11. Establish the identity $\arg(z_1/z_2) = \arg z_1 - \arg z_2$.
12. Describe the set of complex numbers for which $\operatorname{Arg}(1/z) \neq -\operatorname{Arg}(z)$.

13. Show that $\arg(1/z) = -\arg z$.

14. Show that $\arg(z_1\overline{z_2}) = \arg z_1 - \arg z_2$.

15. Show that
(a) $\text{Arg}(z\overline{z}) = 0$ **(b)** $\text{Arg}(z + \overline{z}) = 0$ **when Re** $z > 0$.

16. Let $z \neq z_0$. Show that the polar representation $z - z_0 = \rho(\cos\phi + i\sin\phi)$ can be used to denote the displacement vector from z_0 to z as indicated in Figure 1.10.

17. Let z_1, z_2, and z_3 form the vertices of a triangle as indicated in Figure 1.11. Show that

$$\alpha = \arg\left(\frac{z_2 - z_1}{z_3 - z_1}\right) = \arg(z_2 - z_1) - \arg(z_3 - z_1)$$

is an expression for the angle at the vertex z_1.

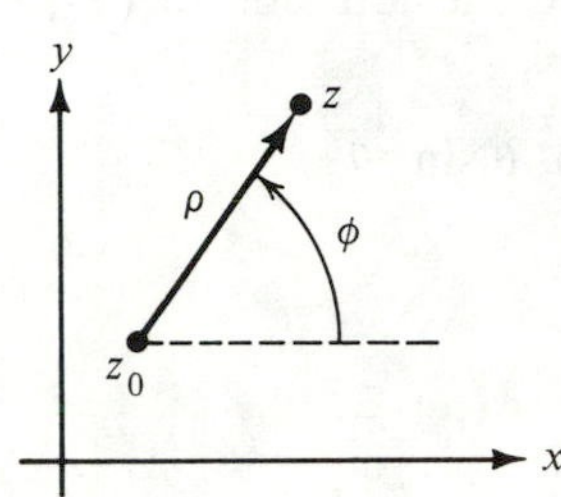

FIGURE 1.10 Accompanies Exercise 16.

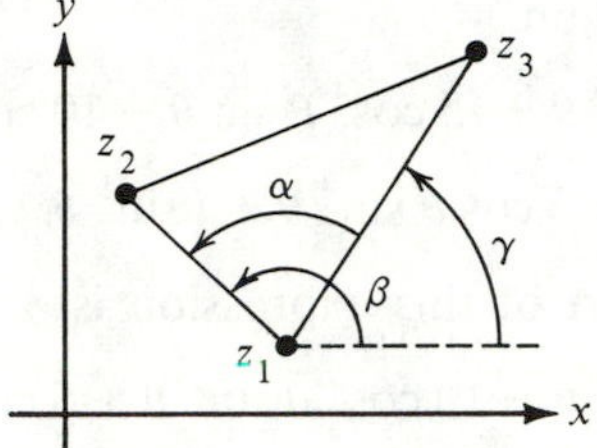

FIGURE 1.11 Accompanies Exercise 17.

1.5 Powers, Roots, and the Quadratic Formula

If we use the polar coordinate representation $z = r(\cos\theta + i\sin\theta)$, then we obtain the formula

(1) $\quad z^2 = r^2(\cos 2\theta + i\sin 2\theta)$.

Using $z^3 = z^2 z$ and the rule for multiplication, we see that

(2) $\quad z^3 = r^3(\cos 3\theta + i\sin 3\theta)$.

Continuing in a similar manner, we can use mathematical induction to establish the general rule

(3) $\quad z^n = r^n(\cos n\theta + i\sin n\theta)$,

which is valid for all integer values of n. The proof is left as an exercise. It is sometimes convenient to express (3) in the exponential form

(4) $\quad z^n = r^n e^{in\theta}$.

If we set $r = 1$ in equation (3), then we obtain

(5) $\quad (\cos\theta + i\sin\theta)^n = \cos n\theta + i\sin n\theta$,

which is known as *De Moivre's Formula*. This important identity is useful in finding powers and roots of complex numbers and in deriving trigonometric identities for $\cos n\theta$ and $\sin n\theta$.

EXAMPLE 1.8 Let us show that $(1+i)^5 = -4-4i$. If we write

$$z = 1 + i = \sqrt{2}\left(\cos\frac{\pi}{4} + i\sin\frac{\pi}{4}\right)$$

and use equation (3) with $n = 5$, then calculation reveals that

$$(1+i)^5 = (\sqrt{2})^5\left(\cos\frac{5\pi}{4} + i\sin\frac{5\pi}{4}\right) = -4-4i.$$

EXAMPLE 1.9 De Moivre's Formula (5) can be used to show that

$$\cos 5\theta = \cos^5\theta - 10\cos^3\theta\sin^2\theta + 5\cos\theta\sin^4\theta.$$

If we let $n = 5$, and use the Binomial Formula to expand the left side of (5), then we obtain

$$\cos^5\theta + i5\cos^4\theta\sin\theta - 10\cos^3\theta\sin^2\theta - 10i\cos^2\theta\sin^3\theta$$
$$+ 5\cos\theta\sin^4\theta + i\sin^5\theta.$$

The real part of this expression is

$$\cos^5\theta - 10\cos^3\theta\sin^2\theta + 5\cos\theta\sin^4\theta.$$

Equating this to the real part of $\cos 5\theta + i\sin 5\theta$ on the right side of (5) establishes the desired result.

Let us show how De Moivre's Formula (5) can be used to find the n solutions to the equation

$$z^n = 1. \tag{6}$$

If we let $z = \cos\theta + i\sin\theta$, identity (5) and equation (6) lead to

$$(\cos\theta + i\sin\theta)^n = \cos n\theta + i\sin n\theta = \cos 0 + i\sin 0 = 1.$$

Equating the real and imaginary parts, we see that $\cos n\theta = \cos 0$ and $\sin n\theta = \sin 0$. The solutions to these equations are $\theta = 2\pi k/n$ where k is an integer. Because of the periodicity of $\sin t$ and $\cos t$, we can display the n distinct solutions to equation (6) as

$$z_k = \cos\frac{2\pi k}{n} + i\sin\frac{2\pi k}{n} \qquad \text{for } k = 0, 1, \ldots, n-1. \tag{7}$$

The solutions (7) are often referred to as the *nth roots of unity*. The value ω_n given by

$$\omega_n = \cos\frac{2\pi}{n} + i\sin\frac{2\pi}{n} \tag{8}$$

is called a *primitive nth root of unity*. By De Moivre's Formula the nth roots of unity are given by

$$1, \omega_n, \omega_n^2, \ldots, \omega_n^{n-1}. \tag{9}$$

Geometrically, the nth roots of unity are equally spaced points that lie on the unit circle $|z| = 1$ and form the vertices of a regular polygon with n sides.

EXAMPLE 1.10 The solutions to the equation

$$z^8 = 1$$

are given by the n values

$$z_k = \cos\frac{2\pi k}{8} + i\sin\frac{2\pi k}{8} \qquad \text{for } k = 0, 1, \ldots, 7.$$

In Cartesian form these solutions are ± 1, $\pm i$, $\pm(\sqrt{2} + i\sqrt{2})/2$, $\pm(\sqrt{2} - i\sqrt{2})/2$ and are illustrated in Figure 1.12.

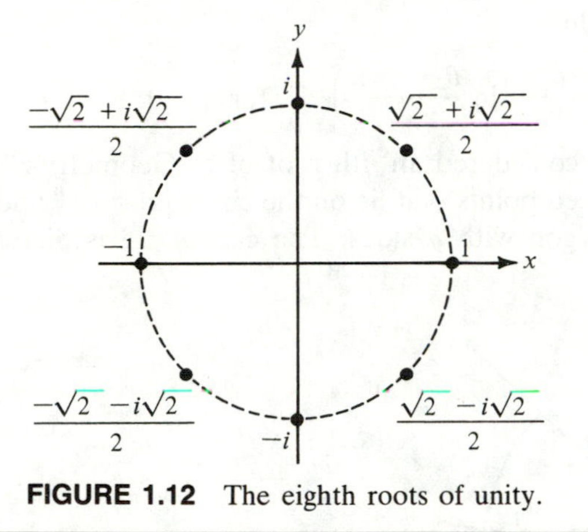

FIGURE 1.12 The eighth roots of unity.

The square roots of the complex number $z = r(\cos\theta + i\sin\theta)$ are given by

$$\text{(10)} \qquad z^{1/2} = \pm r^{1/2}\left(\cos\frac{\theta}{2} + i\sin\frac{\theta}{2}\right) \qquad \text{where } \theta \text{ is an argument of } z.$$

The solutions (10) can easily be verified by direct squaring. Let us recall the trigonometric identities

$$\text{(11)} \qquad \begin{aligned} \cos\frac{\theta}{2} &= \sqrt{\frac{1+\cos\theta}{2}} \qquad \text{for } -\pi < \theta \leqq \pi, \qquad \text{and} \\ \sin\frac{\theta}{2} &= \sqrt{\frac{1-\cos\theta}{2}} \qquad \text{for } 0 \leqq \theta \leqq 2\pi. \end{aligned}$$

If we multiply identities (11) by $r^{1/2}$, the result is

$$r^{1/2}\cos\frac{\theta}{2} = \sqrt{\frac{r + r\cos\theta}{2}} = \sqrt{\frac{|z| + x}{2}} \qquad \text{and}$$

$$r^{1/2}\sin\frac{\theta}{2} = \sqrt{\frac{r - r\cos\theta}{2}} = \sqrt{\frac{|z| - x}{2}}.$$

These can be substituted into (10) to obtain the Cartesian formula

$$\text{(12)} \qquad z^{1/2} = \pm\left[\sqrt{\frac{|z| + x}{2}} + i(\text{sign } y)\sqrt{\frac{|z| - x}{2}}\right]$$

where sign $y = 1$ if $y \geqq 0$ and sign $y = -1$ if $y < 0$.

EXAMPLE 1.11 If $z = 3 + 4i$, then $|z| = 5$, $x = 3$, and $y = 4$. Then we obtain

$$(3+4i)^{1/2} = \pm\left[\sqrt{\frac{5+3}{2}} + i\sqrt{\frac{5-3}{2}}\right] = \pm(2+i).$$

Let n be a positive integer, and let $z = r(\cos\theta + i\sin\theta)$ be a nonzero complex number. Then there are n distinct solutions to the equation

(13) $\quad w^n = z,$

and they are given by the values

(14) $$w_k = r^{1/n}\left(\cos\frac{\theta + 2\pi k}{n} + i\sin\frac{\theta + 2\pi k}{n}\right) \qquad \text{for } k = 0, 1, \ldots, n-1.$$

Each solution in (14) can be considered an nth root of z. Geometrically, the nth roots of z are equally spaced points that lie on the circle $|z| = r^{1/n}$ and form the vertices of a regular polygon with n sides. The case $n = 5$ is pictured in Figure 1.13.

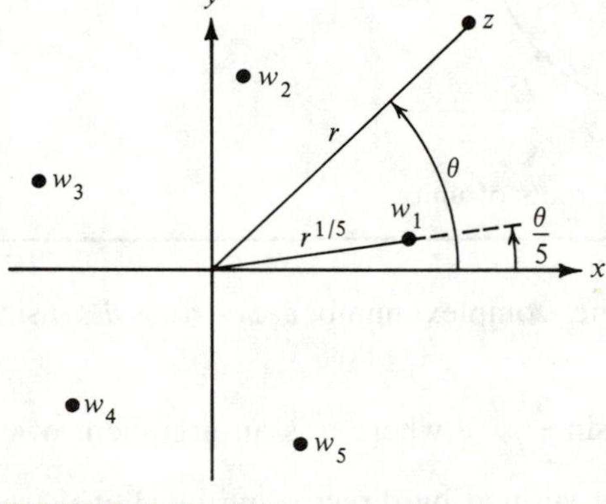

FIGURE 1.13 The solutions to the equation $w^5 = z$.

EXAMPLE 1.12 We can use formula (14) to find the cube roots of $8i = 8(\cos(\pi/2) + i\sin(\pi/2))$. Calculation reveals that the roots are

$$w_k = 2\left(\cos\frac{(\pi/2) + 2\pi k}{3} + i\sin\frac{(\pi/2) + 2\pi k}{3}\right) \qquad \text{for } k = 0, 1, 2$$

The Cartesian forms of the solutions are $w_0 = \sqrt{3} + i$, $w_1 = -\sqrt{3} + i$, and $w_2 = -2i$, as shown in Figure 1.14.

The solutions to the quadratic equation

(15) $\quad Az^2 + Bz + C = 0$

are given by the familiar formula

(16) $$z = \frac{-B \pm (B^2 - 4AC)^{1/2}}{2A}$$

and are valid when the coefficients $A \neq 0$, and B, C are complex numbers.

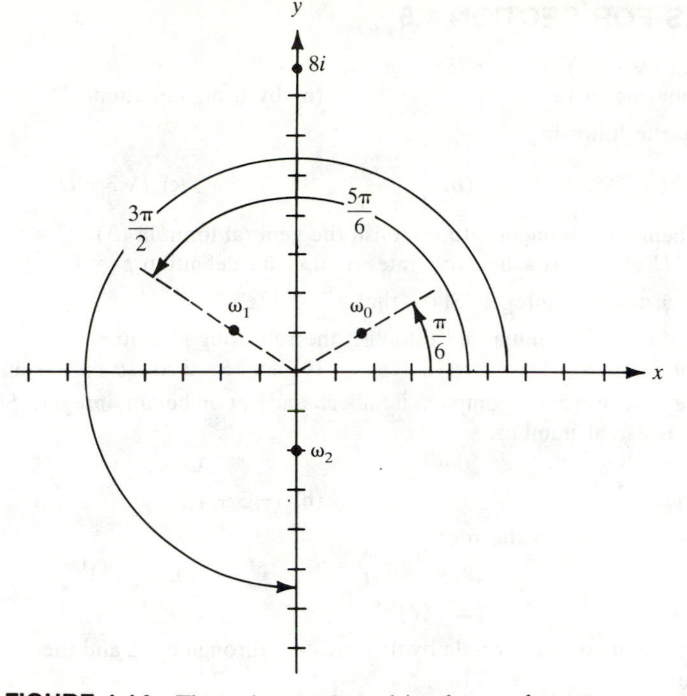

FIGURE 1.14 The point $z = 8i$ and its three cube roots ω_0, ω_1, and ω_2.

EXAMPLE 1.13 The solutions to the quadratic equation

$$z^2 - (3 + 2i)z + 1 + 3i = 0$$

are given by

$$z = \frac{3 + 2i \pm ((-3 - 2i)^2 - 4(1 + 3i))^{1/2}}{2} = \frac{3 + 2i \pm 1^{1/2}}{2}.$$

Therefore the solutions are $1 + i$ and $2 + i$.

EXAMPLE 1.14 Find all the roots of the equation

$$z^3 + (1 - 2i)z^2 + (-1 - 6i)z - 5 = 0$$

given that $z = i$ is a root.

Solution Long division can be used to divide the factor $(z - i)$ into the cubic to obtain

$$(z - i)(z^2 + (1 - i)z - 5i) = z^3 + (1 - 2i)z^2 + (-1 - 6i)z - 5.$$

The quadratic formula can be used to show that the remaining two roots of the quadratic factor are

$$z = \frac{-1 + i \pm (18i)^{1/2}}{2} = \frac{-1 + i \pm (3 + 3i)}{2}$$
$$= 1 + 2i \quad \text{and} \quad -2 - i.$$

EXERCISES FOR SECTION 1.5

1. Show that $(\sqrt{3}+i)^4 = -8 + i8\sqrt{3}$ in two ways:
(a) by squaring twice **(b)** by using equation (3)

2. Calculate the following.
(a) $(1 - i\sqrt{3})^3(\sqrt{3}+i)^2$ **(b)** $\dfrac{(1+i)^3}{(1-i)^5}$ **(c)** $(\sqrt{3}+i)^6$

3. Use mathematical induction to establish the general formula (3). $z^n = r^n(\cos n\theta + i \sin n\theta)$. *Hint*: If n is a negative integer, use the definition $z^n = (z^{-1})^{-n}$.

4. Let n be a negative integer. Show that $z^{-n} = 1/z^n$.

5. Use De Moivre's Formula and establish the following identities.
(a) $\cos 3\theta = \cos^3\theta - 3\cos\theta\sin^2\theta$ **(b)** $\sin 3\theta = 3\cos^2\theta\sin\theta - \sin^3\theta$

6. Let z be any nonzero complex number, and let n be an integer. Show that $z^n + (\bar{z})^n$ is a real number.

7. Find
(a) $(5 - 12i)^{1/2}$ **(b)** $(-3 + 4i)^{1/2}$

For Exercises 8–12, find all the roots.

8. $(-2+2i)^{1/3}$ **9.** $(-64)^{1/4}$ **10.** $(-1)^{1/5}$
11. $(16i)^{1/4}$ **12.** $(8)^{1/6}$

13. Establish the quadratic formula by first dividing through by A and then completing the square.

14. Find the solutions to the equation $z^2 + (1+i)z + 5i = 0$.

15. Solve the equation $(z+1)^3 = z^3$.

16. Find all the roots of the equation $z^3 + z^2 + 3z - 5 = 0$.

17. Show that $1 + i$ is a root of the equation $z^{17} + 2z^{15} - 512 = 0$.

18. Let $P(z) = a_n z^n + a_{n-1}z^{n-1} + \cdots + a_1 z + a_0$ be a polynomial with *real* coefficients $a_n, a_{n-1}, \ldots, a_1, a_0$. If z_1 is a complex root of $P(z)$, show that $\overline{z_1}$ is also a root. *Hint*: Show that $\overline{P(\bar{z}_1)} = P(z_1) = 0$.

19. Find all the roots of the equation $z^4 - 4z^3 + 6z^2 - 4z + 5 = 0$ given that $z_1 = i$ is a root.

20. Let m and n be positive integers that have no common factor. Show that there are n distinct solutions to $w^n = z^m$ and that they are given by

$$w_k = r^{m/n}\left(\cos\frac{m(\theta + 2\pi k)}{n} + i\sin\frac{m(\theta+2\pi k)}{n}\right) \quad \text{for } k = 0, 1, \ldots, n-1.$$

21. Find the three solutions to $z^{3/2} = 4\sqrt{2} + i4\sqrt{2}$.

22. **(a)** If $z \neq 1$, show that $1 + z + z^2 + \cdots + z^n = \dfrac{1 - z^{n+1}}{1 - z}$.

(b) Use part (a) and De Moivre's Formula to derive *Lagrange's identity*:

$$1 + \cos\theta + \cos 2\theta + \cdots + \cos n\theta = \frac{1}{2} + \frac{\sin((n+\frac{1}{2})\theta)}{2\sin(\theta/2)} \quad \text{where } 0 < \theta < 2\pi.$$

23. Let $z_k \neq 1$ be an nth root of unity. Prove that

$$1 + z_k + z_k^2 + \cdots + z_k^{n-1} = 0.$$

24. If $1 = z_0, z_1, z_2, \ldots, z_{n-1}$ are the nth roots of unity, prove that

$$(z - z_1)(z - z_2)\cdots(z - z_{n-1}) = 1 + z + z^2 + \cdots + z^{n-1}.$$

1.6 Curves, Regions, and Domains in the Plane

In this section we investigate some basic ideas concerning sets of points in the plane. The first concept is that of a curve. Let $x = x(t)$ and $y = y(t)$ be two continuous functions of the real parameter t that are defined for values of t in the interval $a \leqq t \leqq b$. Then a *curve* C in the plane that joins the point $(x(a), y(b))$ to the point $(x(b), y(b))$ is given by

(1) $$C\colon z(t) = x(t) + iy(t) \qquad \text{for } a \leqq t \leqq b.$$

We say that C is a curve that goes from the initial point $z(a)$ to the terminal point $z(b)$. For example, if $z_0 = x_0 + iy_0$ and $z_1 = x_1 + iy_1$ are two given points, then the straight line segment joining z_0 to z_1 is

(2) $$C\colon z(t) = [x_0 + (x_1 - x_0)t] + i[y_0 + (y_1 - y_0)t] \qquad \text{for } 0 \leqq t \leqq 1$$

and is pictured in Figure 1.15. One way to derive formula (2) is to use the vector form of a line. A point on the line is $z_0 = x_0 + iy_0$ and the direction of the line is $z_1 - z_0$; hence the line C in (2) is given by

$$C\colon z(t) = z_0 + (z_1 - z_0)t \qquad \text{for } 0 \leqq t \leqq 1.$$

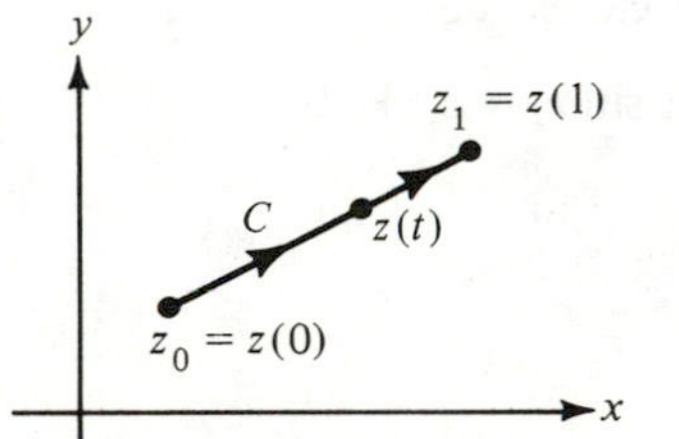

FIGURE 1.15 The straight line segment C joining z_0 to z_1.

A curve C with the property that $z(a) = z(b)$ is said to be a *closed curve*. The line segment (2) is not a closed curve. The curve $x(t) = \sin 2t \cos t$, $y(t) = \sin 2t \sin t$ for $0 \leqq t \leqq 2\pi$ forms the four-leaved rose in Figure 1.16. Observe carefully that as t goes from 0 to $\pi/2$, the point is on leaf 1; from $\pi/2$ to π it is on leaf 2; between π and $3\pi/2$ it is on leaf 3; and finally, for t between $3\pi/2$ and 2π it is on leaf 4. Notice that the curve has crossed over itself at the origin.

Remark In calculus the curve in Figure 1.16 was given the polar coordinate parameterization $r = \sin 2\theta$.

We want to be able to distinguish when a curve does not cross over itself. The curve C is called simple if it does not cross over itself, which is

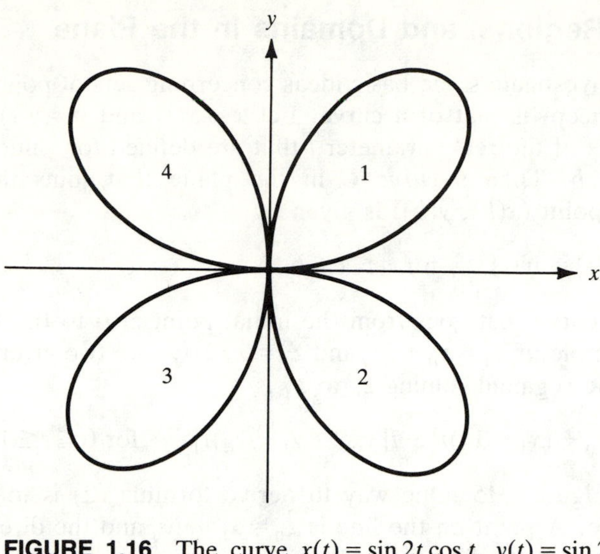

FIGURE 1.16 The curve $x(t) = \sin 2t \cos t$, $y(t) = \sin 2t \sin t$ for $0 \leqq t \leqq 2\pi$, which forms a four-leaved rose.

expressed by requiring that $z(t_1) \neq z(t_2)$ whenever $t_1 \neq t_2$, except possibly when $t_1 = a$ and $t_2 = b$. For example, the circle C with center $z_0 = x_0 + iy_0$ and radius R can be parameterized to form a simple closed curve:

$$C: z(t) = (x_0 + R \cos t) + i(y_0 + R \sin t) = z_0 + \mathrm{R}e^{it} \tag{3}$$

for $0 \leqq t \leqq 2\pi$ as shown in Figure 1.17.

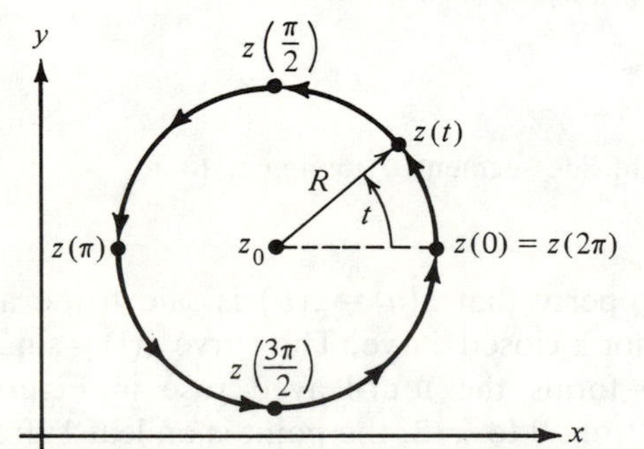

FIGURE 1.17 The simple closed curve $z(t) = z_0 + \mathrm{R}e^{it}$ for $0 \leqq t \leqq 2\pi$.

We need to develop some vocabulary that will help us describe sets of points in the plane. One fundamental idea is that of an *ε-neighborhood* of the point z_0, that is, the set of all points satisfying the inequality

$$|z - z_0| < \varepsilon. \tag{4}$$

This set is the open disk of radius $\varepsilon > 0$ about z_0 shown in Figure 1.18. In particular, the solution sets of the inequalities

$$|z| < 1, \qquad |z - i| < 2, \qquad |z + 1 + 2i| < 3$$

are neighborhoods of the points 0, i, $-1 - 2i$, respectively, of radius 1, 2, 3, respectively.

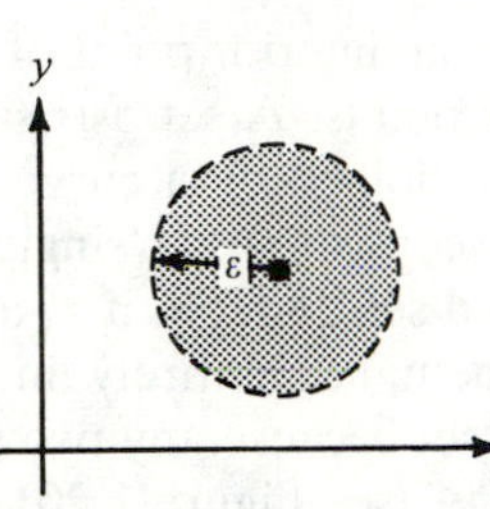

FIGURE 1.18 An ε-neighborhood of the point z_0.

The point z_0 is said to be an *interior point* of the set S provided that there exists an ε-neighborhood of z_0 that contains only points of S; and z_0 is called an *exterior point* of the set S if there exists an ε-neighborhood of z_0 that contains no points of S. If z_0 is neither an interior point nor an exterior point of S, then it is called a *boundary point* of S and has the property that each ε-neighborhood of z_0 contains both points in S and points not in S. The situation is illustrated in Figure 1.19.

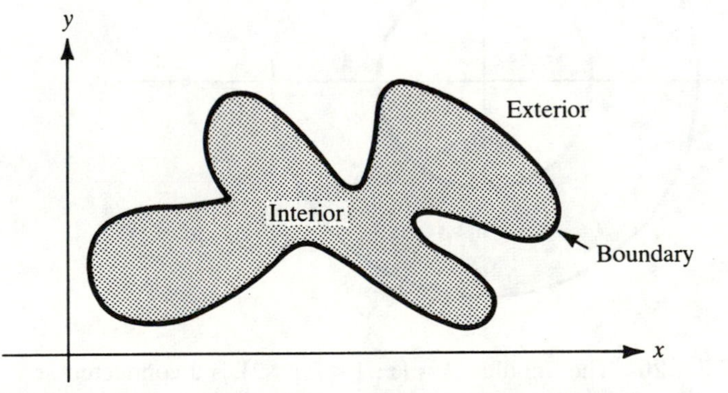

FIGURE 1.19 The interior, exterior, and boundary of a set.

EXAMPLE 1.15 Let $S = \{z: |z| < 1\}$. Find the interior, exterior, and boundary of S.

Solution Let z_0 be a point of S. Then $|z_0| < 1$ so that we can choose $\varepsilon = 1 - |z_0| > 0$. If z lies in the disk $|z - z_0| < \varepsilon$, then

$$|z| = |z_0 + z - z_0| \leqq |z_0| + |z - z_0| < |z_0| + \varepsilon < 1.$$

Hence the ε-neighborhood of z_0 is contained in S, and z_0 is an interior point of S. It follows that the interior of S is the set $\{z: |z|<1\}$.

Similarly, it can be shown that the exterior of S is the set $\{z: |z|>1\}$. The boundary of S is the unit circle $\{z: |z|=1\}$. This is true because if $z_0 = e^{i\theta_0}$ is any point on the circle, then any ε-neighborhood of z_0 will contain the point $(1-\varepsilon/2)e^{i\theta_0}$, which belongs to S, and $(1+\varepsilon/2)e^{i\theta_0}$, which does not belong to S.

A set S is called *open* if every point of S is an interior point of S. A set S is called *closed* if it contains all of its boundary points. A set S is said to be *connected* if every pair of points z_1 and z_2 can be joined by a curve that lies entirely in S. Roughly speaking, a connected set consists of a "single piece." The unit disk $D=\{z: |z|<1\}$ is an open connected set. Indeed, if z_1 and z_2 lie in D, then the straight line segment joining them lies entirely in D. The annulus $A=\{z: 1<|z|<2\}$ is an open connected set because any two points in A can be joined by a curve C that lies entirely in A (see Figure 1.20). The set $B=\{z: |z+2|<1 \text{ or } |z-2|<1\}$ consists of two disjoint disks; hence it is not connected (see Figure 1.21).

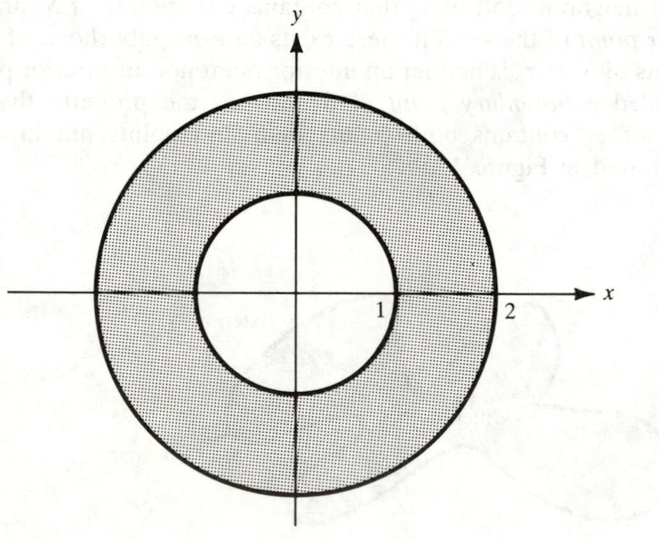

FIGURE 1.20 The annulus $A=\{z: 1<|z|<2\}$ is a connected set.

We call an open connected set a *domain*. For example, the right half-plane $H=\{z: \text{Re } z>0\}$ is a domain. This is true because if $z_0 = x_0 + iy_0$ is any point in H, then we can choose $\varepsilon = x_0$, and the ε-neighborhood of z_0 lies in H. Also, any two points in H can be connected with the line segment between them. The open unit disk $|z|<1$ is also a domain. However, the closed unit disk $|z| \leqq 1$ is not a domain. It should be noted that the term "domain" is a noun and is a kind of set.

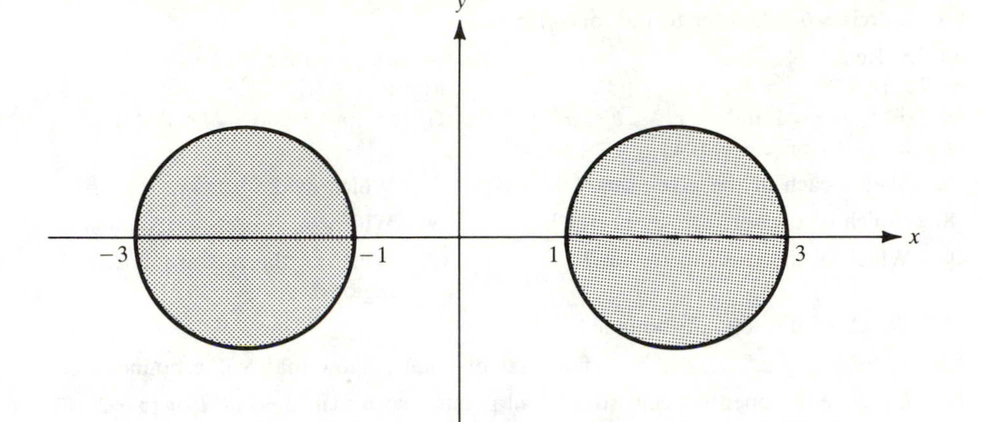

FIGURE 1.21 $B = \{z\colon |z+2| < 1 \text{ or } |z-2| < 1\}$ is not a connected set.

A domain together with some, none, or all of its boundary points is called a *region*. For example, the horizontal strip $\{z\colon 1 < \operatorname{Im} z \leqq 2\}$ is a region. A set that is formed by taking the union of a domain and its boundary is called a *closed region*; that is, the half-plane $\{z\colon x \leqq y\}$ is a closed region. A set is said to be *bounded* if every point can be enveloped by a circle of some finite fixed radius, that is, there exists an $R > 0$ and for each z in S we have $|z| \leqq R$. The rectangle $\{z\colon |x| \leqq 4 \text{ and } |y| \leqq 3\}$ is bounded because it is contained inside the circle $|z| = 5$. A set that cannot be enclosed by a circle is called *unbounded*.

EXERCISES FOR SECTION 1.6

1. Sketch the curve $z(t) = t^2 + 2t + i(t+1)$
(a) for $-1 \leqq t \leqq 0$.
(b) for $1 \leqq t \leqq 2$.
Hint: Use $x = t^2 + 2t$, $y = t + 1$ and eliminate the parameter t.

2. Find a parameterization of the line that
(a) joins the origin to the point $1 + i$.
(b) joins the point i to the point $1 + i$.
(c) joins the point 1 to the point $1 + i$.
(d) joins the point 2 to the point $1 + i$.

3. Find a parameterization of the curve that is a portion of the parabola $y = x^2$ that
(a) joins the origin to the point $2 + 4i$.
(b) joins the point $-1 + i$ to the origin.
(c) joins the point $1 + i$ to the origin.
Hint: For parts (a) and (b), use the parameter $t = x$.

4. Find a parameterization of the curve that is a portion of the circle $|z| = 1$ that joins the point $-i$ to i if
(a) the curve is the right semicircle.
(b) the curve is the left semicircle.

5. Find a parameterization of the curve that is a portion of the circle $|z| = 1$ that joins the point 1 to i if
(a) the parameterization is counterclockwise along the quarter circle.
(b) the parameterization is clockwise.

For Exercises 6–12, refer to the following sets:

(a) $\{z: \text{Re } z > 1\}$.
(b) $\{z: -1 < \text{Im } z \leqq 2\}$.
(c) $\{z: |z - 2 - i| \leqq 2\}$.
(d) $\{z: |z + 3i| > 1\}$.
(e) $\{re^{i\theta}: 0 < r < 1 \text{ and } -\pi/2 < \theta < \pi/2\}$.
(f) $\{re^{i\theta}: r > 1 \text{ and } \pi/4 < \theta < \pi/3\}$.
(g) $\{z: |z| < 1 \text{ or } |z - 4| < 1\}$.

6. Sketch each of the given sets.

7. Which of the sets are open?

8. Which of the sets are connected?

9. Which of the sets are domains?

10. Which of the sets are regions?

11. Which of the sets are closed regions?

12. Which of the sets are bounded?

13. Let $S = \{z_1, z_2, \ldots, z_n\}$ be a finite set of points. Show that S is a bounded set.

14. Let S be the open set consisting of all points z such that $|z + 2| < 1$ or $|z - 2| < 1$. Show that S is not connected.

15. Prove that the neighborhood $|z - z_0| < \varepsilon$ is an open set.

16. Prove that the neighborhood $|z - z_0| < \varepsilon$ is a connected set.

17. Prove that the boundary of the neighborhood $|z - z_0| < \varepsilon$ is the circle $|z - z_0| = \varepsilon$.

2

Complex Functions

2.1 Functions of a Complex Variable

A function f of the complex variable z is a rule that assigns to each value z in a set D one and only one complex value w. We write

(1) $\qquad w = f(z)$

and call w the *image of z under f*. The set D is called the *domain of definition of f*, and the set of all images $R = \{w = f(z) \colon z \in D\}$ is called the *range* of f. Just as z can be expressed by its real and imaginary parts, $z = x + iy$, we write $w = u + iv$ where u and v are the real and imaginary parts of w, respectively. This gives us the representation

(2) $\qquad f(x + iy) = u + iv.$

Since u and v depend on x and y, they can be considered to be real functions of the real variables x and y; that is,

(3) $\qquad u = u(x, y) \qquad \text{and} \qquad v = v(x, y).$

Combining (1), (2), and (3), it is customary to write a complex function f in the form

(4) $\qquad f(z) = f(x + iy) = u(x, y) + iv(x, y).$

Conversely, if $u(x, y)$ and $v(x, y)$ are two given real-valued functions of the real variables x and y, then (4) can be used to define the complex function f.

EXAMPLE 2.1 Write $f(z) = z^4$ in the form $f(z) = u(x, y) + iv(x, y)$.

Solution Using the Binomial Formula, we obtain

$$f(z) = (x + iy)^4 = x^4 + 4x^3iy + 6x^2(iy)^2 + 4x(iy)^3 + (iy)^4$$
$$= (x^4 - 6x^2y^2 + y^4) + i(4x^3y - 4xy^3).$$

EXAMPLE 2.2 Express $f(z) = \overline{z}\,\mathrm{Re}\,z + z^2 + \mathrm{Im}\,z$ in the form $f(z) = u(x, y) + iv(x, y)$.

Solution Using the elementary properties of complex numbers, it follows that

$$f(z) = (x - iy)x + (x^2 - y^2 + i2xy) + y = (2x^2 - y^2 + y) + i(xy).$$

These examples show how to find $u(x, y)$ and $v(x, y)$ when a rule for computing f is given. Conversely, if $u(x, y)$ and $v(x, y)$ are given, then the formulae

$$x = \frac{z + \overline{z}}{2} \qquad \text{and} \qquad y = \frac{z - \overline{z}}{2i}$$

can be used to find a formula for f involving the variables z and $\overline{z}$.

EXAMPLE 2.3 Express $f(z) = 4x^2 + i4y^2$ by a formula involving the variables z and $\overline{z}$.

Solution Calculation reveals that

$$\begin{aligned} f(z) &= 4\left(\frac{z + \overline{z}}{2}\right)^2 + i4\left(\frac{z - \overline{z}}{2i}\right)^2 \\ &= z^2 + 2z\overline{z} + \overline{z}^2 - i(z^2 - 2z\overline{z} + \overline{z}^2) \\ &= (1 - i)z^2 + (2 + 2i)z\overline{z} + (1 - i)\overline{z}^2. \end{aligned}$$

It may be convenient to use $z = re^{i\theta}$ in the expression of a complex function f. This gives us the representation

$$f(re^{i\theta}) = u + iv \tag{5}$$

where u and v are to be considered as real functions of the real variables r and θ. That is,

$$u = u(r, \theta) \qquad \text{and} \qquad v = v(r, \theta). \tag{6}$$

Combining (1), (5), and (6), we obtain the representation

$$f(z) = f(re^{i\theta}) = u(r, \theta) + iv(r, \theta). \tag{7}$$

We remark that the functions u and v defined by equations (3) and (6) are *different*, since (3) involves Cartesian coordinates and (6) involves polar coordinates.

EXAMPLE 2.4 Express $f(z) = z^5 + 4z^2 - 6$ in the polar coordinate form $u(r, \theta) + iv(r, \theta)$.

Solution Using equation (3) of Section 1.5, we obtain

$$f(z) = r^5(\cos 5\theta + i \sin 5\theta) + 4r^2(\cos 2\theta + i\, 2\theta) - 6$$
$$= (r^5 \cos 5\theta + 4r^2 \cos 2\theta - 6) + i(r^5 \sin 5\theta + 4r^2 \sin 2\theta).$$

EXERCISES FOR SECTION 2.1

1. Let $f(z) = x + y + i(x^3y - y^2)$. Find
(a) $f(-1 + 3i)$ **(b)** $f(3i - 2)$

2. Let $f(z) = z^2 + 4z\bar{z} - 5 \operatorname{Re} z + \operatorname{Im} z$. Find
(a) $f(-3 + 2i)$ **(b)** $f(2i - 1)$

3. Find $f(1 + i)$ for the following functions.
(a) $f(z) = z + z^{-2} + 5$ **(b)** $f(z) = \dfrac{1}{z^2 + 1}$

4. Find $f(2i - 3)$ for the following functions.
(a) $f(z) = (z + 3)^3(z - 5i)^2$ **(b)** $f(z) = \dfrac{z + 2 - 3i}{z + 4 - i}$

5. Let $f(z) = z^{21} - 5z^7 + 9z^4$. Use polar coordinates to find
(a) $f(-1 + i)$ **(b)** $f(1 + i\sqrt{3})$

6. Express $f(z) = \bar{z}^2 + (2 - 3i)z$ in the form $u + iv$.

7. Express $f(z) = \dfrac{z + 2 - i}{z - 1 + i}$ in the form $u + iv$.

8. Express $f(z) = z^5 + \bar{z}^5$ in the polar coordinate form $u(r, \theta) + iv(r, \theta)$.

9. Express $f(z) = z^5 + \bar{z}^3$ in the polar coordinate form $u(r, \theta) + iv(r, \theta)$.

10. Let $f(z) = e^x \cos y + ie^x \sin y$. Find
(a) $f(0)$ **(b)** $f(1)$ **(c)** $f(i\pi/4)$
(d) $f(1 + i\pi/4)$ **(e)** $f(i2\pi/3)$ **(f)** $f(2 + i\pi)$

11. Let $f(z) = (1/2)\ln(x^2 + y^2) + i \arctan(y/x)$. Find
(a) $f(1)$ **(b)** $f(1 + i)$ **(c)** $f(\sqrt{3})$
(d) $f(\sqrt{3} + i)$ **(e)** $f(1 + i\sqrt{3})$ **(f)** $f(3 + 4i)$

12. Let $f(z) = r^2 \cos 2\theta + ir^2 \sin 2\theta$ where $z = re^{i\theta}$. Find
(a) $f(1)$ **(b)** $f(2e^{i\pi/4})$
(c) $f(\sqrt{2}e^{i\pi/3})$ **(d)** $f(\sqrt{3}e^{i7\pi/6})$

13. Let $f(z) = \ln r + i\theta$ where $r = |z|$, $\theta = \operatorname{Arg} z$. Find
(a) $f(1)$ **(b)** $f(1 + i)$
(c) $f(-2)$ **(d)** $f(-\sqrt{3} + i)$

14. A line that carries a charge of $2\pi\varepsilon_0$ per unit length is perpendicular to the z-plane and passes through the point z_0. The electric field intensity $\mathbf{E}(z)$ at the point z varies inversely as the distance from z_0 and is directed along the line from z_0 to z. Show that

$$\mathbf{E}(z) = \frac{1}{\bar{z} - \bar{z}_0}$$

15. Suppose that three positively charged rods carry a charge of $2\pi\varepsilon_0$ per unit length and pass through the three points 0, $1 - i$, and $1 + i$. Use the result of Exercise 14 and show that $\mathbf{E}(z) = 0$ at the points $z = (2/3) \pm i(\sqrt{2}/3)$.

16. Suppose that a positively charged rod carrying a charge of $2\pi\varepsilon_0$ passes through the point 0 and that positively charged rods carrying a charge of $4\pi\varepsilon_0$ pass through the points $2 + i$ and $-2 + i$. Use the result of Exercise 14 and show that $\mathbf{E}(z) = 0$ at the points $z = \pm\frac{4}{5} + i\frac{3}{5}$.

2.2 Transformations and Linear Mappings

We now take our first look at the geometric interpretation of a complex function. If D is the domain of definition of the real-valued functions $u(x, y)$ and $v(x, y)$, then the system of equations

$$(1) \qquad u = u(x, y) \quad \text{and} \quad v = v(x, y)$$

describes a transformation or mapping from D in the xy plane into the uv plane. Therefore the function

$$(2) \qquad w = f(z) = u(x, y) + iv(x, y)$$

can be considered as a mapping or transformation from the set D in the z plane onto the range R in the w plane. This is illustrated in Figure 2.1.

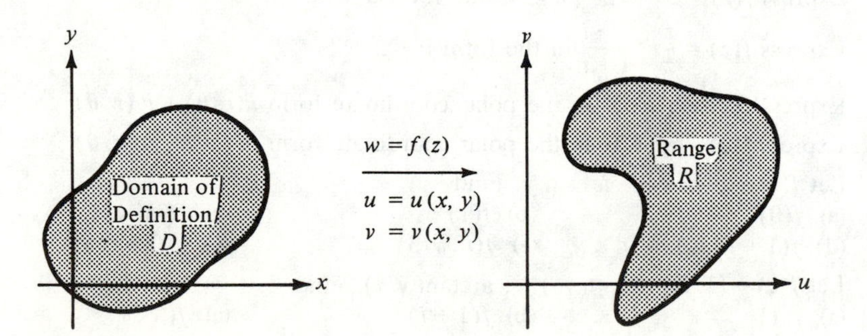

FIGURE 2.1 The mapping $w = f(z)$.

If A is a subset of the domain of definition D, then the set $B = \{w = f(z)\colon z \in A\}$ is called the *image* of the set A, and f is said to *map A onto B*. The image of a single point is a single point, and the image of the entire domain D is the range R. The mapping $w = f(z)$ is said to be *from A into S* if the image of A is contained in S. The inverse image of a point w is the set of all points z in D such that $w = f(z)$. The inverse image of a point may be one point, several points, or none at all. If the latter case occurs, then the point w is not in the range of f.

The function f is said to be *one-to-one* if it maps distinct points $z_1 \neq z_2$ onto distinct points $f(z_1) \neq f(z_2)$. If $w = f(z)$ maps the set A one-to-one and onto the set B, then for each w in B there exists exactly one point z in A such that $w = f(z)$. Then loosely speaking, we can solve the equation $w = f(z)$ by solving for z as a function of w. That is, the *inverse function* $z = g(w)$ can be found, and the following equations hold:

$$(3) \qquad \begin{aligned} g(f(z)) &= z \qquad \text{for all } z \text{ in } A \qquad \text{and} \\ f(g(w)) &= w \qquad \text{for all } w \text{ in } B. \end{aligned}$$

Conversely, if $w = f(z)$ and $z = g(w)$ are functions that map A into B and B into A, respectively, and equations (3) hold, then $w = f(z)$ maps the set A one-to-one and onto the set B. The one-to-one property is easy to show, for if we have $f(z_1) = f(z_2)$, then $g(f(z_1)) = g(f(z_2))$; and using (3), we obtain $z_1 = z_2$. To show that f is onto, we must show that each point w in B is the image of some point in A. If $w \in B$, then $z = g(w)$ lies in A and $f(g(w)) = w$, and we conclude that f is a one-to-one mapping from A onto B.

We observe that if f is a one-to-one mapping from D onto R and if A is a subset of D, then f is a one-to-one mapping from A onto its image B. One can also show that if $\xi = f(z)$ is a one-to-one mapping from A onto S, and $w = g(\xi)$ is a one-to-one mapping from S onto B, then the composition mapping $w = g(f(z))$ is a one-to-one mapping from A onto B.

It is useful to find the image B of a specified set A under a given mapping $w = f(z)$. The set A is usually described with an equation or inequality involving x and y. A chain of equivalent statements can be constructed that lead to a description of the set B in terms of an equation or an inequality involving u and v.

EXAMPLE 2.5 Show that the function $f(z) = iz$ maps the line $y = x + 1$ onto the line $v = -u - 1$.

Solution We can write f in the Cartesian form $u + iv = f(z) = i(x + iy) = -y + ix$, and see that the transformation can be given by the equations $u = -y$ and $v = x$. We can substitute these into the equation $y = x + 1$ to obtain $-u = v + 1$, which can be written as $v = -u - 1$.

We now turn our attention to the investigation of some elementary mappings. Let $B = a + ib$ denote a fixed complex number. Then the transformation

$$(4) \qquad w = T(z) = z + B = x + a + i(y + b)$$

is a one-to-one mapping of the z plane onto the w plane and is called a *translation*. This transformation can be visualized as a rigid translation whereby the point z is displaced through the vector $a + ib$ to its new position $w = T(z)$.

The inverse mapping is given by

(5) $z = T^{-1}(w) = w - B = u - a + i(v - b)$

and shows that T is a one-to-one mapping from the z plane onto the w plane. The effect of a translation is pictured in Figure 2.2.

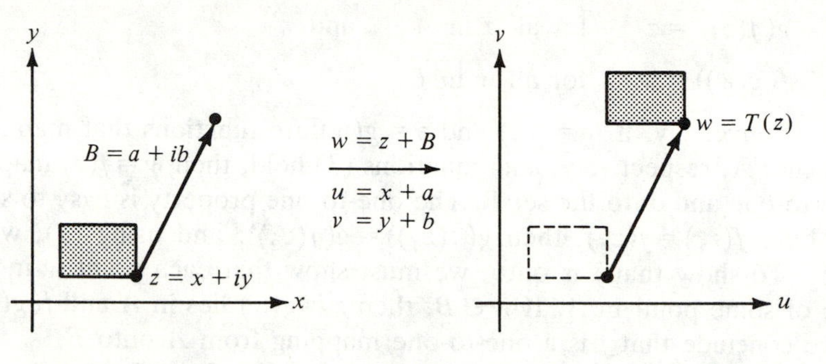

FIGURE 2.2 The translation $w = T(z) = z + B = x + a + i(y + b)$.

Let α be a fixed real number. Then the transformation

(6) $w = R(z) = ze^{i\alpha} = re^{i\theta}e^{i\alpha} = re^{i(\theta+\alpha)}$

is a one-to-one mapping of the z plane onto the w plane and is called a *rotation*. It can be visualized as a rigid rotation whereby the point z is rotated about the origin through an angle α to its new position $w = R(z)$. If we use polar coordinates $w = \rho e^{i\phi}$ in the w plane, then the inverse mapping is given by

(7) $z = R^{-1}(w) = we^{-i\alpha} = \rho e^{i\phi}e^{-i\alpha} = \rho e^{i(\phi-\alpha)}$.

This shows that R is a one-to-one mapping of the z plane onto the w plane. The effect of rotation is pictured in Figure 2.3.

Let $K > 0$ be a fixed positive real number. Then the transformation

(8) $w = S(z) = Kz = Kx + iKy$

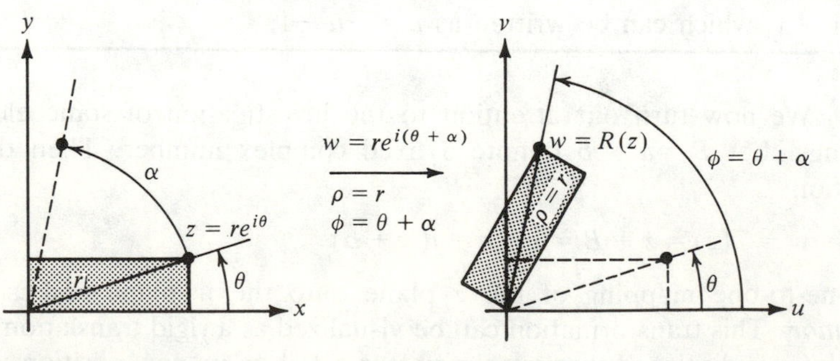

FIGURE 2.3 The rotation $w = R(z) = re^{i(\theta+\alpha)}$.

is a one-to-one mapping of the z plane onto the w plane and is called a *magnification*. If $K > 1$, it has the effect of stretching the distance between points by the factor K. If $K < 1$, then it reduces the distance between points by the factor K. The inverse transformation is given by

$$(9) \qquad z = S^{-1}(w) = \frac{1}{K} w = \frac{1}{K} u + i \frac{1}{K} v$$

and shows that S is one-to-one mapping from the z plane onto the w plane. The effect of magnification is shown in Figure 2.4.

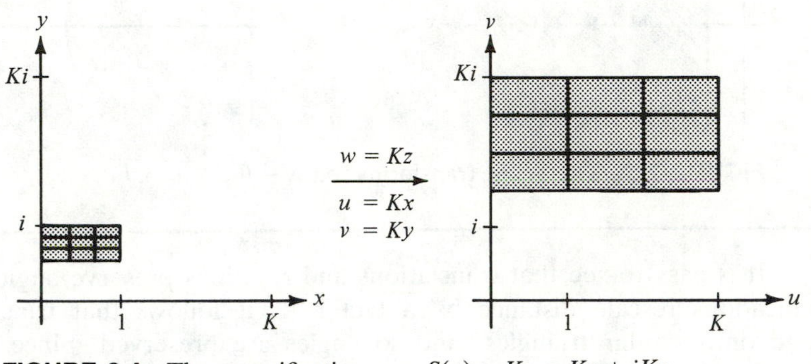

FIGURE 2.4 The magnification $w = S(z) = Kz = Kx + iKy$.

Let $A = Ke^{i\alpha}$ and $B = a + ib$ where $K > 0$ is a positive real number. Then the transformation

$$(10) \qquad w = W(z) = Az + B$$

is a one-to-one mapping of the z plane onto the w plane and is called a *linear transformation*. It can be considered as the composition of a rotation, a magnification, and a translation. It has the effect of rotating the plane through an angle $\alpha = \operatorname{Arg} A$, followed by a magnification by the factor $K = |A|$, followed by a translation by the vector $B = a + ib$. The inverse mapping is given by

$$(11) \qquad z = W^{-1}(w) = \frac{1}{A} w - \frac{B}{A}$$

and shows that W is a one-to-one mapping from the z plane onto the w plane.

EXAMPLE 2.6 Show that the linear transformation $w = iz + i$ maps the right half-plane $\operatorname{Re} z > 1$ onto the upper half-plane $\operatorname{Im} w > 2$.

Solution We can write $w = f(z)$ in Cartesian form $u + iv = i(x + iy) + i = -y + i(x + 1)$ and see that the transformation can be given by the equations $u = -y$ and $v = x + 1$. The substitution $x = v - 1$ can be used in the inequality

$\text{Re } z = x > 1$ to see that the image values must satisfy $v - 1 > 1$ or $v > 2$, which is the upper half-plane $\text{Im } w > 2$. The effect of the transformation $w = f(z)$ is a rotation of the plane through the angle $\alpha = \pi/2$ followed by a translation by the vector $B = i$ and is illustrated in Figure 2.5.

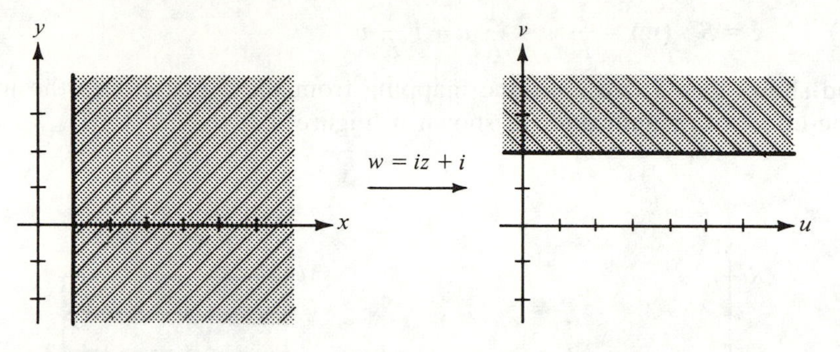

FIGURE 2.5 The linear transformation $w = f(z) = iz + i$.

It is easy to see that translations and rotations preserve angles. Since magnifications rescale distance by a factor K, it follows that triangles are mapped onto similar triangles, and so angles are preserved. Since a linear transformation can be considered as a composition of a rotation, a magnification, and a translation, it follows that linear transformations preserve angles. Consequently, any geometric object is mapped onto an object that is similar to the original object; hence linear transformations can be called *similarity mappings*.

EXAMPLE 2.7 Show that the image of the open disk $|z + 1 + i| < 1$ under the linear transformation $w = (3 - 4i)z + 6 + 2i$ is the open disk $|w + 1 - 3i| < 5$.

Solution The inverse transformation is given by

$$z = \frac{w - 6 - 2i}{3 - 4i},$$

and this substitution can be used to show that the image points must satisfy the inequality

$$\left| \frac{w - 6 - 2i}{3 - 4i} + 1 + i \right| < 1.$$

Multiplying both sides by $|3 - 4i| = 5$ results in

$$|w - 6 - 2i + (1 + i)(3 - 4i)| < 5,$$

which can be simplified to obtain the inequality

$$|w + 1 - 3i| < 5.$$

Hence the disk with center $-1-i$ and radius 1 is mapped one-to-one and onto the disk with center $-1+3i$ and radius 5 as pictured in Figure 2.6.

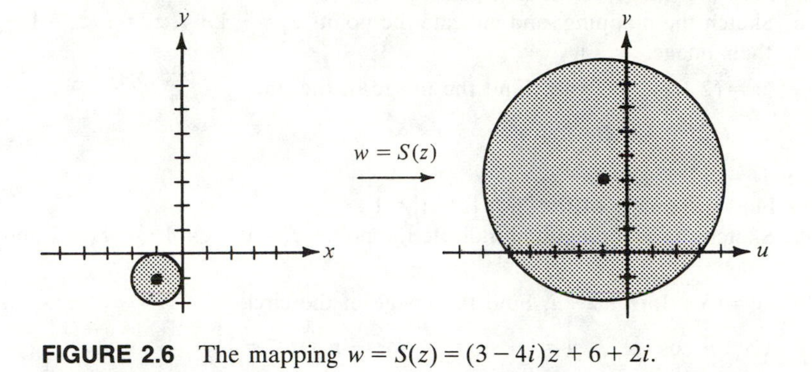

FIGURE 2.6 The mapping $w = S(z) = (3-4i)z + 6 + 2i$.

EXAMPLE 2.8 Show that the image of the right half-plane $\text{Re}\, z > 1$ under the linear transformation $w = (-1+i)z - 2 + 3i$ is the half-plane $v > u + 7$.

Solution The inverse transformation is given by

$$z = \frac{w + 2 - 3i}{-1+i} = \frac{u + 2 + i(v-3)}{-1+i},$$

which can be expressed in the component form

$$x + iy = \frac{-u + v - 5}{2} + i\,\frac{-u - v + 1}{2}.$$

The substitution $x = (-u + v - 5)/2$ can be used in the inequality $\text{Re}\, z = x > 1$ to see that the image points must satisfy $(-u + v - 5)/2 > 1$. This can be simplified to yield the inequality $v > u + 7$. The mapping is illustrated in Figure 2.7.

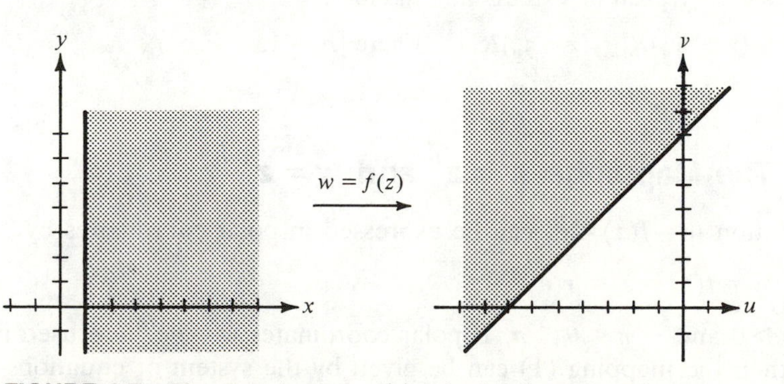

FIGURE 2.7 The mapping $w = f(z) = (-1+i)z - 2 + 3i$.

EXERCISES FOR SECTION 2.2

1. Let $w = (1 - i)z + 1 - 2i$.
(a) Find the image of the half-plane $\text{Im } z > 1$.
(b) Sketch the mapping, and indicate the points $z_1 = -1 + i$, $z_2 = i$, $z_3 = 1 + i$ and their images w_1, w_2, w_3.

2. Let $w = (2 + i)z - 3 + 4i$. Find the image of the line

$$x = t, 2y = 1 - 2t \qquad \text{for } -\infty < t < \infty.$$

3. Let $w = (3 + 4i)z - 2 + i$.
(a) Find the image of the disk $|z - 1| < 1$.
(b) Sketch the mapping, and indicate the points $z_1 = 0$, $z_2 = 1 - i$, $z_3 = 2$ and their images.

4. Let $w = (3 + 4i)z - 2 + i$. Find the image of the circle

$$x = 1 + \cos t, \qquad y = 1 + \sin t \qquad \text{for } -\pi < t \leqq \pi.$$

5. Let $w = (2 + i)z - 2i$. Find the triangle onto which the triangle with vertices $z_1 = -2 + i$, $z_2 = -2 + 2i$, and $z_3 = 2 + i$ is mapped.

6. Find the linear transformation $w = f(z)$ that maps the points $z_1 = 2$ and $z_2 = -3i$ onto the points $w_1 = 1 + i$ and $w_2 = 1$, respectively.

7. Find the linear transformation $w = S(z)$ that maps the circle $|z| = 1$ onto the circle $|w - 3 + 2i| = 5$ and satisfies the condition $S(-i) = 3 + 3i$.

8. Find the linear transformation $w = f(z)$ that maps the triangle with vertices $-4 + 2i$, $-4 + 7i$, and $1 + 2i$ onto the triangle with vertices 1, 0, and $1 + i$.

9. Let $S(z) = Kz$ where $K > 0$ is a postive real constant. Show that $|S(z_1) - S(z_2)| = K|z_1 - z_2|$, and interpret this result geometrically.

10. Give a proof that the image of a circle under a linear transformation is a circle. *Hint*: Let the given circle have the parameterization $x = x_0 + R\cos t$, $y = y_0 + R\sin t$.

11. Prove that the composition of two linear transformations is a linear transformation.

12. Show that a linear transformation that maps the circle $|z - z_0| = R_1$ onto the circle $|w - w_0| = R_2$ can be expressed in the form

$$A(w - w_0)R_1 = (z - z_0)R_2 \qquad \text{where } |A| = 1.$$

2.3 The Mappings $w = z^n$ and $w = z^{1/n}$

The function $w = f(z) = z^2$ can be expressed in polar coordinates by

$$(1) \qquad w = f(z) = z^2 = r^2 e^{i2\theta}$$

where $r > 0$ and $-\pi < \theta \leqq \pi$. If polar coordinates $w = \rho e^{i\phi}$ are used in the w plane, then the mapping (1) can be given by the system of equations

$$(2) \qquad \rho = r^2 \qquad \text{and} \qquad \phi = 2\theta,$$

which give the modulus of the image as a function of the modulus of z and the argument of the image as a function of the argment of z. If we consider the wedge-shaped set A given by $A = \{re^{i\theta}: r > 0 \text{ and } -\pi/4 < \theta < \pi 4\}$, then the image of A under the mapping f is the right half-plane $\rho > 0$, $-\pi/2 < \phi < \pi/2$. Since the argument of the product zz is twice the argument of z, we say that f doubles angles at the origin. Points that lie on the ray $r > 0$, $\theta = \alpha$ are mapped onto points that lie on the ray $\rho > 0$, $\phi = 2\alpha$.

If the domain of definition D for $f(z) = z^2$ is restricted to be the set

(3) $$D = \left\{ re^{i\theta}: r > 0 \quad \text{and} \quad \frac{-\pi}{2} < \theta < \frac{\pi}{2} \right\},$$

then the image of D under the mapping $w = z^2$ consists of all points in the w plane (except the point $w = 0$) and all the points that lie along the negative u axis. The inverse mapping of f is

(4) $$z = f^{-1}(w) = w^{1/2} = \rho^{1/2} e^{i\phi/2} \quad \text{where } \rho > 0 \text{ and } -\pi < \phi < \pi.$$

The function $f^{-1}(w) = w^{1/2}$ in (4) is called the *principal square root function* and shows that f is one-to-one when its domain is restricted by (3). The mappings $w = z^2$ and $z = w^{1/2}$ are illustrated in Figure 2.8.

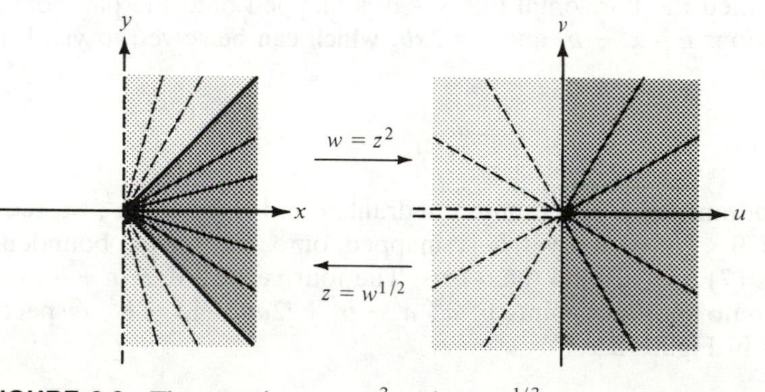

FIGURE 2.8 The mappings $w = z^2$ and $z = w^{1/2}$.

Since $f(-z) = (-z)^2 = z^2$, we see that the image of the left half-plane Re $z < 0$ under the mapping $w = z^2$ is the w plane slit along the negative u axis as indicated in Figure 2.9. We will discuss the inverse of this portion of the mapping in Section 4.1, where we will discuss branches.

Other useful properties of the mapping $w = z^2$ can be investigated if we use the Cartesian form

(5) $$w = f(z) = z^2 = x^2 - y^2 + i2xy$$

and the resulting system of equations

(6) $$u = x^2 - y^2 \quad \text{and} \quad v = 2xy.$$

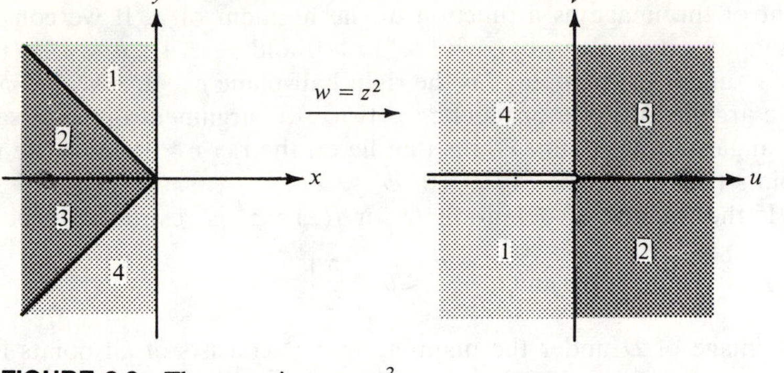

FIGURE 2.9 The mapping $w = z^2$.

If $a > 0$, then the vertical line $x = a$ is mapped onto the parabola given by the equations $u = a^2 - y^2$ and $v = 2ay$, which can be solved to yield the single equation

$$(7) \qquad u = a^2 - \frac{v^2}{4a^2}.$$

If $b > 0$, then the horizontal line $y = b$ is mapped onto the parabola given by the equations $u = x^2 - b^2$ and $v = 2xb$, which can be solved to yield the single equation

$$(8) \qquad u = -b^2 + \frac{v^2}{4b^2}.$$

Since quadrant I is mapped onto quadrants I and II by $w = z^2$, we see that the rectangle $0 < x < a$, $0 < y < b$ is mapped onto the region bounded by the parabolas (7) and (8) and the u axis. The four vertices 0, a, $a + ib$, and ib are mapped onto the four points 0, a^2, $a^2 - b^2 + i2ab$, and $-b^2$, respectively, as indicated in Figure 2.10.

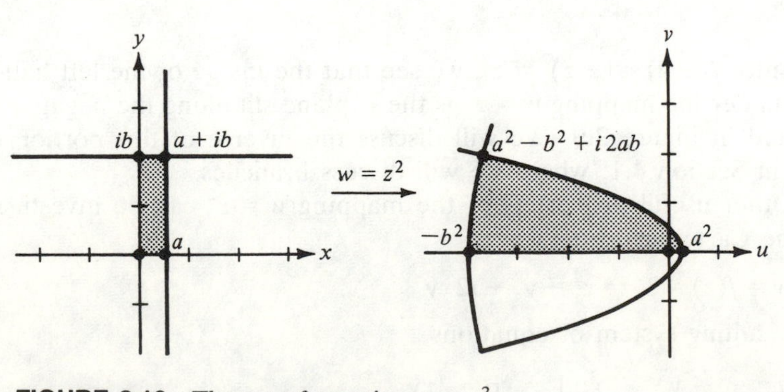

FIGURE 2.10 The transformation $w = z^2$.

The mapping $w = z^{1/2}$ can be expressed in polar form

$$(9) \qquad w = f(z) = z^{1/2} = r^{1/2}e^{i\theta/2}$$

where the domain of definition D for f is restricted to be $r > 0$, $-\pi < \theta \leqq \pi$. If polar coordinates $w = \rho e^{i\phi}$ are used in the w plane, then (9) can be represented by the system

$$(10) \qquad \rho = r^{1/2} \qquad \text{and} \qquad \phi = \frac{\theta}{2}.$$

From (10) we see that the argument of the image is half the argument of z and that the modulus of the image is the square root of the modulus of z. Points that lie on the ray $r > 0$, $\theta = \alpha$ are mapped onto the ray $\rho > 0$, $\phi = \alpha/2$. The image of the z plane (with the point $z = 0$ deleted) consists of the right half-plane $\text{Re}\, w > 0$ together with the positive v axis, and the mapping is pictured in Figure 2.11.

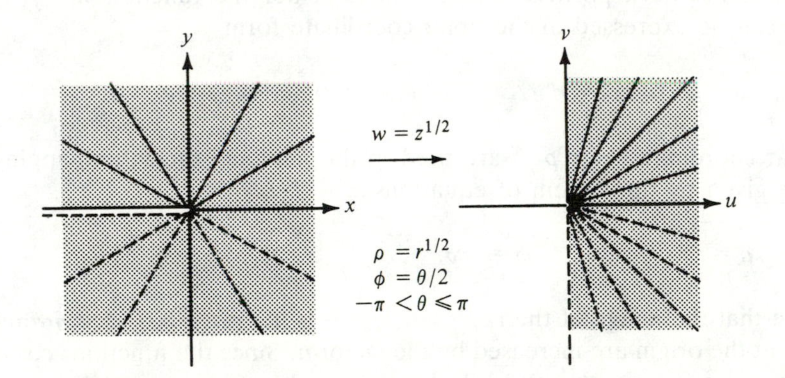

FIGURE 2.11 The mapping $w = z^{1/2}$.

The mapping $w = z^{1/2}$ can be studied through our knowledge about its inverse mapping $z = w^2$. If we use the Cartesian formula

$$(11) \qquad z = w^2 = u^2 - v^2 + i2uv,$$

then the mapping $z = w^2$ is given by the system of equations

$$(12) \qquad x = u^2 - v^2 \qquad \text{and} \qquad y = 2uv.$$

Let $a > 0$. Then the system (12) can be used to see that the right half-plane $\text{Re}\, z = x > a$ is mapped onto the region in the right half-plane satisfying $u^2 - v^2 > a$ and lies to the right of the hyperbola $u^2 - v^2 = a$. If $b > 0$, then the system (12) can be used to see that the upper half-plane $\text{Im}\, z = y > b$ is mapped onto the region in quadrant I satisfying $2uv > b$ and lies above the hyperbola $2uv = b$. The situation is illustrated in Figure 2.12.

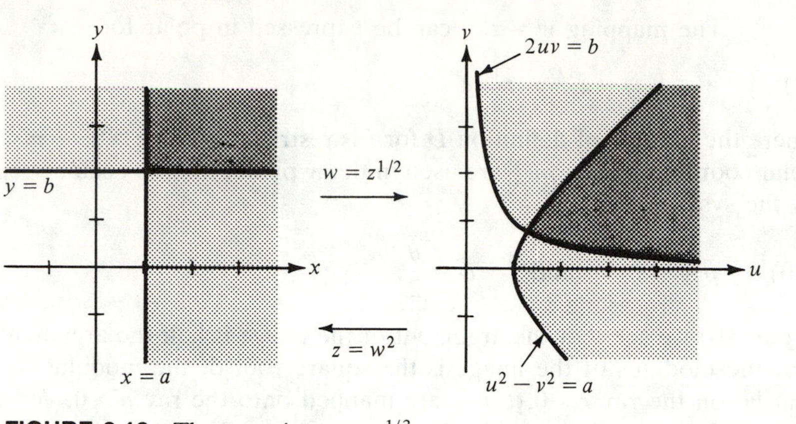

FIGURE 2.12 The mapping $w = z^{1/2}$.

Let n be a positive integer and consider the function $w = f(z) = z^n$, which can be expressed in the polar coordinate form

$$(13) \qquad w = f(z) = z^n = r^n e^{in\theta} \qquad \text{where } r > 0 \text{ and } -\pi < \theta \leqq \pi.$$

If polar coordinates $w = \rho e^{i\phi}$ are used in the w plane, then the mapping (13) can be given by the system of equations

$$(14) \qquad \rho = r^n \qquad \text{and} \qquad \phi = n\theta.$$

We see that the image of the ray $r > 0$, $\theta = \alpha$ is the ray $\rho > 0$, $\phi = n\alpha$ and that angles at the origin are increased by the factor n. Since the functions $\cos n\theta$ and $\sin n\theta$ are periodic with period $2\pi/n$, we see that f is in general an n-to-one function; that is, n points in the z plane are mapped onto each point in the w plane (except $w = 0$). If the domain of definition D of f in (13) is restricted to be

$$(15) \qquad D = \left\{ re^{i\theta} : r > 0, \frac{-\pi}{n} < \theta \leqq \frac{\pi}{n} \right\},$$

then the image of D under the mapping $w = f(z) = z^n$ consists of all points in the w plane (except the origin $w = 0$), and the inverse function is given by

$$(16) \qquad z = f^{-1}(w) = w^{1/n} = \rho^{1/n} e^{i\phi/n} \qquad \text{where } \rho > 0 \text{ and } -\pi < \phi \leqq \pi.$$

The function $f^{-1}(w) = w^{1/n}$ is called the *principal nth root function* and shows that f is one-to-one when its domain is restricted by (15). The mappings $w = z^n$ and $z = w^{1/n}$ are shown in Figure 2.13. Other branches of f will be discussed in Chapter 4.

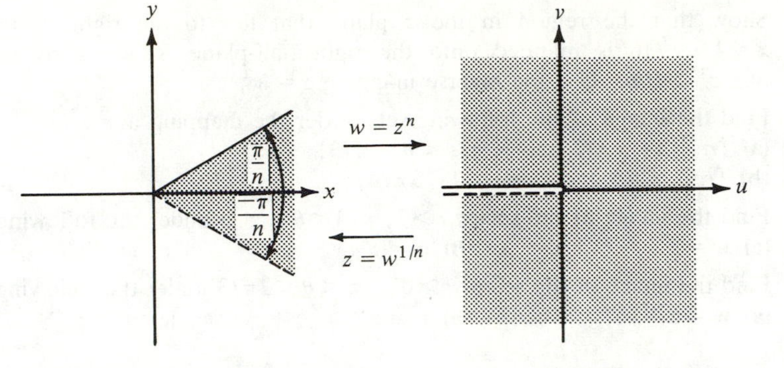

FIGURE 2.13 The mappings $w = z^n$ and $z = w^{1/n}$.

EXERCISES FOR SECTION 2.3

1. Show that the image of the horizontal line $y = 1$ under the mapping $w = z^2$ is the parabola $u = v^2/4 - 1$.

2. Show that the image of the vertical line $x = 2$ under the mapping $w = z^2$ is the parabola $u = 4 - v^2/16$.

3. Find the image of the rectangle $0 < x < 2$, $0 < y < 1$ under the mapping $w = z^2$. Sketch the mapping.

4. Find the image of the triangle with vertices 0, 2, and $2 + 2i$ under the mapping $w = z^2$. Sketch the mapping.

5. Show that the infinite strip $1 < x < 2$ is mapped onto the region that lies between the parabolas $u = 1 - v^2/4$ and $u = 4 - v^2/16$ by the mapping $w = z^2$.

6. For what values of z does $(z^2)^{1/2} = z$ hold if the principal value of the square root is to be used?

7. Sketch the set of points satisfying the following relations.
(a) $\text{Re}(z^2) > 4$ **(b)** $\text{Im}(z^2) > 6$

8. Show that the region in the right half-plane that lies to the right of the hyperbola $x^2 - y^2 = 1$ is mapped onto the right half-plane $\text{Re}\, w > 1$ by the mapping $w = z^2$.

9. Show that the image of the line $x = 4$ under the mapping $w = z^{1/2}$ is the right branch of the hyperbola $u^2 - v^2 = 4$.

10. Find the image of the following sets under the mapping $w = z^{1/2}$.
(a) $\{re^{i\theta} : r > 1 \text{ and } \pi/3 < \theta < \pi/2\}$
(b) $\{re^{i\theta} : 1 < r < 9 \text{ and } 0 < \theta < 2\pi/3\}$
(c) $\{re^{i\theta} : r < 4 \text{ and } -\pi < \theta < \pi/2\}$

11. Find the image of the right half-plane $\text{Re}\, z > 1$ under the mapping $w = z^2 + 2z + 1 = (z + 1)^2$.

12. Show that the infinite strip $2 < y < 6$ is mapped onto the region in the first quadrant that lies between the hyperbolas $uv = 1$ and $uv = 3$ by the mapping $w = z^{1/2}$.

13. Find the image of the region in the first quadrant that lies between the hyperbolas $xy = \frac{1}{2}$ and $xy = 4$ under the mapping $w = z^2$.

14. Show that the region in the z plane that lies to the right of the parabola $x = 4 - y^2/16$ is mapped onto the right half-plane $\text{Re } w > 2$ by the mapping $w = z^{1/2}$. *Hint*: Use the inverse mapping $z = w^2$.

15. Find the image of the following sets under the mapping $w = z^3$.
    (a) $\{re^{i\theta} : 1 < r < 2 \text{ and } -\pi/4 < \theta < \pi/3\}$
    (b) $\{re^{i\theta} : r > 3 \text{ and } 2\pi/3 < \theta < 3\pi/4\}$

16. Find the image of the sector $r > 2$, $\pi/4 < \theta < \pi/3$ under the following mappings.
    (a) $w = z^3$ **(b)** $w = z^4$ **(c)** $w = z^6$

17. Find the image of the sector $r > 0$, $-\pi < \theta < 2\pi/3$ under the following mappings.
    (a) $w = z^{1/2}$ **(b)** $w = z^{1/3}$ **(c)** $w = z^{1/4}$

2.4 Limits and Continuity

Let $u = u(x, y)$ be a real-valued function of the two real variables x and y. We say that u has the *limit* u_0 as (x, y) approaches (x_0, y_0) provided that the value of $u(x, y)$ gets close to the value u_0 as (x, y) gets close to (x_0, y_0). We write

$$(1) \qquad \lim_{(x,y)\to(x_0,y_0)} u(x, y) = u_0.$$

That is, u has the limit u_0 as (x, y) approaches (x_0, y_0) if and only if $|u(x, y) - u_0|$ can be made arbitrarily small by making both $|x - x_0|$ and $|y - y_0|$ small. This is like the definition of limit for functions of one variable, except that there are two variables instead of one. Since (x, y) is a point in the xy plane, and the distance between (x, y) and (x_0, y_0) is $\sqrt{(x - x_0)^2 + (y - y_0)^2}$, we can give a precise definition of limit as follows. To each number $\varepsilon > 0$, there corresponds a number $\delta > 0$ such that

$$(2) \qquad |u(x, y) - u_0| < \varepsilon \qquad \text{whenever } 0 < \sqrt{(x - x_0)^2 + (y - y_0)^2} < \delta.$$

EXAMPLE 2.9 If $u(x, y) = x^3/(x^2 + y^2)$, then

$$(3) \qquad \lim_{(x,y)\to(0,0)} u(x, y) = 0.$$

Solution If $x = r\cos\theta$ and $y = r\sin\theta$, then

$$u(x, y) = \frac{r^3\cos^3\theta}{r^2\cos^2\theta + r^2\sin^2\theta} = r\cos^3\theta.$$

Since $\sqrt{(x - 0)^2 + (y - 0)^2} = r$, we see that

$$|u(x, y) - 0| = r|\cos^3\theta| < \varepsilon \qquad \text{whenever } 0 < \sqrt{x^2 + y^2} = r < \varepsilon.$$

Hence for any $\varepsilon > 0$, inequality (2) is satisfied for $\delta = \varepsilon$; that is, $u(x, y)$ has the limit $u_0 = 0$ as (x, y) approaches $(0, 0)$.

The value u_0 of the limit must *not* depend on how (x, y) approaches (x_0, y_0). So it follows that $u(x, y)$ must approach the value u_0 when (x, y) approaches (x_0, y_0) along any curve that ends at the point (x_0, y_0). Conversely, if we can find two curves C_1 and C_2 that end at (x_0, y_0) along which $u(x, y)$ approaches the two distinct values u_1 and u_2, respectively, then $u(x, y)$ *does not* have a limit as (x, y) approaches (x_0, y_0).

EXAMPLE 2.10 The function $u(x, y) = xy/(x^2 + y^2)$ *does not* have a limit as (x, y) approaches $(0, 0)$. If we let (x, y) approach $(0, 0)$ along the x axis, then

$$\lim_{(x,0)\to(0,0)} u(x, 0) = \lim_{(x,0)\to(0,0)} \frac{(x)(0)}{x^2 + 0^2} = 0.$$

But if we let (x, y) approach $(0, 0)$ along the line $y = x$, then

$$\lim_{(x,x)\to(0,0)} u(x, x) = \lim_{(x,x)\to(0,0)} \frac{(x)(x)}{x^2 + x^2} = \frac{1}{2}.$$

Since the two values are different, the value of the limit is dependent on how (x, y) approaches $(0, 0)$. We conclude that $u(x, y)$ does not have a limit as (x, y) approaches $(0, 0)$.

Let $f(z)$ be a complex function of the complex variable z that is defined for all values of z in some neighborhood of z_0, except perhaps at the point z_0. We say that f has the *limit* w_0 as z approaches z_0, provided that the value $f(z)$ gets close to the value w_0 as z gets close to z_0; and we write

$$\lim_{z\to z_0} f(z) = w_0. \tag{4}$$

Since the distance between the points z and z_0 can be expressed by $|z - z_0|$, we can give a precise definition of the limit (4). For each positive number $\varepsilon > 0$, there exists a $\delta > 0$ such that

$$|f(z) - w_0| < \varepsilon \quad \text{whenever } 0 < |z - z_0| < \delta. \tag{5}$$

Geometrically, this says that for each ε-neighborhood $|w - w_0| < \varepsilon$ of the point w_0 there is a deleted δ-neighborhood $0 < |z - z_0| < \delta$ of z_0 such that the image of each point in the δ-neighborhood, except perhaps z_0, lies in the ε-neighborhood of w_0. The image of the δ-neighborhood does not have to fill up the entire ε-neighborhood; but if z approaches z_0 along a curve that ends at z_0, then $w = f(z)$ approaches w_0. The situation is illustrated in Figure 2.14.

If we consider $w = f(z)$ as a mapping from the z plane into the w plane and think about the previous geometric interpretation of a limit, then we are led to conclude that the limit of a function f should be determined by the limits of its real and imaginary parts u and v. This will also give us a tool for computing limits.

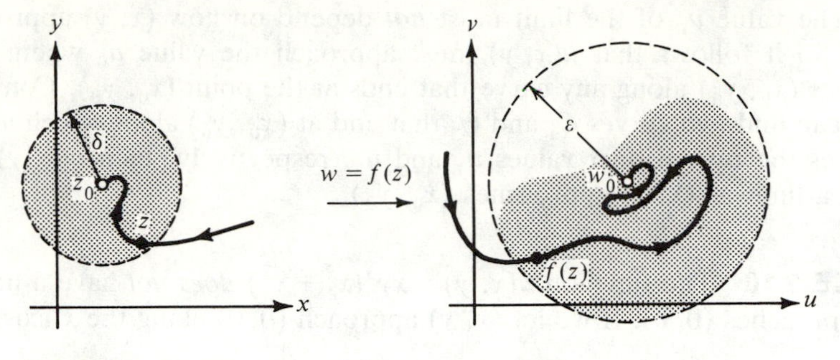

FIGURE 2.14 The limit $f(z) \to w_0$ as $z \to z_0$.

Theorem 2.1 *Let $f(z) = u(x, y) + iv(x, y)$ be a complex function that is defined in some neighborhood of z_0, except perhaps at the point $z_0 = x_0 + iy_0$. Then*

$$\lim_{z \to z_0} f(z) = w_0 = u_0 + iv_0 \tag{6}$$

if and only if

$$\lim_{(x,y)\to(x_0,y_0)} u(x, y) = u_0 \quad \textit{and} \quad \lim_{(x,y)\to(x_0,y_0)} v(x, y) = v_0. \tag{7}$$

Proof Let us first assume that statement (6) is true, and show that (7) is true. According to the definition of limit, for each $\varepsilon > 0$, there corresponds a $\delta > 0$ such that

$$|f(z) - w_0| < \varepsilon \qquad \text{whenever } 0 < |z - z_0| < \delta.$$

Since $f(z) - w_0 = u(x, y) - u_0 + i(v(x, y) - v_0)$, we can use (2) of Sec. 1.3 to conclude that

$$|u(x, y) - u_0| \leqq |f(z) - w_0| \qquad \text{and} \qquad |v(x, y) - v_0| \leqq |f(z) - w_0|.$$

It now follows that $|u(x, y) - u_0| < \varepsilon$ and $|v(x, y) - v_0| < \varepsilon$ whenever $0 < |z - z_0| < \delta$ so that (7) is true.

Conversely, now let us assume that (7) is true. Then for each $\varepsilon > 0$, there exists $\delta_1 > 0$ and $\delta_2 > 0$ so that

$$|u(x, y) - u_0| < \frac{\varepsilon}{2} \qquad \text{whenever } 0 < |z - z_0| < \delta_1 \qquad \text{and}$$

$$|v(x, y) - v_0| < \frac{\varepsilon}{2} \qquad \text{whenever } 0 < |z - z_0| < \delta_2,$$

Let δ be chosen to be the minimum of the two values δ_1 and δ_2. Then we can use the triangle inequality

$$|f(z) - w_0| \leqq |u(x, y) - u_0| + |v(x, y) - v_0|$$

to conclude that

$$|f(z) - w_0| < \frac{\varepsilon}{2} + \frac{\varepsilon}{2} = \varepsilon \qquad \text{whenever } 0 < |z - z_0| < \delta.$$

Hence the truth of (7) implies the truth of (6), and the proof is complete.

For example, $\lim_{z \to 1+i}(z^2 - 2z + 1) = -1$. To show this result, we let

$$f(z) = z^2 - 2z + 1 = x^2 - y^2 - 2x + 1 + i(2xy - 2y).$$

Computing the limits for u and v, we obtain

$$\lim_{(x,y)\to(1,1)} u(x, y) = 1 - 1 - 2 + 1 = -1 \qquad \text{and}$$

$$\lim_{(x,y)\to(1,1)} v(x, y) = 2 - 2 = 0.$$

So Theorem 2.1 implies that $\lim_{z \to 1+i} f(z) = -1$.

Limits of complex functions are formally the same as in the case of real functions, and the sum, difference, product, and quotient of functions have limits given by the sum, difference, product, and quotient of the respective limits. We state this result as a theorem and leave the proof as an exercise.

Theorem 2.2 *Let* $\lim_{z\to z_0} f(z) = A$ *and* $\lim_{z \to z_0} g(z) = B$. *Then*

(8) $$\lim_{z\to z_0} (f(z) \pm g(z)) = A \pm B.$$

(9) $$\lim_{z\to z_0} f(z)g(z) = AB.$$

(10) $$\lim_{z\to z_0} \frac{f(z)}{g(z)} = \frac{A}{B} \qquad \textit{where } B \neq 0.$$

Let $u(x, y)$ be a real-valued function of the two real variables x and y. We say that u is *continuous* at the point (x_0, y_0) if the following three conditions are satisfied:

(11) $$\lim_{(x,y)\to(x_0,y_0)} u(x, y) \text{ exists.}$$

(12) $$u(x_0, y_0) \text{ exists.}$$

(13) $$\lim_{(x,y)\to(x_0,y_0)} u(x, y) = u(x_0, y_0).$$

Condition (13) actually contains conditions (11) and (12), since the existence of the quantity on each side of the equation there is implicitly understood to exist. For example, if $u(x, y) = x^3/(x^2 + y^2)$ when $(x, y) \neq (0, 0)$ and if $u(0, 0) = 0$, then we have already seen that $u(x, y) \to 0$ as $(x, y) \to (0, 0)$ so that (11), (12), and (13) are satisfied. Hence $u(x, y)$ is continuous at $(0, 0)$.

Let $f(z)$ be a complex function of the complex variable z that is defined for all values of z in some neighborhood of z_0. We say that f is *continuous* at z_0 if the following three conditions are satisfied:

(14) $\lim_{z \to z_0} f(z)$ exists.

(15) $f(z_0)$ exists.

(16) $\lim_{z \to z_0} f(z) = f(z_0)$.

A complex function f is continuous if and only if its real and imaginary parts u and v are continuous, and the proof of this fact is an immediate consequence of Theorem 2.1. Continuity of complex functions is formally the same as in the case of real functions, and the sum, difference, and product of continuous functions are continuous; their quotient is continuous at points where the denominator is not zero. These results are summarized by the following theorems, and the proofs are left as exercises.

Theorem 2.3 *Let $f(z) = u(x, y) + iv(x, y)$ be defined in some neighborhood of z_0. Then f is continuous at $z_0 = x_0 + iy_0$ if and only if u and v are continuous at (x_0, y_0).*

Theorem 2.4 *Suppose that f and g are continuous at the point z_0. Then the following functions are continuous at z_0:*

(17) *Their sum $f(z) + g(z)$.*

(18) *Their difference $f(z) - g(z)$.*

(19) *Their product $f(z)g(z)$.*

(20) *Their quotient $\dfrac{f(z)}{g(z)}$ provided that $g(z_0) \neq 0$.*

(21) *Their composition $f(g(z))$ provided that $f(z)$ is continuous in a neighborhood of the point $g(z_0)$.*

Let us use Theorem 2.2 to show that the polynomial P of degree n given by $P(z) = a_0 + a_1 z + a_2 z^2 + \cdots + a_n z^n$ is continuous at each point z_0. We observe that if a_0 is the constant function, then $\lim_{z \to z_0} a_0 = a_0$; and if $a_1 \neq 0$, then we can use definition (5) with $f(z) = a_1 z$ and the choice $\delta = \varepsilon / |a_1|$ to prove that $\lim_{z \to z_0} a_1 z = a_1 z_0$. Then using property (9) and mathematical induction, we obtain

(22) $\lim_{z \to z_0} a_k z^k = a_k z_0^k \quad$ for $k = 0, 1, 2, \ldots, n$.

Property (8) can be extended to a finite sum of terms, and we can use the result of (22) to obtain

$$\lim_{z \to z_0} P(z) = \lim_{z \to z_0} \left(\sum_{k=0}^{n} a_k z^k \right) = \sum_{k=0}^{n} a_k z_0^k = P(z_0). \tag{23}$$

Since conditions (14), (15), and (16) are satisfied, we can conclude that P is continuous at z_0.

One technique for computing limits is the use of (20). Let P and Q be polynomials. If $Q(z_0) \neq 0$, then

$$\lim_{z \to z_0} \frac{P(z)}{Q(z)} = \frac{P(z_0)}{Q(z_0)}.$$

Another technique involves factoring polynomials. If both $P(z_0) = 0$ and $Q(z_0) = 0$, then P and Q can be factored $P(z) = (z - z_0)P_1(z)$ and $Q(z) = (z - z_0)Q_1(z)$. If $Q_1(z_0) \neq 0$, then the limit is given by

$$\lim_{z \to z_0} \frac{P(z)}{Q(z)} = \lim_{z \to z_0} \frac{(z - z_0)P_1(z)}{(z - z_0)Q_1(z)} = \frac{P_1(z_0)}{Q_1(z_0)}.$$

For example, let us show that

$$\lim_{z \to 1+i} \frac{z^2 - 2i}{z^2 - 2z + 2} = 1 - i.$$

Here P and Q can be factored in the form

$$P(z) = (z - 1 - i)(z + 1 + i) \qquad \text{and} \qquad Q(z) = (z - 1 - i)(z - 1 + i)$$

so that the limit is obtained by the calculation

$$\lim_{z \to 1+i} \frac{z^2 - 2i}{z^2 - 2z + 2} = \lim_{z \to 1+i} \frac{(z - 1 - i)(z + 1 + i)}{(z - 1 - i)(z - 1 + i)}$$
$$= \lim_{z \to 1+i} \frac{z + 1 + i}{z - 1 + i} = 1 - i.$$

The question of factoring a polynomial will be discussed in Section 5.6, and the question about finding limits of quotients will be discussed in more detail in Section 3.1.

EXERCISES FOR SECTION 2.4

1. Find $\lim_{z \to 2+i} (z^2 - 4z + 2 + 5i)$.

2. Find $\lim_{z \to i} \frac{z^2 + 4z + 2}{z + 1}$.

3. Find $\lim_{z \to i} \frac{z^4 - 1}{z - i}$.

4. Find $\lim_{z \to 1+i} \frac{z^2 + z - 2 + i}{z^2 - 2z + 1}$.

5. Find $\lim_{z \to 1+i} \frac{z^2 + z - 1 - 3i}{z^2 - 2z + 2}$ by factoring.

6. Show that $\lim_{z \to 0} \frac{x^2}{z} = 0$.

7. State why $\lim_{z \to z_0} (e^x \cos y + ix^2 y) = e^{x_0} \cos y_0 + ix_0^2 y_0$.

8. State why $\lim_{z \to z_0} (\ln(x^2 + y^2) + iy) = \ln(x_0^2 + y_0^2) + iy_0$ provided that $|z_0| \neq 0$.

9. Show that $\lim_{z \to 0} \dfrac{|z|^2}{z} = 0$.

10. Let $f(z) = \dfrac{z^2}{|z|^2} = \dfrac{x^2 - y^2 + i2xy}{x^2 + y^2}$.

(a) Find $\lim_{z \to 0} f(z)$ as $z \to 0$ along the line $y = x$.
(b) Find $\lim_{z \to 0} f(z)$ as $z \to 0$ along the line $y = 2x$.
(c) Find $\lim_{z \to 0} f(z)$ as $z \to 0$ along the parabola $y = x^2$.
(d) What can you conclude about the limit of $f(z)$ as $z \to 0$?

11. Let $f(z) = \bar{z}/z$. Show that $f(z)$ does not have a limit as $z \to 0$.

12. Does $u(x, y) = (x^3 - 3xy^2)/(x^2 + y^2)$ have a limit as $(x, y) \to (0, 0)$?

13. Let $f(z) = z^{1/2} = r^{1/2}(\cos(\theta/2) + i\sin(\theta/2))$ where $r > 0$ and $-\pi < \theta \leqq \pi$. Use the polar form of z and show that

(a) $f(z) \to i$ as $z \to -1$ along the upper semicircle $r = 1$, $0 < \theta < \pi$.
(b) $f(z) \to -i$ as $z \to -1$ along the lower semicircle $r = 1$, $-\pi < \theta < 0$.

14. Does $\lim_{z \to -4} \operatorname{Arg} z$ exist? Why?

15. Determine where the following functions are continuous.

(a) $z^4 - 9z^2 + iz - 2$ **(b)** $\dfrac{z + 1}{z^2 + 1}$ **(c)** $\dfrac{z^2 + 6z + 5}{z^2 + 3z + 2}$

(d) $\dfrac{z^4 + 1}{z^2 + 2z + 2}$ **(e)** $\dfrac{x + iy}{x - 1}$ **(f)** $\dfrac{x + iy}{|z| - 1}$

16. Let $f(z) = (z \operatorname{Re} z)/|z|$ when $z \neq 0$, and let $f(0) = 0$. Show that $f(z)$ is continuous for all values of z.

17. Let $f(z) = xe^y + iy^2e^{-x}$. Show that $f(z)$ is continuous for all values of z.

18. Let $f(z) = (x^2 + iy^2)/|z|^2$ when $z \neq 0$, and let $f(0) = 1$. Show that $f(z)$ is not continuous at $z_0 = 0$.

19. Let $f(z) = \operatorname{Re} z/|z|$ when $z \neq 0$, and let $f(0) = 1$. Is $f(z)$ continuous at the origin?

20. Let $f(z) = (\operatorname{Re} z)^2/|z|$ when $z \neq 0$, and let $f(0) = 1$. Is $f(z)$ continuous at the origin?

21. Let $f(z) = z^{1/2} = r^{1/2}(\cos(\theta/2) + i\sin(\theta/2))$ where $r > 0$ and $-\pi < \theta \leqq \pi$. Show that $f(z)$ is discontinuous at each point along the negative x axis.

22. Let $f(z) = \ln|z| + i \operatorname{Arg} z$ where $-\pi < \operatorname{Arg} z \leqq \pi$. Show that $f(z)$ is discontinuous at $z_0 = 0$ and at each point along the negative x axis.

23. Let A and B be complex constants. Use Theorem 2.1 to prove that $\lim_{z \to z_0} (Az + B) = Az_0 + B$.

24. Let $\Delta z = z - z_0$. Show that $\lim_{z \to z_0} f(z) = w_0$ if and only if $\lim_{\Delta z \to 0} f(z_0 + \Delta z) = w_0$.

25. Let $|g(z)| \leqq M$ and $\lim_{z \to z_0} f(z) = 0$. Show that $\lim_{z \to z_0} f(z)g(z) = 0$.

26. Establish identity (8). **27.** Establish identity (9). **28.** Establish identity (10).

29. Let $f(z)$ be continuous for all values of z.
(a) Show that $g(z) = f(\overline{z})$ is continuous for all z.
(b) Show that $h(z) = \overline{f(z)}$ is continuous for all z.

30. Establish the results of (17) and (18).

31. Establish the result (19).

32. Establish the result (20).

33. Establish the result (21).

2.5 The Reciprocal Transformation $w = 1/z$ (Prerequisite for Section 8.2)

The mapping $w = 1/z$ is called the *reciprocal transformation* and maps the z plane one-to-one and onto the w plane except for the point $z = 0$, which has no image, and the point $w = 0$, which has no preimage or inverse image. Since $z\overline{z} = |z|^2$, we can express the reciprocal transformation as a composition:

$$(1) \qquad w = \overline{Z} \qquad \text{and} \qquad Z = \frac{z}{|z|^2}.$$

The transformation $Z = z/|z|^2$ on the right side of (1) is called the *inversion mapping* with respect to the unit circle $|z| = 1$. It has the property that a nonzero point z is mapped onto the point Z such that

$$(2) \qquad |Z||z| = 1 \qquad \text{and} \qquad \arg Z = \arg z.$$

Hence it maps points inside the circle $|z| = 1$ onto points outside the circle $|Z| = 1$, and conversely. Any point of unit modulus is mapped onto itself. The inversion mapping is illustrated in Figure 2.15.

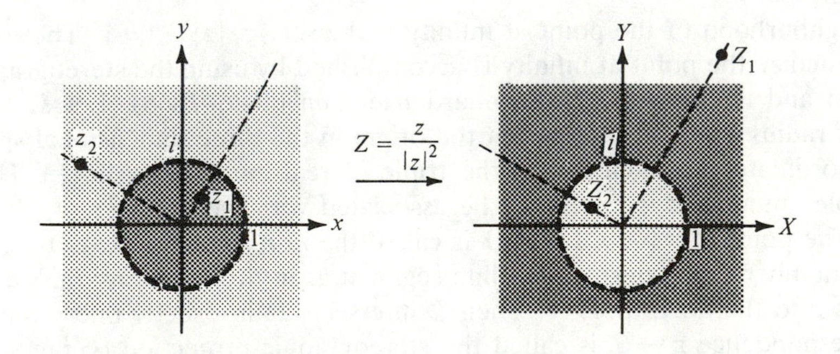

FIGURE 2.15 The inversion mapping.

The geometric description of the reciprocal transformation is now evident from the composition given in (1). It is an inversion followed by a reflection through the x axis. If we use the polar coordinate form

(3) $$w = \rho e^{i\phi} = \frac{1}{r} e^{-i\theta} \qquad \text{where } z = re^{i\theta},$$

then we see that the ray $r > 0$, $\theta = \alpha$ is mapped one-to-one and onto the ray $\rho > 0$, $\phi = -\alpha$. Also, points that lie inside the circle $|z| = 1$ are mapped onto points that lie outside the circle $|w| = 1$, and vice versa. The situation is illustrated in Figure 2.16.

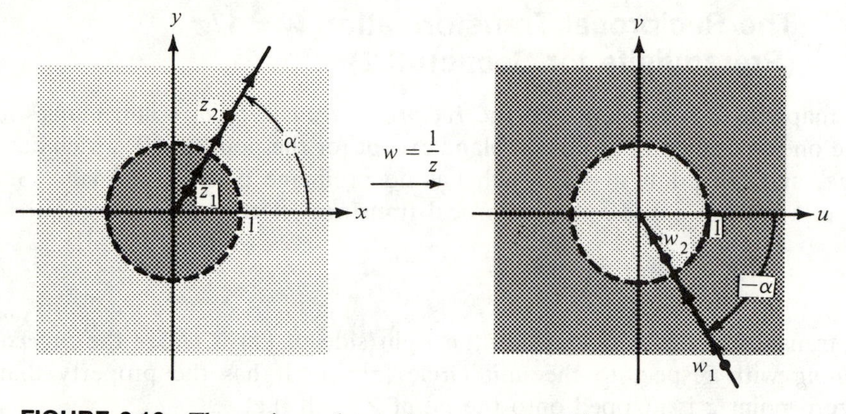

FIGURE 2.16 The reciprocal transformation $w = 1/z$.

It is convenient to extend the system of complex numbers by joining to it an "ideal" point denoted by ∞ and called the *point at infinity*. This new set is called the extended complex plane. The point ∞ has the property that

(4) $$\lim_{n\to\infty} z_n = \infty \qquad \text{if and only if } \lim_{n\to\infty} |z_n| = \infty.$$

An ε-neighborhood of the point at infinity is the set $\{z: |z| > 1/\varepsilon\}$. The usual way to visualize the point at infinity is accomplished by using the stereographic projection and is attributed to Bernhard Riemann (1826–1866). Let $\mathbf{\Omega}$ be a sphere of radius 1 that is centered at the origin in the three-dimensional space where coordinates are denoted by the triple of real numbers (x, y, ξ). Here the complex number $z = x + iy$ will be associated with the point $(x, y, 0)$.

The point $\mathcal{N} = (0, 0, 1)$ on $\mathbf{\Omega}$ is called the *north pole* of $\mathbf{\Omega}$. Let z be a complex number, and consider the line segment L in three-dimensional space that joins z to the north pole $\mathcal{N}$. Then L intersects $\mathbf{\Omega}$ in exactly one point $\mathscr{Z}$. The correspondence $z \leftrightarrow \mathscr{Z}$ is called the stereographic projection of the complex z plane onto the *Riemann sphere* $\mathbf{\Omega}$. A point $z = x + iy$ of unit modulus will correspond to $\mathscr{Z} = (x, y, 0)$. If z has modulus greater than 1, then $\mathscr{Z}$ will lie in the upper hemisphere where $\xi > 0$. If z has modulus less than 1, then $\mathscr{Z}$ will lie in the lower hemisphere where $\xi < 0$. The complex number $z = 0$ corresponds to the *south pole* $\mathcal{S} = (0, 0, -1)$. It is easy to visualize that $z \to \infty$ if and only if $\mathscr{Z} \to \mathcal{N}$. Hence $\mathcal{N}$ corresponds to the "ideal" point at infinity. The situation is shown in Figure 2.17.

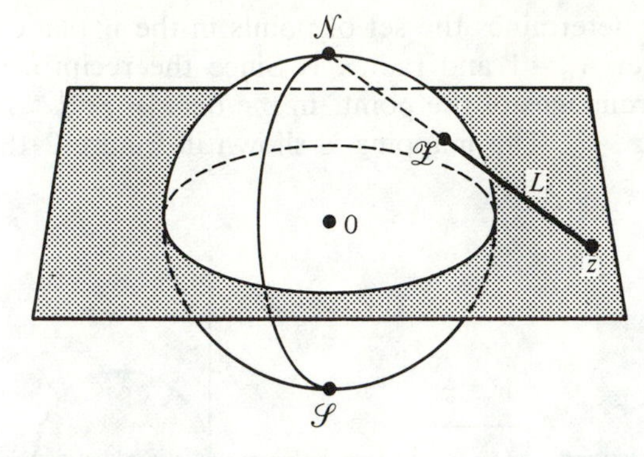

FIGURE 2.17 The Riemann sphere.

Let us reconsider the mapping $w = 1/z$. Let us assign the images $w = \infty$ and $w = 0$ to the points $z = 0$ and $z = \infty$, respectively. The reciprocal transformation can now be written as

$$(5) \qquad w = f(z) = \begin{cases} 1/z & \text{when } z \neq 0,\ z \neq \infty, \\ 0 & \text{when } z = \infty, \\ \infty & \text{when } z = 0. \end{cases}$$

It is easy to see that the transformation $w = f(z)$ is a one-to-one mapping of the extended complex z plane onto the extended complex w plane. Using property (4) of the point at infinity, it is easy to show that f is a continuous mapping from the extended z plane onto the extended w plane. The details are left for the reader.

EXAMPLE 2.11 Show that the image of the right half-plane $\operatorname{Re} z > \frac{1}{2}$, under the mapping $w = 1/z$, is the disk $|w - 1| < 1$.

Solution The inverse mapping $z = 1/w$ can be written as

$$(6) \qquad x + iy = z = \frac{1}{w} = \frac{u - iv}{u^2 + v^2}.$$

Equating the real and imaginary parts in equation (6), we obtain the equations

$$(7) \qquad x = \frac{u}{u^2 + v^2} \qquad \text{and} \qquad y = \frac{-v}{u^2 + v^2}.$$

The requirement that $x > \frac{1}{2}$ forces the image values to satisfy the inequality

$$(8) \qquad \frac{u}{u^2 + v^2} > \frac{1}{2}.$$

It is easy to manipulate inequality (8) to obtain

$$(9) \qquad u^2 - 2u + 1 + v^2 < 1,$$

which is an inequality that determines the set of points in the w plane that lie inside the circle with center $w_0 = 1$ and radius 1. Since the reciprocal transformation is one-to-one, preimages of the points in the disk $|w-1|<1$ will lie in the right half-plane $\text{Re}\, z > \frac{1}{2}$. The mapping is shown in Figure 2.18.

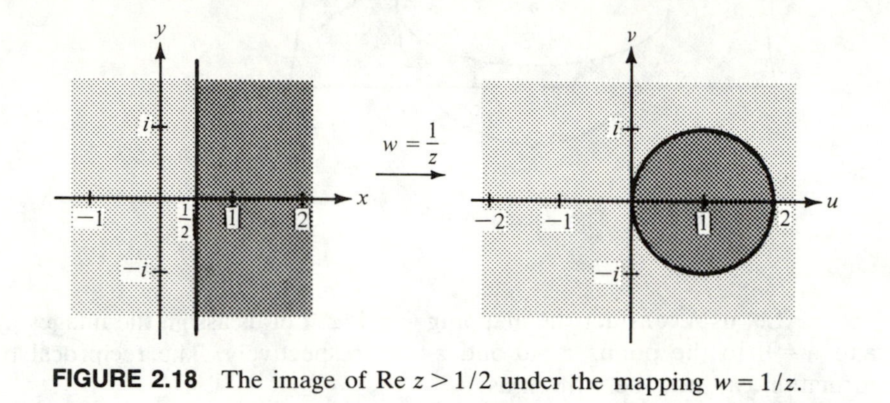

FIGURE 2.18 The image of $\text{Re}\, z > 1/2$ under the mapping $w = 1/z$.

EXAMPLE 2.12 Find the image of the portion of the right half-plane $\text{Re}\, z > \frac{1}{2}$ that lies inside the circle $|z - \frac{1}{2}| < 1$ under the transformation $w = 1/z$.

Solution Using the result of Example 2.11, we need only find the image of the disk $|z-\frac{1}{2}|<1$ and intersect it with the disk $|w-1|<1$. To start with, we can express the disk $|z-\frac{1}{2}|<1$ by the inequality

$$(10) \qquad x^2 + y^2 - x < \tfrac{3}{4}.$$

We can use the identities in (7) to show that the image values of points satisfying inequality (10) must satisfy the inequality

$$(11) \qquad \frac{1}{u^2+v^2} - \frac{u}{u^2+v^2} < \frac{3}{4}.$$

Inequality (11) can now be manipulated to yield

$$(\tfrac{4}{3})^2 < (u + \tfrac{2}{3})^2 + v^2,$$

which is an inequality that determines the set of points in the w plane that lie exterior to the circle $|w + \frac{2}{3}| = \frac{4}{3}$. Therefore the image is the crescent-shaped region illustrated in Figure 2.19.

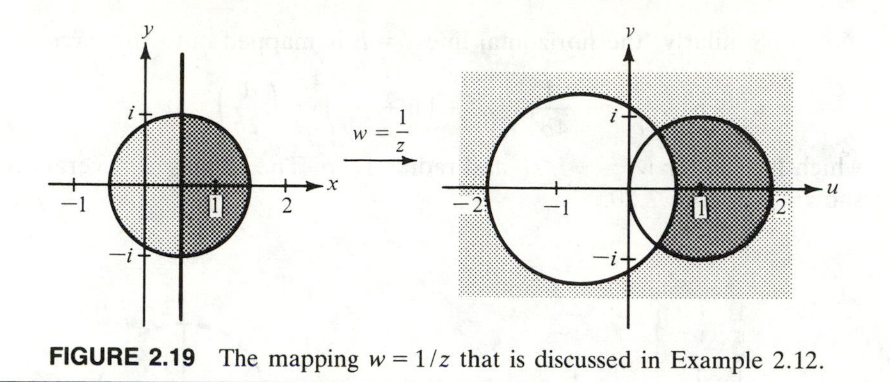

FIGURE 2.19 The mapping $w = 1/z$ that is discussed in Example 2.12.

Let us consider the equation

$$A(x^2 + y^2) + Bx + Cy + D = 0 \tag{12}$$

where A, B, C, and D are real numbers. Then equation (12) represents either a circle or a line, depending on whether $A \neq 0$ or $A = 0$, respectively. If we use polar coordinates, then equation (12) has the form

$$Ar^2 + r(B \cos \theta + C \sin \theta) + D = 0. \tag{13}$$

Using the polar coordinate form of the reciprocal transformation given in (3), we find that the image of the curve in (13) can be expressed by the equation

$$A + \rho(B \cos \phi - C \sin \phi) + D\rho^2 = 0, \tag{14}$$

which represents either a circle or a line, depending on whether $D \neq 0$ or $D = 0$, respectively. Therefore we have shown that the reciprocal transformation $w = 1/z$ carries the class of lines and circles onto itself.

EXAMPLE 2.13 Find the images of the vertical lines $x = a$ and the horizontal lines $y = b$ under the mapping $w = 1/z$.

Solution The image of the line $x = 0$ is the line $u = 0$; that is, the y axis is mapped onto the v axis. Similarly, the x axis is mapped onto the u axis.

If $a \neq 0$, then using (7), we see that the vertical line $x = a$ is mapped onto the circle

$$\frac{u}{u^2 + v^2} = a. \tag{15}$$

It is easy to manipulate the equation in (15) to obtain

$$u^2 - \frac{1}{a}u + \frac{1}{4a^2} + v^2 = \left(u - \frac{1}{2a}\right)^2 + v^2 = \left(\frac{1}{2a}\right)^2,$$

which is the equation of a circle in the w plane with center $w_0 = 1/2a$ and radius $1/2a$.

Similarly, the horizontal line $y = b$ is mapped onto the circle

$$u^2 + v^2 + \frac{1}{b}v + \frac{1}{4b^2} = u^2 + \left(v + \frac{1}{2b}\right)^2 = \left(\frac{1}{2b}\right)^2,$$

which has center $w_0 = -i/2b$ and radius $1/2b$. The images of several lines are shown in Figure 2.20.

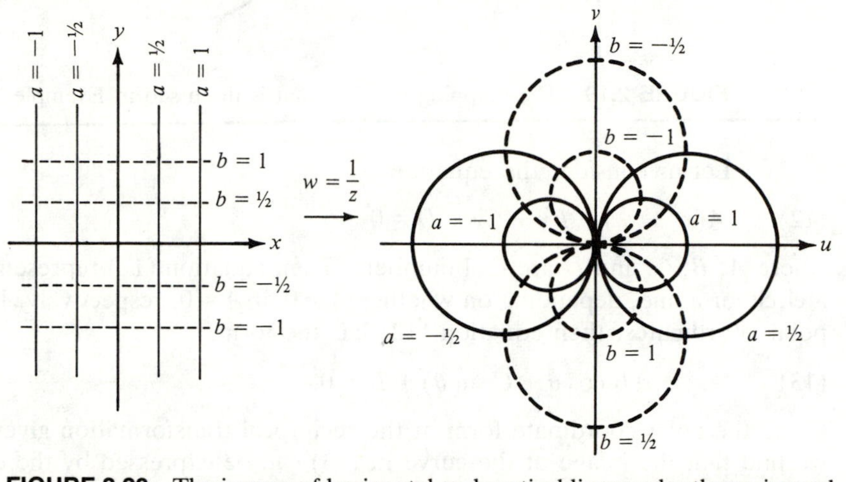

FIGURE 2.20 The images of horizontal and vertical lines under the reciprocal transformation.

EXERCISES FOR SECTION 2.5

For Exercises 1–8, find the image of the given circle or line under the reciprocal transformation $w = 1/z$.

1. The horizontal line $\operatorname{Im} z = \frac{1}{5}$.

2. The circle $|z + i/2| = \frac{1}{2}$.

3. The vertical line $\operatorname{Re} z = -3$.

4. The circle $|z + 2| = 1$.

5. The line $2x + 2y = 1$.

6. The circle $|z - i/2| = 1$.

7. The circle $|z - \frac{3}{2}| = 1$.

8. The circle $|z + 1 - i| = 2$.

9. **(a)** Show that $\lim_{z\to\infty} (1/z) = 0$. **(b)** Show that $\lim_{z\to 0} (1/z) = \infty$.

10. Show that the reciprocal transformation $w = 1/z$ maps the vertical strip $0 < x < \frac{1}{2}$ onto the region in the right half-plane $\operatorname{Re} w > 0$ that lies outside the circle $|w - 1| = 1$.

11. Find the image of the disk $|z + 2i/3| < \frac{4}{3}$ under the reciprocal transformation.

12. Show that the reciprocal transformation maps the disk $|z - 1| < 2$ onto the region that lies exterior to the circle $|w + \frac{1}{3}| = \frac{2}{3}$.

13. Find the image of the half-plane $y > \frac{1}{2} - x$ under the mapping $w = 1/z$.

14. Show that the half-plane $y < x - \frac{1}{2}$ is mapped onto the disk $|w - 1 - i| < \sqrt{2}$ by the reciprocal transformation.

15. Find the image of the quadrant $x > 1$, $y > 1$ under the mapping $w = 1/z$.

16. Show that the transformation $w = 2/z$ maps the disk $|z - i| < 1$ onto the lower half-plane $\text{Im } w < -1$.

17. Show that the transformation $w = (2 - z)/z = -1 + 2/z$ maps the disk $|z - 1| < 1$ onto the right half-plane $\text{Re } w > 0$.

18. Show that the parabola $2x = 1 - y^2$ is mapped onto the cardioid $\rho = 1 + \cos \phi$ by the reciprocal transformation.

19. Limits involving ∞. The function $f(z)$ is said to have the limit L as z approaches ∞, and we write

$$\lim_{z \to \infty} f(z) = L$$

if for every $\varepsilon > 0$ there exists an $R > 0$ so that

$$|f(z) - L| < \varepsilon \qquad \text{whenever } |z| > R.$$

Use the above definition to prove that

$$\lim_{z \to \infty} \frac{z + 1}{z - 1} = 1.$$

Analytic and Harmonic Functions

3.1 Differentiable Functions

Let f be a complex function that is defined at all points in some neighborhood of z_0. The *derivative of f at z_0* is written $f'(z_0)$ and is defined by the equation

$$(1) \qquad f'(z_0) = \lim_{z \to z_0} \frac{f(z) - f(z_0)}{z - z_0}$$

provided that the limit exists. The function f is said to be *differentiable* at z_0 if it has a derivative at z_0. If we write $\Delta z = z - z_0$, then definition (1) can be expressed in the form

$$(2) \qquad f'(z_0) = \lim_{\Delta z \to 0} \frac{f(z_0 + \Delta z) - f(z_0)}{\Delta z}.$$

If we let $w = f(z)$ and $\Delta w = f(z) - f(z_0)$, then the notation dw/dz for the derivative is expressed by

$$(3) \qquad f'(z_0) = \frac{dw}{dz} = \lim_{\Delta z \to 0} \frac{\Delta w}{\Delta z}.$$

EXAMPLE 3.1 If $f(z) = z^3$, then we can use definition (1) to show that $f'(z) = 3z^2$.

Solution Calculation reveals that

$$f'(z_0) = \lim_{z \to z_0} \frac{z^3 - z_0^3}{z - z_0} = \lim_{z \to z_0} \frac{(z - z_0)(z^2 + z_0 z + z_0^2)}{z - z_0} = 3z_0^2.$$

The subscript on z_0 can be dropped to obtain the general formula $f'(z) = 3z^2$.

We must pay careful attention to the complex value Δz in equation (3), since the value of the limit must be independent of the manner in which $\Delta z \to 0$. If we can find two curves that end at z_0 along which $\Delta w/\Delta z$ approaches distinct values, then $\Delta w/\Delta z$ does *not* have a limit as $\Delta z \to 0$ and f does *not* have a derivative at z_0.

EXAMPLE 3.2 The function $w = f(z) = \overline{z} = x - iy$ is nowhere differentiable.

Solution To show this, we choose two approaches to the point $z_0 = x_0 + iy_0$ and compute limits of the difference quotients. First, we approach z_0 along a line parallel to the x axis:

$$\lim_{z \to z_0} \frac{f(x_0 + \Delta x + iy_0) - f(x_0 + iy_0)}{x_0 + \Delta x + iy_0 - x_0 - iy_0} = \lim_{\Delta x \to 0} \frac{x_0 + \Delta x - iy_0 - x_0 + iy_0}{\Delta x}$$

$$= \lim_{\Delta x \to 0} \frac{\Delta x}{\Delta x} = 1.$$

Second, we approach z_0 along a line parallel to the y axis:

$$\lim_{z \to z_0} \frac{f(x_0 + iy_0 + i\Delta y) - f(x_0 + iy_0)}{x_0 + iy_0 + i\Delta y - x_0 - iy_0} = \lim_{\Delta y \to 0} \frac{x_0 - iy_0 - i\Delta y - x_0 + iy_0}{i\Delta y}$$

$$= \lim_{\Delta y \to 0} \frac{-i\Delta y}{i\Delta y} = -1.$$

Since the limits along the two approaches are different, there is no computable limit for the right side of equation (1). Therefore $f(z) = \overline{z}$ is not differentiable at the point z_0. Since z_0 was arbitrary, $f(z)$ is nowhere differentiable.

Our definition of derivative for complex functions is formally the same as for real functions and is the natural extension from real variables to complex variables. The basic differentiation formulas follow identically as in the case of real functions, and we obtain the same rules for differentiating powers, sums, products, quotients, and compositions of functions. The proof of the differentiation formulas are easily established by using the limit theorems.

Let C denote a complex constant. From definition (1) and the technique exhibited in the solution to Example 3.1, the following are easily established, just as they were in the real case:

(4) $\quad \dfrac{d}{dz} C = 0 \quad$ and

(5) $\quad \dfrac{d}{dz} z^n = nz^{n-1} \quad$ where n is a positive integer.

Furthermore, the rules for finding derivatives of combinations of two differentiable functions f and g are identical to those developed in calculus:

(6) $\quad \dfrac{d}{dz} [Cf(z)] = Cf'(z),$

(7) $\quad \dfrac{d}{dz} [f(z) + g(z)] = f'(z) + g'(z),$

(8) $\quad \dfrac{d}{dz} [f(z)g(z)] = f'(z)g(z) + f(z)g'(z),$

(9) $$\frac{d}{dz}\frac{f(z)}{g(z)} = \frac{f'(z)g(z) - f(z)g'(z)}{[g(z)]^2} \qquad \text{provided that } g(z) \neq 0$$

(10) $$\frac{d}{dz} f(g(z)) = f'(g(z))g'(z).$$

Important particular cases of (9) and (10), respectively, are

(11) $$\frac{d}{dz}\frac{1}{z^n} = \frac{-n}{z^{n+1}} \qquad \text{for } z \neq 0 \text{ and where } n \text{ is a positive integer.}$$

(12) $$\frac{d}{dz}[f(z)]^n = n[f(z)]^{n-1}f'(z) \qquad \text{where } n \text{ is a positive integer.}$$

EXAMPLE 3.3 If we use equation (12) with $f(z) = z^2 + i2z + 3$ and $f'(z) = 2z + 2i$, then we see that

$$\frac{d}{dz}(z^2 + i2z + 3)^4 = 8(z^2 + i2z + 3)^3(z + i).$$

Several proofs involving complex functions rely on the continuity of functions. The following result shows that a differentiable function is a continuous function.

Theorem 3.1 *If f is differentiable at z_0, then f is continuous at z_0.*

Proof Since f is differentiable at z_0, from (1), we obtain

$$\lim_{z \to z_0} \frac{f(z) - f(z_0)}{z - z_0} = f'(z_0).$$

Using the multiplicative property of limits given by (19) in Section 2.4, we see that

$$\begin{aligned}\lim_{z \to z_0} [f(z) - f(z_0)] &= \lim_{z \to z_0} \frac{f(z) - f(z_0)}{z - z_0}(z - z_0) \\ &= \lim_{z \to z_0} \frac{f(z) - f(z_0)}{z - z_0} \lim_{z \to z_0}(z - z_0) \\ &= f'(z_0) \cdot 0 = 0.\end{aligned}$$

Hence $\lim_{z \to z_0} f(z) = f(z_0)$, and f is continuous at z_0.

Using Theorem 3.1, we are able to establish formula (8). Letting $h(z) = f(z)g(z)$ and using definition (1), we write

$$h'(z_0) = \lim_{z \to z_0} \frac{h(z) - h(z_0)}{z - z_0} = \lim_{z \to z_0} \frac{f(z)g(z) - f(z_0)g(z_0)}{z - z_0}.$$

If we add and subtract the term $f(z_0)g(z)$ in the numerator, then we can regroup the last term and obtain

$$h'(z_0) = \lim_{z\to z_0} \frac{f(z)g(z) - f(z_0)g(z)}{z - z_0} + \lim_{z\to z_0} \frac{f(z_0)g(z) - f(z_0)g(z_0)}{z - z_0}$$

$$= \lim_{z\to z_0} \frac{f(z) - f(z_0)}{z - z_0} \lim_{z\to z_0} g(z) + f(z_0) \lim_{z\to z_0} \frac{g(z) - g(z_0)}{z - z_0}.$$

Using definition (1) for derivative and the continuity of g, we obtain $h'(z_0) = f'(z_0)g(z_0) + f(z_0)g'(z_0)$. Hence formula (8) is established. The proofs of the other formulas are left as exercises.

The rule for differentiating a polynomial can be extended to complex variables. Let $P(z)$ be a polynomial of degree n:

$$(13) \qquad P(z) = a_0 + a_1 z + a_2 z^2 + \cdots + a_n z^n.$$

Then mathematical induction can be used with formulas (5) and (7) to obtain the derivative of (13):

$$(14) \qquad P(z) = a_1 + a_2 z + a_3 z^2 + \cdots + a_n z^{n-1}.$$

The proof is left as an exercise.

Properties of limits and derivatives can be used to establish L'Hôpital's Rule, which we state as follows.

Theorem 3.2 (L'Hôpital's Rule) *Let f and g be differentiable at z_0. If $f(z_0) = 0$ and $g(z_0) = 0$, but $g'(z_0) \neq 0$, then*

$$\lim_{z\to z_0} \frac{f(z)}{g(z)} = \frac{f'(z_0)}{g'(z_0)}.$$

Proof Since $f(z_0) = g(z_0) = 0$, we can write

$$\lim_{z\to z_0} \frac{f(z)}{g(z)} = \lim_{z\to z_0} \frac{f(z) - f(z_0)}{g(z) - g(z_0)} = \lim_{z\to z_0} \frac{(f(z) - f(z_0))/(z - z_0)}{(g(z) - g(z_0))/(z - z_0)} = \frac{f'(z_0)}{g'(z_0)}.$$

Limits involving $0/0$ are easy to find using L'Hôpital's Rule.

EXAMPLE 3.4 Let us show that

$$\lim_{z\to 1+i} \frac{z^2 + z - 1 - 3i}{z^2 - 2z + 2} = 1 - \frac{3}{2} i.$$

If we let $f(z) = z^2 + z - 1 + 3i$ and $g(z) = z^2 - 2z + 2$, then $f(1 + i) = 0$, $g(1 + i) = 0$, and $g'(1 + i) = 2i \neq 0$. Computation reveals that $f'(1 + i) = 3 + 2i$; hence the limit is given by

$$\lim_{z\to 1+i} \frac{z^2 + z - 1 - 3i}{z^2 - 2z + 2} = \frac{f'(1 + i)}{g'(1 + i)} = \frac{3 + 2i}{2i} = 1 - \frac{3}{2} i.$$

EXERCISES FOR SECTION 3.1

1. Find the derivatives of the following functions.

(a) $f(z) = 5z^3 - 4z^2 + 7z - 8$

(b) $g(z) = (z^2 - iz + 9)^5$

(c) $h(z) = \dfrac{2z+1}{z+2}$ for $z \neq -2$

(d) $F(z) = (z^2 + (1-3i)z + 1)(z^4 + 3z^2 + 5i)$

2. Use definition (1), and show that $\dfrac{d}{dz}\dfrac{1}{z} = \dfrac{-1}{z^2}$.

3. If f is differentiable for all z, then we say that f is an *entire* function. If f and g are entire functions, decide which of the following are entire functions.

(a) $[f(z)]^3$ **(b)** $f(z)g(z)$ **(c)** $f(z)/g(z)$

(d) $f(1/z)$ **(e)** $f(z-1)$ **(f)** $f(g(z))$

4. Use definition (1) to establish (5).

5. Let P be a polynomial of degree n given by $P(z) = a_0 + a_1 z + \cdots + a_n z^n$. Show that $P'(z) = a_1 + 2a_2 z + \cdots + na_n z^{n-1}$.

6. Let P be a polynomial of degree 2, given by

$$P(z) = (z - z_1)(z - z_2)$$

where z_1 and z_2 are distinct. Show that

$$\frac{P'(z)}{P(z)} = \frac{1}{z - z_1} + \frac{1}{z - z_2}.$$

7. Use L'Hôpital's Rule to find the following limits.

(a) $\displaystyle\lim_{z \to i} \frac{z^4 - 1}{z - i}$ **(b)** $\displaystyle\lim_{z \to 1+i} \frac{z^2 - iz - 1 - i}{z^2 - 2z + 2}$

(c) $\displaystyle\lim_{z \to -i} \frac{z^6 + 1}{z^2 + 1}$ **(d)** $\displaystyle\lim_{z \to 1+i} \frac{z^4 + 4}{z^2 - 2z + 2}$

(e) $\displaystyle\lim_{z \to 1+i\sqrt{3}} \frac{z^6 - 64}{z^3 + 8}$ **(f)** $\displaystyle\lim_{z \to -1+i\sqrt{3}} \frac{z^9 - 512}{z^3 - 8}$

8. Let f be differentiable at z_0. Show that there exists a function $\eta(z)$, such that

$$f(z) = f(z_0) + f'(z_0)(z - z_0) + \eta(z)(z - z_0) \quad \text{where } \eta(z) \to 0 \text{ as } z \to z_0.$$

9. Show that $\dfrac{d}{dz} z^{-n} = -nz^{-n-1}$ where n is a positive integer.

10. Establish the identity

$$\frac{d}{dz} f(z)g(z)h(z) = f'(z)g(z)h(z) + f(z)g'(z)h(z) + f(z)g(z)h'(z).$$

11. Show that the function $f(z) = \bar{z} + 2z$ is nowhere differentiable.

12. Show that the function $f(z) = |z|^2$ is differentiable only at the point $z_0 = 0$. *Hint*: To show that f is *not* differentiable at $z_0 \neq 0$, choose horizontal and vertical lines through the point z_0, and show that $\Delta w/\Delta z$ approaches two distinct values as $\Delta z \to 0$.

13. Establish identity (4).

14. Establish identity (7).

15. Establish identity (9).

16. Establish identity (10).

17. Establish identity (12).

18. Consider the differentiable function $f(z) = z^3$ and the two points $z_1 = 1$ and $z_2 = i$. Show that there does not exist a point c on the line $y = 1 - x$ between 1 and i such that

$$\frac{f(z_2) - f(z_1)}{z_2 - z_1} = f'(c).$$

This shows that the Mean Value Theorem for derivatives does not extend to complex functions.

3.2 The Cauchy-Riemann Equations

Let $f(z) = u(x, y) + iv(x, y)$ be a complex function that is differentiable at the point z_0. Then it is natural to seek a formula for computing $f'(z_0)$ in terms of the partial derivatives of $u(x, y)$ and $v(x, y)$. If we investigate this idea, then it is easy to find the required formula; but we will find that there are special conditions that must be satisfied before it can be used. In addition, we will discover two important equations relating the partial derivatives of u and v, which were discovered independently by the French mathematician A. L. Cauchy and the German mathematician G. F. B. Riemann.

Theorem 3.3 (Cauchy-Riemann Equations) *Let $f(z) = u(x, y) + iv(x, y)$ be differentiable at the point $z_0 = x_0 + iy_0$. Then the partial derivatives of u and v exist at the point (x_0, y_0) and satisfy the equations*

$$(1) \qquad u_x(x_0, y_0) = v_y(x_0, y_0) \quad \textit{and} \quad u_y(x_0, y_0) = -v_x(x_0, y_0).$$

Proof We shall choose horizontal and vertical lines that pass through the point (x_0, y_0) and compute the limiting values of $\Delta w/\Delta z$ along these lines. Equating the two resulting limits will result in (1). For the horizontal approach to z_0 we set $z = x + iy_0$ and obtain

$$f'(z_0) = \lim_{(x, y_0) \to (x_0, y_0)} \frac{f(x + iy_0) - f(x_0 + iy_0)}{x + iy_0 - (x_0 + iy_0)}$$

$$= \lim_{x \to x_0} \frac{u(x, y_0) - u(x_0, y_0) + i[v(x, y_0) - v(x_0, y_0)]}{x - x_0}$$

$$= \lim_{x \to x_0} \frac{u(x, y_0) - u(x_0, y_0)}{x - x_0} + i \lim_{x \to x_0} \frac{v(x, y_0) - v(x_0, y_0)}{x - x_0}.$$

The last limits are recognized to be the partial derivatives of u and v with respect to x, and we obtain

$$(2) \qquad f'(z_0) = u_x(x_0, y_0) + iv_x(x_0, y_0).$$

Along the vertical approach to z_0, we have $z = x_0 + iy$. Calculation reveals that

$$f'(z_0) = \lim_{(x_0,y)\to(x_0,y_0)} \frac{f(x_0 + iy) - f(x_0 + iy_0)}{x_0 + iy - (x_0 + iy_0)}$$

$$= \lim_{y\to y_0} \frac{u(x_0, y) - u(x_0, y_0) + i[v(x_0, y) - v(x_0, y_0)]}{i(y - y_0)}$$

$$= \lim_{y\to y_0} \frac{v(x_0, y) - v(x_0, y_0)}{y - y_0} - i \lim_{y\to y_0} \frac{u(x_0, y) - u(x_0, y_0)}{y - y_0}.$$

The last limits are recognized to be the partial derivatives of u and v with respect to y, and we obtain

$$(3) \qquad f'(z_0) = v_y(x_0, y_0) - iu_y(x_0, y_0).$$

Since f is differentiable at z_0, the limits given by equations (2) and (3) must be equal. If we equate the real and imaginary parts in (2) and (3), then the result is (1), and the proof is complete.

At this stage we may be tempted to use equation (2) or (3) to compute $f'(z_0)$, but we cannot do this without further investigations.

EXAMPLE 3.5 The function defined by

$$f(z) = \frac{(\bar{z})^2}{z} = \frac{x^3 - 3xy^2}{x^2 + y^2} + i\frac{y^3 - 3x^2y}{x^2 + y^2}$$

when $z \neq 0$ and $f(0) = 0$ is *not* differentiable at the point $z_0 = 0$. However, the Cauchy-Riemann equations (1) hold true at $(0, 0)$. To verify this, we must use limits to calculate the partial derivatives at $(0, 0)$. Indeed,

$$u_x(0, 0) = \lim_{x\to 0} \frac{u(x, 0) - u(0, 0)}{x - 0} = \frac{\lim\limits_{x\to 0} \dfrac{x^3 - 0}{x^2 + 0}}{x} = 1.$$

In a similar fashion, one can show that

$$u_y(0, 0) = 0, \qquad v_x(0, 0) = 0, \qquad \text{and} \qquad v_y(0, 0) = 1.$$

Hence the Cauchy-Riemann equations hold at the point $(0, 0)$.

We now show that f is *not* differentiable at $z_0 = 0$. If we let z approach 0 along the x axis, then

$$\lim_{(x,0)\to(0,0)} \frac{f(x + 0i) - f(0)}{x + 0i - 0} = \lim_{x\to 0} \frac{x - 0}{x - 0} = 1.$$

But if we let z approach 0 along the line given by the parametric equations $x = t$ and $y = t$, then

$$\lim_{(t,t)\to(0,0)} \frac{f(t + it) - f(0)}{t + it - 0} = \lim_{t\to 0} \frac{-t - it}{t + it} = -1.$$

Since the two limits are distinct, we concude that f is not differentiable at the origin.

The next theorem will give us sufficient conditions under which we can use equations (2) and/or (3) to compute the derivative $f'(z_0)$. They are referred to as the *Cauchy-Riemann conditions* for differentiability.

Theorem 3.4 (Sufficient Conditions) *Let $f(z) = u(x, y) + iv(x, y)$ be a continuous function that is defined in some neighborhood of the point $z_0 = x_0 + iy_0$. If all the partial derivatives u_x, u_y, v_x, and v_y are continuous at the point (x_0, y_0) and if the Cauchy-Riemann equations*

$$u_x(x_0, y_0) = v_y(x_0, y_0) \quad \text{and} \quad u_y(x_0, y_0) = -v_x(x_0, y_0)$$

hold, then f is differentiable at z_0, and the derivative $f'(z_0)$ can be computed with either formula (2) or (3).

We postpone the proof of Theorem 3.4 until the end of this section. Let us consider its utility.

EXAMPLE 3.6 The function $f(z) = e^{-y} \cos x + ie^{-y} \sin x$ is differentiable for all z, and its derivative is $f'(z) = -e^{-y} \sin x + ie^{-y} \cos x$. To show this, we first write $u(x, y) = e^{-y} \cos x$ and $v(x, y) = e^{-y} \sin x$ and compute the partial derivatives:

$$u_x(x, y) = v_y(x, y) = -e^{-y} \sin x \quad \text{and}$$

$$v_x(x, y) = -u_y(x, y) = e^{-y} \cos x.$$

We see that u, v, u_x, u_y, v_x, and v_y are all continuous functions and that the Cauchy-Riemann equations hold for all values of (x, y). Hence using equation (2), we write

$$f'(z) = u_x(x, y) + iv_x(x, y) = -e^{-y} \sin x + ie^{-y} \cos x.$$

The Cauchy-Riemann conditions are particularly useful in determining the set of points for which a function f is differentiable.

EXAMPLE 3.7 The function $f(z) = x^3 + 3xy^2 + i(y^3 + 3x^2y)$ is differentiable only at points that lie on the coordinate axes.

Solution To show this, we write $u(x, y) = x^3 + 3xy^2$ and $v(x, y) = y^3 + 3x^2y$ and compute the partial derivatives:

$$u_x(x, y) = 3x^2 + 3y^2, \qquad v_y(x, y) = 3x^2 + 3y^2,$$

$$u_y(x, y) = 6xy, \qquad v_x(x, y) = 6xy.$$

Here u, v, u_x, u_y, v_x, v_y are all continuous, and $u_x(x, y) = v_y(x, y)$ holds for all (x, y). But $u_y(x, y) = -v_x(x, y)$ if and only if $6xy = -6xy$, which is equivalent to $12xy = 0$. Therefore the Cauchy-Riemann equations hold only when $x = 0$ or $y = 0$, and according to Theorem 3.4, f is differentiable only at points that lie on the coordinate axes.

When polar coordinates (r, θ) are used to locate points in the plane, then it is convenient to use expression (7) of Section 2.1 for a complex function; that is,

$$f(z) = f(re^{i\theta}) = u(r, \theta) + iv(r, \theta).$$

In this case, u and v are real functions of the real variables r and θ. The polar form of the Cauchy-Riemann equations and a formula for finding $f'(z)$ in terms of the partial derivatives of $u(r, \theta)$ and $v(r, \theta)$ are given in the following result.

Theorem 3.5 (Polar Form) *Let $f(z) = u(r, \theta) + iv(r, \theta)$ be a continuous function that is defined in some neighborhood of the point $z_0 = r_0 e^{i\theta_0}$. If all the partial derivatives u_r, u_θ, v_r, and v_θ are continuous at the point (r_0, θ_0) and if the Cauchy-Riemann equations*

$$(4) \qquad u_r(r_0, \theta_0) = \frac{1}{r_0} v_\theta(r_0, \theta_0) \quad \textit{and} \quad v_r(r_0, \theta_0) = \frac{-1}{r_0} u_\theta(r_0, \theta_0)$$

hold, then f is differentiable at z_0, and the derivative $f'(z_0)$ can be computed by either of the following formulas:

$$(5) \qquad f'(z_0) = e^{-i\theta_0}[u_r(r_0, \theta_0) + iv_r(r_0, \theta_0)] \quad \textit{or}$$

$$(6) \qquad f'(z_0) = \frac{1}{r_0} e^{-i\theta_0}[v_\theta(r_0, \theta_0) - iu_\theta(r_0, \theta_0)].$$

EXAMPLE 3.8 If f is given by

$$f(z) = z^{1/2} = r^{1/2} \cos \frac{\theta}{2} + ir^{1/2} \sin \frac{\theta}{2}$$

where the domain is restricted to be $r > 0$ and $-\pi < \theta < \pi$, then the derivative is given by

$$f'(z) = \frac{1}{2z^{1/2}} = \frac{1}{2} r^{-1/2} \cos \frac{\theta}{2} - i \frac{1}{2} r^{-1/2} \sin \frac{\theta}{2}$$

where $r > 0$ and $-\pi < \theta < \pi$.

Solution To show this, we write

$$u(r, \theta) = r^{1/2} \cos \frac{\theta}{2} \quad \text{and} \quad v(r, \theta) = r^{1/2} \sin \frac{\theta}{2}.$$

Here

$$u_r(r, \theta) = \frac{1}{r} v_\theta(r, \theta) = \frac{1}{2} r^{-1/2} \cos \frac{\theta}{2} \quad \text{and}$$

$$v_r(r, \theta) = \frac{-1}{r} u_\theta(r, \theta) = \frac{1}{2} r^{-1/2} \sin \frac{\theta}{2}.$$

Using these results in identity (5), we obtain

$$f'(z_0) = e^{-i\theta}\left(\frac{1}{2}\, r^{-1/2} \cos \frac{\theta}{2} + i\, \frac{1}{2}\, r^{-1/2} \sin \frac{\theta}{2}\right)$$

$$= e^{-i\theta}\left(\frac{1}{2}\, r^{-1/2} e^{i\theta/2}\right) = \frac{1}{2}\, r^{-1/2} e^{-i\theta/2} = \frac{1}{2z^{1/2}}.$$

Proof of Theorem 3.4 Let $\Delta z = \Delta x + i\Delta y$ and $\Delta w = \Delta u + i\Delta v$, and let Δz be chosen small enough that z lies in the ε-neighborhood of z_0 in which the hypotheses holds true. We will show that $\Delta w/\Delta z$ approaches the limit given in equation (2) as Δz approaches zero. The difference Δu can be written as

$$\Delta u = u(x_0 + \Delta x, y_0 + \Delta y) - u(x_0, y_0).$$

If we add and subtract the term $u(x_0, y_0 + \Delta y)$, then the result is

$$(7) \qquad \Delta u = [u(x_0 + \Delta x, y_0 + \Delta y) - u(x_0, y_0 + \Delta y)] + [u(x_0, y_0 + \Delta y) - u(x_0, y_0)].$$

Since the partial derivatives u_x and u_y exist, the Mean Value Theorem for real functions of two variables implies that a value x^* exists between x_0 and $x_0 + \Delta x$ such that the first term in brackets on the right side of (7) can be written as

$$(8) \qquad u(x_0 + \Delta x, y_0 + \Delta y) - u(x_0, y_0 + \Delta y) = u_x(x^*, y_0 + \Delta y)\Delta x.$$

Furthermore, since u_x and u_y are continuous at (x_0, y_0), there exists a quantity ε_1 such that

$$(9) \qquad u_x(x^*, y_0 + \Delta y) = u_x(x_0, y_0) + \varepsilon_1$$

where $\varepsilon_1 \to 0$ as $x^* \to x_0$ and $\Delta y \to 0$. Since $\Delta x \to 0$ forces $x^* \to x_0$, we can use the equation

$$(10) \qquad u(x_0 + \Delta x, y_0 + \Delta y) - u(x_0, y_0 + \Delta y) = [u_x(x_0, y_0) + \varepsilon_1]\Delta x$$

where $\varepsilon_1 \to 0$ as $\Delta x \to 0$ and $\Delta y \to 0$. Similarly, there exists a quantity ε_2 such that the second term in brackets on the right side of (7) satisfies the equation

$$(11) \qquad u(x_0, y_0 + \Delta y) - u(x_0, y_0) = [u_y(x_0, y_0) + \varepsilon_2]\Delta y$$

where $\varepsilon_2 \to 0$ as $\Delta x \to 0$ and $\Delta y \to 0$.

Combining (10) and (11), we obtain

$$(12) \qquad \Delta u = (u_x + \varepsilon_1)\Delta x + (u_y + \varepsilon_2)\Delta y$$

where the partial derivatives u_x and u_y are evaluated at the point (x_0, y_0) and ε_1 and ε_2 tend to zero as Δx and Δy both tend to zero. Similarly, the change Δv is related to the changes Δx and Δy by the equation

$$(13) \qquad \Delta v = (v_x + \varepsilon_3)\Delta x + (v_y + \varepsilon_4)\Delta y$$

where the partial derivatives v_x and v_y are evaluated at the point (x_0, y_0) and ε_3 and ε_4 tend to zero as Δx and Δy both tend to zero. Combining (12) and (13), we have

$$(14) \qquad \Delta w = u_x \Delta x + u_y \Delta y + i(v_x \Delta x + v_y \Delta y) + \varepsilon_1 \Delta x + \varepsilon_2 \Delta y + i(\varepsilon_3 \Delta x + \varepsilon_4 \Delta y).$$

The Cauchy-Riemann equations can be used in equation (14) to obtain

$$\Delta w = u_x \Delta x - v_x \Delta y + i(v_x \Delta x + u_x \Delta y) + \varepsilon_1 \Delta x + \varepsilon_2 \Delta y + i(\varepsilon_3 \Delta x + \varepsilon_4 \Delta y).$$

Now the terms can be rearranged to yield

$$(15) \qquad \Delta w = u_x[\Delta x + i\Delta y] + iv_x[\Delta x + i\Delta y] + \varepsilon_1 \Delta x + \varepsilon_2 \Delta y + i(\varepsilon_3 \Delta x + \varepsilon_4 \Delta y).$$

Since $\Delta z = \Delta x + i\Delta y$, we can divide both sides of equation (15) by Δz and take the limit as $\Delta z \to 0$:

$$(16) \qquad \lim_{\Delta z \to 0} \frac{\Delta w}{\Delta z} = u_x + iv_x + \lim_{\Delta z \to 0} \left[\frac{\varepsilon_1 \Delta x}{\Delta z} + \frac{\varepsilon_2 \Delta y}{\Delta z} + i \frac{\varepsilon_3 \Delta x}{\Delta z} + i \frac{\varepsilon_4 \Delta y}{\Delta z} \right].$$

Using the property of ε_1 mentioned in (9), we have

$$\lim_{\Delta z \to 0} \left| \frac{\varepsilon_1 \Delta x}{\Delta z} \right| = \lim_{\Delta z \to 0} |\varepsilon_1| \left| \frac{\Delta x}{\Delta z} \right| \leqq \lim_{\Delta z \to 0} |\varepsilon_1| = 0.$$

Similarly, the limits of the other quantities in (16) involving ε_2, ε_3, ε_4 are zero. Therefore the limit in (16) becomes

$$\lim_{\Delta z \to 0} \frac{\Delta w}{\Delta z} = f'(z_0) = u_x(x_0, y_0) + iv_x(x_0, y_0),$$

and the proof of the theorem is complete.

EXERCISES FOR SECTION 3.2

1. Use the Cauchy-Riemann conditions to show that the following functions are differentiable for all z, and find $f'(z)$.
(a) $f(z) = iz + 4i$ **(b)** $f(z) = z^3$
(c) $f(z) = -2(xy + x) + i(x^2 - 2y - y^2)$

2. Let $f(z) = e^x \cos y + ie^x \sin y$. Show that both $f(z)$ and $f'(z)$ are differentiable for all z.

3. Find the constants a and b such that $f(z) = (2x - y) + i(ax + by)$ is differentiable for all z.

4. Show that $f(z) = (y + ix)/(x^2 + y^2)$ is differentiable for all $z \neq 0$.

5. Show that $f(z) = e^{2xy}[\cos(y^2 - x^2) + i \sin(y^2 - x^2)]$ is differentiable for all z.

6. Use the Cauchy-Riemann conditions to show that the following functions are nowhere differentiable.
(a) $f(z) = \overline{z}$ **(b)** $g(z) = z + \overline{z}$
(c) $h(z) = e^y \cos x + ie^y \sin x$

7. Let $f(z) = |z|^2$. Show that f is differentiable at the point $z_0 = 0$ but is not differentiable at any other point.

8. Show that the function $f(z) = x^2 + y^2 + i2xy$ has a derivative only at points that lie on the x axis.

9. Let f be a differentiable function. Establish the identity $|f'(z)|^2 = u_x^2 + v_x^2 = u_y^2 + v_y^2$.

10. Let $f(z) = (\ln r)^2 - \theta^2 + i2\theta \ln r$ where $r > 0$ and $-\pi < \theta \leqq \pi$. Show that f is differentiable for $r > 0$, $-\pi < \theta < \pi$, and find $f'(z)$.

11. Let f be differentiable at $z_0 = r_0 e^{i\theta_0}$. Let z approach z_0 along the ray $r > 0$, $\theta = \theta_0$, and use definition (1) of Section 3.1 to show that equation (5) of Section 3.2 holds.

12. A vector field $\mathbf{F}(z) = U(x, y) + iV(x, y)$ is said to be *irrotational* if $U_y(x, y) = V_x(x, y)$. It is said to be *solenoidal* if $U_x(x, y) = -V_y(x, y)$. If $f(z)$ is an analytic function, show that $\mathbf{F}(z) = f(z)$ is both irrotational and solenoidal.

13. The polar form of the Cauchy-Riemann equations.

(a) Use the coordinate transformation

$$x = r\cos\theta \quad \text{and} \quad y = r\sin\theta$$

and the chain rules

$$u_r = u_x \frac{\partial x}{\partial r} + u_y \frac{\partial y}{\partial r} \quad \text{and} \quad u_\theta = u_x \frac{\partial x}{\partial \theta} + u_y \frac{\partial y}{\partial \theta} \quad \text{etc.}$$

to prove that

$$u_r = u_x \cos\theta + u_y \sin\theta \quad \text{and} \quad u_\theta = -u_x r\sin\theta + u_y r\cos\theta \quad \text{and}$$

$$v_r = v_x \cos\theta + v_y \sin\theta \quad \text{and} \quad v_\theta = -v_x r\sin\theta + v_y r\cos\theta.$$

(b) Use the results of part (a) to prove that

$$ru_r = v_\theta \quad \text{and} \quad rv_r = -u_\theta.$$

3.3 Analytic Functions and Harmonic Functions

It is seldom of interest to study functions that are differentiable at a single point. Complex functions that have a derivative at all points in a neighborhood of z_0 deserve further study. In Chapter 6 we will learn that if the complex function f can be represented by a Taylor series at z_0, then it must be differentiable in some neighborhood of z_0. The function f is said to be *analytic at* z_0 if its derivative exists at each point z in some neighborhood of z_0. If f is analytic at each point in the region R, then we say that f is analytic on R. If f is analytic on the whole complex plane, then f is said to be *entire*.

Points of nonanalyticity are called *singular points*. They are important for certain applications in physics and engineering.

EXAMPLE 3.9 The function $f(z) = x^2 + y^2 + i2xy$ is nowhere analytic.

Solution We identify the functions $u(x, y) = x^2 + y^2$ and $v(x, y) = 2xy$. The equation $u_x = v_y$ becomes $2x = 2x$, which holds everywhere. But the equation $u_y = -v_x$ becomes $2y = -2y$, which holds only when $y = 0$. Thus $f(x)$

is differentiable only at points that lie on the x axis. However, for any point $z_0 = x_0 + 0i$ on the x axis and any δ-neighborhood of z_0, the point $z_1 = x_0 + i\delta/2$ is a point where f is not differentiable. Therefore f is not differentiable in any full neighborhood of z_0, and consequently it is not analytic at z_0.

We have seen that polynomial functions have derivatives at all points in the complex plane; hence polynomials are entire functions. The function $f(z) = e^x \cos y + ie^x \sin y$ has a derivative at all points z, and it is an entire function.

The results in Section 3.2 show that an analytic function must be continuous and must satisfy the Cauchy-Riemann equations. Conversely, if the Cauchy-Riemann conditions hold at all points in a neighborhood of z_0, then f is analytic at z_0. Using properties of derivatives, we see that the sum, difference, and product of two analytic functions are analytic functions. Similarly, the quotient of two analytic functions is analytic, provided that the function in the denominator is not zero. The chain rule can be used to show that the composition $g(f(z))$ of two analytic functions f and g is analytic, provided that g is analytic in a domain that contains the range of f.

The function $f(z) = 1/z$ is analytic for all $z \neq 0$; and if $P(z)$ and $Q(z)$ are polynomials, then their quotient $P(z)/Q(z)$ is analytic at all points where $Q(z) \neq 0$. The square root function is more complicated. If

$$(1) \qquad f(z) = z^{1/2} = r^{1/2} \cos \frac{\theta}{2} + ir^{1/2} \sin \frac{\theta}{2} \quad \text{where } r > 0 \text{ and } -\pi < \theta \leqq \pi$$

then f is analytic at all points except $z_0 = 0$ and except at points that lie along the negative x axis. The function $f(z) = z^{1/2}$ defined by equation (1) is not continuous at points that lie along the negative x axis, and for this reason it is not analytic there.

Let $\phi(x, y)$ be a real-valued function of the two real variables x and y. The partial differential equation

$$(2) \qquad \phi_{xx}(x, y) + \phi_{yy}(x, y) = 0$$

is known as *Laplace's equation* and is sometimes referred to as the potential equation. If ϕ, ϕ_x, ϕ_y, ϕ_{xx}, ϕ_{xy}, ϕ_{yx}, and ϕ_{yy} are all continuous and if $\phi(x, y)$ satisfies Laplace's equation, then $\phi(x, y)$ is called a *harmonic function*. Harmonic functions are important in the areas of applied mathematics, engineering, and mathematical physics. They are used to solve problems involving steady state temperatures, two-dimensional electrostatics, and ideal fluid flow. An important result for our studies is the fact that if $f(z) = u(x, y) + iv(x, y)$ is an analytic function, then both u and v are harmonic functions. In Chapter 9 we will see how complex variable techniques can be used to solve some problems involving harmonic functions.

Theorem 3.6 *Let $f(z) = u(x, y) + iv(x, y)$ be an analytic function in the domain D. If all second-order partial derivatives of u and v are continuous, then both u and v are harmonic functions in D.*

Proof Since f is analytic, u and v satisfy the Cauchy-Riemann equations

$$(3) \qquad u_x = v_y \qquad \text{and} \qquad u_y = -v_x.$$

If we differentiate both sides of the equations in (3) with respect to x, we obtain

$$(4) \qquad u_{xx} = v_{yx} \qquad \text{and} \qquad u_{yx} = -v_{xx}.$$

Similarly, if we differentiate both sides of the equations in (3) with respect to y, then we obtain

$$(5) \qquad u_{xy} = v_{yy} \qquad \text{and} \qquad u_{yy} = -v_{xy}.$$

Since the partial derivatives u_{xy}, u_{yx}, v_{xy}, and v_{yx} are all continuous, a theorem from the calculus of real functions states that the mixed partial derivatives are equal; that is,

$$(6) \qquad u_{xy} = u_{yx} \qquad \text{and} \qquad v_{xy} = v_{yx}.$$

If we use (4), (5), and (6), then it follows that $u_{xx} + u_{yy} = v_{yx} - v_{xy} = 0$, and $v_{xx} + v_{yy} = -u_{yx} + u_{xy} = 0$. Therefore both u and v are harmonic functions.

Remark for Theorem 3.6 Corollary 5.2 in Chapter 5 will show that if $f(z)$ is analytic, then all the partial derivatives of u and v are continuous. Hence Theorem 3.6 holds for all analytic functions.

On the other hand, if we are given a function $u(x, y)$ that is harmonic in the domain D and if we can find another harmonic function $v(x, y)$ where their first partial derivatives satisfy the Cauchy-Riemann equations throughout D, then we say that $v(x, y)$ is the *harmonic conjugate* of $u(x, y)$. It then follows that the function $f(z) = u(x, y) + iv(x, y)$ is analytic in D.

EXAMPLE 3.10 If $u(x, y) = x^2 - y^2$, then $u_{xx}(x, y) + u_{yy}(x, y) = 2 - 2 = 0$; hence u is a harmonic function. We find that $v(x, y) = 2xy$ is also a harmonic function and that

$$u_x = v_y = 2x \qquad \text{and} \qquad u_y = -v_x = -2y.$$

Therefore v is the harmonic conjugate of u, and the function f given by

$$f(z) = x^2 - y^2 + i2xy = z^2$$

is an analytic function.

Harmonic functions are easily constructed from known analytic functions.

EXAMPLE 3.11 The function $f(z) = z^3 = x^3 - 3xy^2 + i(3x^2y - y^3)$ is analytic for all values of $z = x + iy$, and we see that

$$u(x, y) = x^3 - 3xy^2 \quad \text{and} \quad v(x, y) = 3x^2y - y^3$$

are harmonic functions. Indeed, v is the harmonic conjugate of u.

Complex variable techniques can be used to show that certain combinations of harmonic functions are harmonic. For example, if v is the harmonic conjugate of u, then their product $\phi(x, y) = u(x, y)v(x, y)$ is a harmonic function. This can be verified directly by computing the partial derivatives and showing that (2) holds true, but the details are tedious.

If we use complex variable techniques, we can start with the fact that $f(z) = u(x, y) + iv(x, y)$ is an analytic function. Then we observe that the square of f is also an analytic function and is given by $[f(z)]^2 = [u(x, y)]^2 - [v(x, y)]^2 + i2u(x, y)v(x, y)$. Hence the imaginary part of f^2 is $2u(x, y)v(x, y)$ and is a harmonic function. Since a constant multiple of a harmonic function is harmonic, it follows that ϕ is harmonic. It is left as an exercise to show that if u_1 and u_2 are two harmonic functions that are not related in the above fashion, then their product need not be harmonic.

Theorem 3.7 (Construction of a Conjugate) *Let $u(x, y)$ be harmonic in an ε-neighborhood of the point (x_0, y_0). Then there exists a conjugate harmonic function $v(x, y)$ defined in this neighborhood, and $f(z) = u(x, y) + iv(x, y)$ is an analytic function.*

Proof The harmonic function u and its conjugate harmonic function v will satisfy the Cauchy-Riemann equations $u_x = v_y$ and $u_y = -v_x$. We can construct $v(x, y)$ in a two-step process. First integrate v_y (which is equal to u_x) with respect to y:

$$(7) \qquad v(x, y) = \int u_x(x, y)\,dy + C(x)$$

where $C(x)$ is a function of x alone (that is, the partial derivative of $C(x)$ with respect to y is zero). Second, we are able to find $C'(x)$ by differentiating (7) with respect to x and replacing v_x with $-u_y$ on the left side:

$$(8) \qquad -u_y(x, y) = \frac{d}{dx} \int u_x(x, y)\,dy + C'(x).$$

As a matter of concern, all terms except those involving x in (8) will cancel, and a formula for $C'(x)$ involving x alone will be revealed. Now elementary integration of the single-variable function $C'(x)$ can be used to discover $C(x)$.

This technique is a practical method for constructing $v(x, y)$. Notice that both $u_x(x, y)$ and $u_y(x, y)$ are used in the process.

EXAMPLE 3.12 Show that $u(x, y) = xy^3 - x^3y$ is a harmonic function and find the conjugate harmonic function $v(x, y)$.

Solution The first partial derivatives are

$$(9) \qquad u_x(x, y) = y^3 - 3x^2y \quad \text{and} \quad u_y(x, y) = 3xy^2 - x^3.$$

To verify that u is harmonic, we use the second partial derivatives and see that $u_{xx}(x, y) + u_{yy}(x, y) = -6xy + 6xy = 0$, which implies that u is harmonic. To construct $v(x, y)$, we start with (7) and the first equation in (9) to get

$$(10) \qquad v(x, y) = \int [y^3 - 3x^2y]\,dy + C(x) = \frac{1}{4}y^4 - \frac{3}{2}x^2y^2 + C(x).$$

Differentiate the left and right sides of (10) with respect to x and use $-u_y(x, y) = v_x(x, y)$ and (9) on the left side to get

$$(11) \qquad -3xy^2 + x^3 = 0 - 3xy^2 + C'(x).$$

Cancel the terms involving both x and y in (11) and discover that

$$(12) \qquad C'(x) = x^3.$$

Integrate (12) and get $C(x) = \frac{1}{4}x^4 + C$, where C is a constant. Hence the harmonic conjugate of u is

$$v(x, y) = \frac{1}{4}x^4 - \frac{3}{2}x^2y^2 + \frac{1}{4}y^4 + C.$$

EXAMPLE 3.13 Let f be an analytic function in the domain D. If $|f(z)| \equiv K$ where K is a constant, then f is constant in D.

Solution Suppose that $K = 0$. Then $|f(z)|^2 = 0$, and hence $u^2 + v^2 = 0$. It follows that both $u \equiv 0$ and $v \equiv 0$, and therefore $f(z) \equiv 0$ in D.

Now suppose that $K \neq 0$; then we can differentiate the equation $u^2 + v^2 = K^2$ partially with respect to x and then with respect to y to obtain the system of equations

$$(13) \qquad 2uu_x + 2vv_x = 0 \quad \text{and} \quad 2uu_y + 2vv_y = 0.$$

The Cauchy-Riemann equations can be used in (13) to express the system in the form

$$(14) \qquad uu_x - vu_y = 0 \quad \text{and} \quad vu_x + uu_y = 0.$$

Treating u and v as coefficients, we easily solve (14) for the unknowns u_x and u_y:

$$u_x = \frac{\begin{vmatrix} 0 & -v \\ 0 & u \end{vmatrix}}{\begin{vmatrix} u & -v \\ v & u \end{vmatrix}} = \frac{0}{u^2 + v^2} = 0 \quad \text{and} \quad u_y = \frac{\begin{vmatrix} u & 0 \\ v & 0 \end{vmatrix}}{\begin{vmatrix} u & -v \\ v & u \end{vmatrix}} = \frac{0}{u^2 + v^2} = 0.$$

A theorem from the calculus of real functions states that the conditions $u_x \equiv 0$ and $u_y \equiv 0$ together imply that $u(x, y) \equiv c_1$ where c_1 is a constant. Similarly, we find that $v(x, y) \equiv c_2$, and therefore $f(z) \equiv c_1 + ic_2$.

Harmonic functions are solutions to many physical problems. Applications include two-dimensional models of heat flow, electrostatics, and fluid flow. For example, let us see how harmonic functions are used to study fluid flows. We must assume that an incompressible and frictionless fluid flows over the complex plane and that all cross sections in planes parallel to the complex plane are the same. Situations such as this occur when fluid is flowing in a deep channel. The velocity vector at the point (x, y) is

$$\text{(15)} \qquad \mathbf{V}(x, y) = p(x, y) + iq(x, y)$$

and is illustrated in Figure 3.1.

FIGURE 3.1 The vector field $\mathbf{V}(x, y) = p(x, y) + iq(x, y)$, which can be considered as a fluid flow.

The assumptions that the flow is irrotational and has no sources or sinks implies that both the curl and divergence vanish, that is, $q_x - p_y = 0$ and $p_x + q_y = 0$. Hence p and q obey the equations

$$\text{(16)} \qquad p_x(x, y) = -q_y(x, y) \quad \text{and} \quad p_y(x, y) = q_x(x, y).$$

The equations in (16) are similar to the Cauchy-Riemann equations and permit us to define a special complex function:

$$\text{(17)} \qquad f(z) = u(x, y) + iv(x, y) = p(x, y) - iq(x, y).$$

Here we have $u_x = p_x$, $u_y = p_y$, $v_x = -q_x$, and $v_y = -q_y$. Now (16) can be used to obtain the Cauchy-Riemann equations for $f(z)$:

$$\text{(18)} \qquad \begin{aligned} u_x(x, y) &= p_x(x, y) = -q_y(x, y) = v_y(x, y), \\ u_y(x, y) &= p_y(x, y) = q_x(x, y) = -v_x(x, y). \end{aligned}$$

Therefore the function $f(z)$ defined in (17) is analytic, and the fluid flow (15) is the conjugate of an analytic function, that is,

$$\text{(19)} \qquad \mathbf{V}(x, y) = \overline{f(z)}.$$

In Chapter 5 we will prove that every analytic function $f(z)$ has an analytic antiderivative $F(z)$; hence we are justified to write

$$F(z) = \phi(x, y) + i\psi(x, y) \qquad \text{where} \qquad F'(z) = f(z). \tag{20}$$

Observe that $\phi(x, y)$ is a harmonic function. If we use the vector interpretation of a complex number, then the gradient of $\phi(x, y)$ can be written as follows:

$$\text{grad}\,\phi(x, y) = \phi_x(x, y) + i\phi_y(x, y). \tag{21}$$

The Cauchy-Riemann equations applied to $F(z)$ give us $\phi_y = -\psi_x$ and equation (21) becomes

$$\text{grad}\,\phi(x, y) = \phi_x(x, y) - i\psi_x(x, y) = \overline{\phi_x(x, y) + i\psi_x(x, y)}. \tag{22}$$

Theorem 3.3 says that $\phi_x(x, y) + i\psi_x(x, y) = F'(z)$, which can be substituted in (22) to obtain

$$\text{grad}\,\phi(x, y) = \overline{F'(z)}. \tag{23}$$

Now use $F'(z) = f(z)$ in (23) to conclude that $\phi(x, y)$ is the scalar potential function for the fluid flow (19), that is,

$$\mathbf{V}(x, y) = \text{grad}\,\phi(x, y). \tag{24}$$

The curves $\phi(x, y) = \text{constant}$ are called *equipotentials*. The curves $\psi(x, y) = \text{constant}$ are called *streamlines* and describe paths of fluid flow. In Chapter 9 we will see that the family of equipotentials is orthogonal to the family of streamlines (see Figure 3.2).

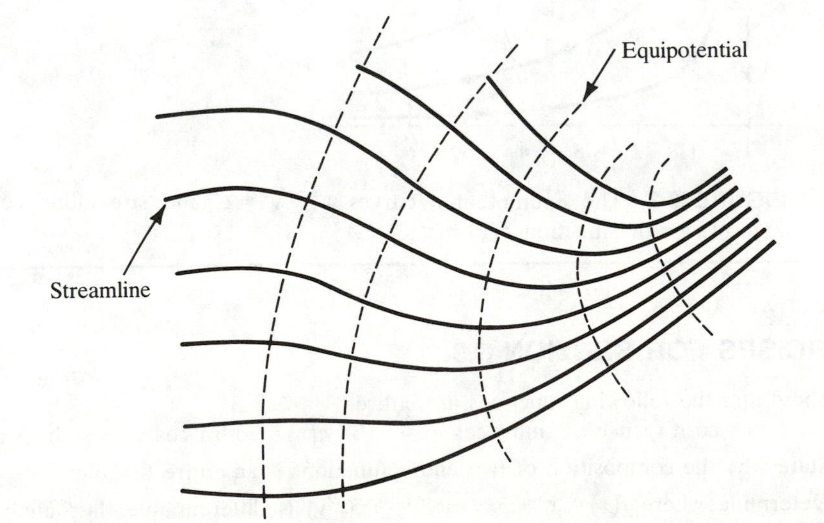

FIGURE 3.2 The families of orthogonal curves $\{\phi(x, y) = \text{constant}\}$ and $\{\psi(x, y) = \text{constant}\}$ for the function $F(z) = \phi(x, y) + i\psi(x, y)$.

EXAMPLE 3.14 Show that the harmonic function $\phi(x, y) = x^2 - y^2$ is the scalar potential function for the fluid flow

$$\mathbf{V}(x, y) = 2x - i2y.$$

Solution The fluid flow can be written as

$$\mathbf{V}(x, y) = \overline{f(z)} = \overline{2x + i2y} = \overline{2z}.$$

The antiderivative of $f(z) = 2z$ is $F(z) = z^2$, and the real part of $F(z)$ is the desired harmonic function:

$$\phi(x, y) = \text{Re } F(z) = \text{Re}(x^2 - y^2 + i2xy) = x^2 - y^2.$$

Observe that the hyperbolas $\phi(x, y) = x^2 - y^2 = C$ are the equipotential curves, and the hyperbolas $\psi(x, y) = 2xy = C$ are the streamline curves; these curves are orthogonal, as is shown in Figure 3.3.

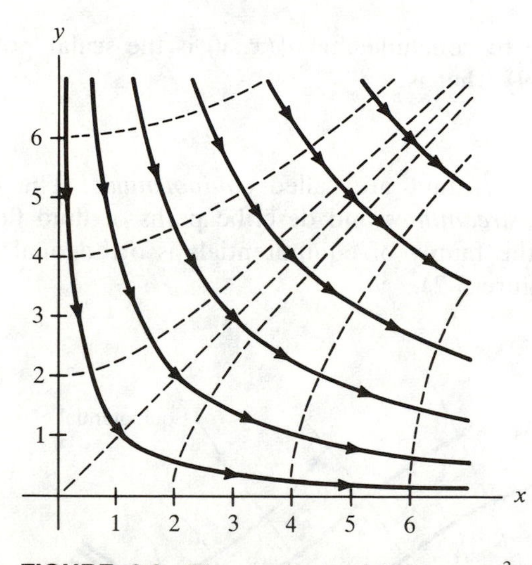

FIGURE 3.3 The equipotential curves $x^2 - y^2 = C$ and streamline curves $2xy = C$ for the function $F(z) = z^2$.

EXERCISES FOR SECTION 3.3

1. Show that the following functions are entire.
   (a) $f(z) = \cosh x \sin y - i \sinh x \cos y$ (b) $g(z) = \cosh x \cos y + i \sinh x \sin y$
2. State why the composition of two entire functions is an entire function.
3. Determine where $f(z) = x^3 + 3xy^2 + i(y^3 + 3x^2y)$ is differentiable. Is f analytic? Why?
4. Determine where $f(z) = 8x - x^3 - xy^2 + i(x^2y + y^3 - 8y)$ is differentiable. Is f analytic? Why?

5. Let $f(z) = x^2 - y^2 + i2|xy|$.
(a) Where does f have a derivative? **(b)** Where is f analytic?

6. Show that $u(x, y) = e^x \cos y$ and $v(x, y) = e^x \sin y$ are harmonic for all values of (x, y).

7. Let $u(x, y) = \ln(x^2 + y^2)$ for $(x, y) \neq (0, 0)$. Compute the partial derivatives of u, and verify that u satisfies Laplace's equation.

8. Let a, b, and c be real constants. Determine a relation among the coefficients that will guarantee that the functions $\phi(x, y) = ax^2 + bxy + cy^2$ is harmonic.

9. Does an analytic function $f(z) = u(x, y) + iv(x, y)$ exist for which $v(x, y) = x^3 + y^3$? Why?

10. Find the analytic function $f(z) = u(x, y) + iv(x, y)$ given the following.
(a) $u(x, y) = y^3 - 3x^2y$ **(b)** $u(x, y) = \sin y \sinh x$
(c) $v(x, y) = e^y \sin x$ **(d)** $v(x, y) = \sin x \cosh y$

11. Let $v(x, y) = \arctan(y/x)$ for $x \neq 0$. Compute the partial derivatives of v, and verify that v satisfies Laplace's equation.

12. Let $u(x, y)$ be harmonic. Show that $U(x, y) = u(x, -y)$ is harmonic. *Hint*: Use the chain rule for differentiation of real functions.

13. Let $u_1(x, y) = x^2 - y^2$ and $u_2(x, y) = x^3 - 3xy^2$. Show that u_1 and u_2 are harmonic functions and that their product $u_1(x, y)u_2(x, y)$ is not a harmonic function.

14. Let v be the harmonic conjugate of u. Show that $-u$ is the harmonic conjugate of v.

15. Let v be the harmonic conjugate of u. Show that $h = u^2 - v^2$ is a harmonic function.

16. Suppose that v is the harmonic conjugate of u and that u is the harmonic conjugate of v. Show that u and v must be constant functions.

17. Let f be an analytic function in the domain D. If $f'(z) = 0$ for all z in D, then show that f is constant in D.

18. Let f and g be analytic functions in the domain D. If $f'(z) = g'(z)$ for all z in D, then show that $f(z) = g(z) + C$ where C is a complex constant.

19. Let f be a nonconstant analytic function in the domain D. Show that the function $g(z) = \overline{f(z)}$ is *not* analytic in D.

20. Let $f(z) = f(re^{i\theta}) = \ln r + i\theta$ where $r > 0$ and $-\pi < \theta < \pi$. Show that f is analytic in the domain indicated and that $f'(z) = 1/z$.

21. Let $f(z) = f(re^{i\theta}) = u(r, \theta) + iv(r, \theta)$ be analytic in a domain D that does not contain the origin. Use the polar form of the Cauchy-Riemann equations $u_\theta = -rv_r$ and $v_\theta = ru_r$, and differentiate them with respect to θ and then with respect to r. Use the results to establish the *polar form of Laplace's equation*:

$$r^2u_{rr}(r, \theta) + ru_r(r, \theta) + u_{\theta\theta}(r, \theta) = 0$$

22. Use the polar form of Laplace's equation given in Exercise 21 to show that $u(r, \theta) = r^n \cos n\theta$ and $v(r, \theta) = r^n \sin n\theta$ are harmonic functions.

23. Use the polar form of Laplace's equation given in Exercise 21 to show that

$$u(r, \theta) = \left(r + \frac{1}{r}\right)\cos\theta \qquad \text{and} \qquad v(r, \theta) = \left(r - \frac{1}{r}\right)\sin\theta$$

are harmonic functions.

24. Let f be an analytic function in the domain D. Show that if $\text{Re}(f(z)) \equiv 0$ at all points in D, then f is constant in D.

25. Assume that $F(z) = \phi(x, y) + i\psi(x, y)$ is analytic in the domain D and that $F'(z) \neq 0$ in D. Consider the families of level curves $\{\phi(x, y) = \text{constant}\}$ and $\{\psi(x, y) = \text{constant}\}$, which are the equipotentials and streamlines for the fluid flow $\mathbf{V}(x, y) = \overline{F'(z)}$. Prove that the two families of curves are orthogonal. *Hint*: Suppose that (x_0, y_0) is a point common to the two curves $\phi(x, y) = c_1$ and $\psi(x, y) = c_2$. Take the gradients of ϕ and ψ, and show that the normals to the curves are perpendicular.

26. The function $F(z) = 1/z$ is used to determine a field known as a dipole. Express $F(z)$ in the form $F(z) = \phi(x, y) + i\psi(x, y)$ and sketch the equipotentials $\phi = 1$, $1/2$, $1/4$ and the streamlines $\psi = 1, 1/2, 1/4$.

27. The logarithmic function will be introduced in Chapter 4. Let $F(z) = \log z = \ln|z| + i \arg z$. Here we have $\phi(x, y) = \ln|z|$ and $\psi(x, y) = \arg z$. Sketch the equipotentials $\phi = 0, \ln 2, \ln 3, \ln 4$ and the streamlines $\psi = k\pi/8$ for $k = 0, 1, \ldots, 7$.

4

Elementary Functions

In this chapter we study various elementary functions that are encountered in calculus. We have already seen how polynomials, quotients of polynomials, and the principal *n*th root function are defined when the real variable is replaced with a complex variable. We will now show how to arrive at natural definitions for the complex functions corresponding to the exponential, logarithmic, trigonometric, and hyperbolic functions. These complex functions will be extensions of the real variable functions and will be consistent with the definitions of the latter. We first turn our attention to a problem mentioned earlier.

4.1 Branches of Functions

In Section 2.3 we defined the principal square root function and investigated some of its properties. We left some unanswered questions concerning the choices of square roots. We now look into this problem because it is similar to situations involving other elementary functions.

In our definition of a function in Section 2.1 we specified that each value of the independent variable in the domain is mapped onto one and only one value of the dependent variable. As a result, one often talks about a "single-valued function," which emphasizes the "only one" part of the definition and allows us to distinguish such functions from multiple-valued functions, which we now introduce.

Let $w = f(z)$ denote a function whose domain is the set D and whose range is the set R. If w is a value in the range, then there is an associated "inverse function" $z = g(w)$ that assigns to each value w the value (or values) of z in D for which the equation $f(z) = w$ holds true. But unless f takes on the value w at most once in D, then the "inverse function" g is necessarily many-valued, and we say that g is a *multivalued function*. For example, the inverse of the function $w = f(z) = z^2$ is the "square root function" $z = g(w) = w^{1/2}$. We see that for each value z other than $z = 0$ the two points z and $-z$ are mapped onto the same point $w = f(z)$; hence g is in general a two-valued function.

The study of limits, continuity, and derivatives loses all meaning if an arbitrary or ambiguous assignment of function values is made. For this reason we did not allow multivalued functions to be considered when we defined these concepts. When working with "inverse functions," it is necessary to carefully specify one of the many possible inverse values when constructing an inverse function. The idea is the same as determining implicit functions in calculus. If the values of a function f are determined by an equation that they satisfy rather than by an explicit formula, then we say that the function is defined implicitly or that f is an *implicit function*. In the theory of complex variables we study a similar concept.

Let $w = f(z)$ be a multiple-valued function. A *branch* of f is any single-valued function f_0 that is analytic in *some* domain and, at each point z in the domain, assigns one of the values of $f(z)$.

EXAMPLE 4.1 Let us consider some branches of the two-valued square root function $f(z) = z^{1/2}$. Let us use the principal square root function and define

$$(1) \qquad f_1(z) = r^{1/2} \cos \frac{\theta}{2} + i r^{1/2} \sin \frac{\theta}{2},$$

where we require that $r > 0$ and $-\pi < \theta \leqq \pi$. Then f_1 is a branch of f. We can find other branches of the square root function. For example, let

$$(2) \qquad f_2(z) = r^{1/2} \cos \frac{\theta + 2\pi}{2} + i r^{1/2} \sin \frac{\theta + 2\pi}{2}$$

where $r > 0$ and $-\pi < \theta \leqq \pi$.

If we use the identities $\cos((\theta + 2\pi)/2) = -\cos(\theta/2)$ and $\sin((\theta + 2\pi)/2) = -\sin(\theta/2)$, then we see that

$$f_2(z) = -r^{1/2} \cos \frac{\theta}{2} - i r^{1/2} \sin \frac{\theta}{2} = -f_1(z),$$

so f_1 and f_2 can be thought of as "plus" and "minus" square root functions.

The negative x axis is called a *branch cut* for the functions f_1 and f_2. It is characterized by the fact that each point on the branch cut is a point of discontinuity for both functions f_1 and f_2.

EXAMPLE 4.2 To show that the function f_1 is discontinuous along the negative x axis, we let $r_0 e^{\pm i\pi} = z_0$ denote a negative real number. Now we compute the limit of $f_1(z)$ as z approaches z_0 through the upper half-plane $\operatorname{Im} z > 0$ and the limit of $f_1(z)$ as z approaches z_0 through the lower half-plane

$\text{Im}\, z < 0$. In polar coordinates these limits are given by

$$\lim_{(r,\theta)\to(r_0,\pi)} f_1(re^{i\theta}) = \lim_{(r,\theta)\to(r_0,\pi)} r^{1/2}\left(\cos\frac{\theta}{2} + i\sin\frac{\theta}{2}\right) = ir_0^{1/2} \quad \text{and}$$

$$\lim_{(r,\theta)\to(r_0,-\pi)} f_1(re^{i\theta}) = \lim_{(r,\theta)\to(r_0,-\pi)} r^{1/2}\left(\cos\frac{\theta}{2} + i\sin\frac{\theta}{2}\right) = -ir_0^{1/2}.$$

Since the two limits are distinct, the function f_1 is discontinuous at z_0. Likewise, f_2 is discontinuous at z_0. The mappings $w = f_1(z)$ and $w = f_2(z)$ and the branch cut are illustrated in Figure 4.1.

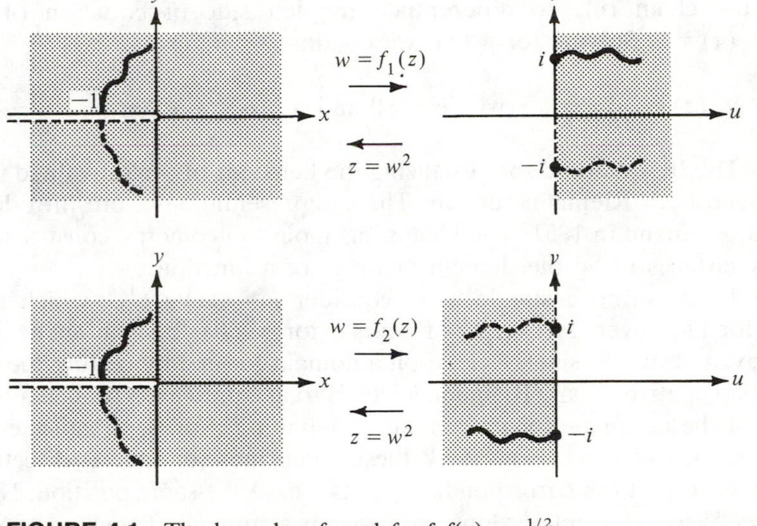

FIGURE 4.1 The branches f_1 and f_2 of $f(z) = z^{1/2}$.

Other branches of the square root function can be constructed by specifying that an argument of z given by $\theta = \arg z$ is to lie in the interval $\alpha < \theta \leqq \alpha + 2\pi$. Then the branch f_α is given by

$$(3) \qquad f_\alpha(z) = r^{1/2}\cos\frac{\theta}{2} + ir^{1/2}\sin\frac{\theta}{2} \qquad \text{where } r > 0 \text{ and } \alpha < \theta \leqq \alpha + 2\pi.$$

The branch cut for f_α is the ray $r \geqq 0$, $\theta = \alpha$, which includes the origin. The point $z = 0$, common to all branch cuts for the multivalued function, is called a *branch point*. The mapping $w = f_\alpha(z)$ and its branch cut are illustrated in Figure 4.2.

The derivative of the square root function can be obtained by using implicit differentiation. If f_α is given by equation (3), then

$$(4) \qquad [f_\alpha(z)]^2 = z.$$

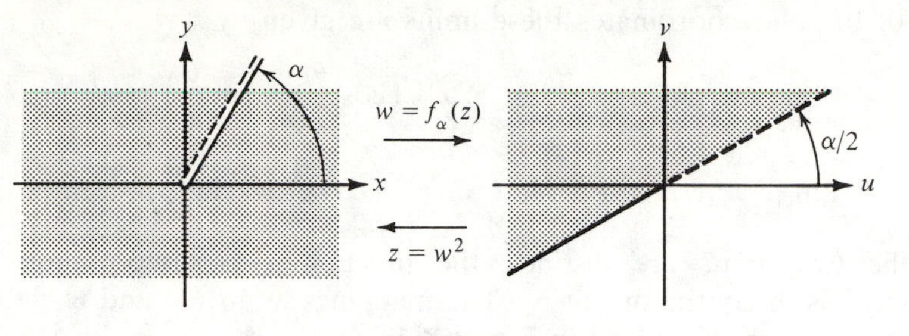

FIGURE 4.2 The branch f_α of $f(z) = z^{1/2}$.

Using the chain rule to differentiate the left side of equation (4) yields $2f_\alpha(z)f'_\alpha(z) = 1$. Solving for $f'_\alpha(z)$, we obtain

$$(5) \qquad f'_\alpha(z) = \frac{1}{2f_\alpha(z)} \qquad \text{where } r > 0 \text{ and } \alpha < \theta < \alpha + 2\pi.$$

The best method for visualizing the behavior of a multivalued function is provided by a Riemann surface. These representations were introduced by G. F. B. Riemann in 1851. The idea is ingenious, a geometric construction that permits surfaces to be the domain or range of a function.

To illustrate a simple case, consider $w = f(z) = z^{1/2}$, which has two values for any given z (except, of course, for $z = 0$). Each branch $f_1(z)$ and $f_2(z)$, given above, is single-valued on a domain formed by cutting the z-plane along the negative x axis. Riemann's idea is to assign to each branch a different replica of the cut plane for its domain of definition. For $f_1(z)$, call the domain S_1; for $f_2(z)$, call it S_2. Now stack these cut planes (or sheets) directly upon each other so that the corresponding points z have the same position. Now join the sheet S_1 to the sheet S_2 along the two cuts as follows. The edge of S_1 in the upper half-plane is joined to the edge of the sheet S_2 in the lower half-plane. Similarly, the edge of the sheet S_1 in the lower half-plane is joined to the edge of S_2 in the upper half-plane. The two planes are regarded as being "glued" together along these two distinct edges and form the Riemann surface for $w = f(z) = z^{1/2}$. Although this surface cannot be accurately drawn or photographed, an attempt is made to visualize it in Figure 4.3.

The points $a_1 = -1 - i$, $b_1 = 1 - i$, $c_1 = 1 + i$ and $d_1 = -1 + i$ lie on the sheet S_1, and the points $a_2 = -1 - i$, $b_2 = 1 - i$, $c_2 = 1 + i$ and $d_2 = -1 + i$ lie on S_2. The two sheets have the point $z = 0$ in common. Now let us describe how to wind around $z = 0$ on the Riemann surface.

Start at a_1 on S_1 and move along the line segments $\overparen{a_1b_1}$, $\overparen{b_1c_1}$ and $\overparen{c_1d_1}$, which form three sides of a square lying entirely on S_1. We cannot move from d_1 to a_1 on S_1 because the sheet is cut along the negative x-axis. Instead, we move along the sheet S_1 from d_1 to the negative x axis, where we cross over onto S_2 and proceed along this sheet of the surface to a_2. Now continue on S_2 and move along the line segments $\overparen{a_2b_2}$, $\overparen{b_2c_2}$, and $\overparen{c_2d_2}$, which form three sides of a different square that lies entirely on S_2. Again, we must cross over a cut in

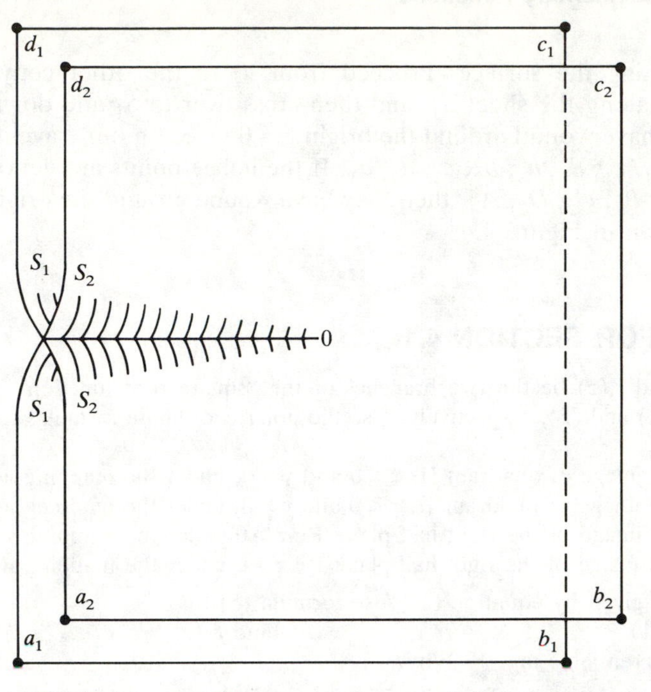

FIGURE 4.3 The Riemann surface consisting of two sheets S_1 and S_2 that is the domain for the mapping $w = f(z) = z^{1/2}$.

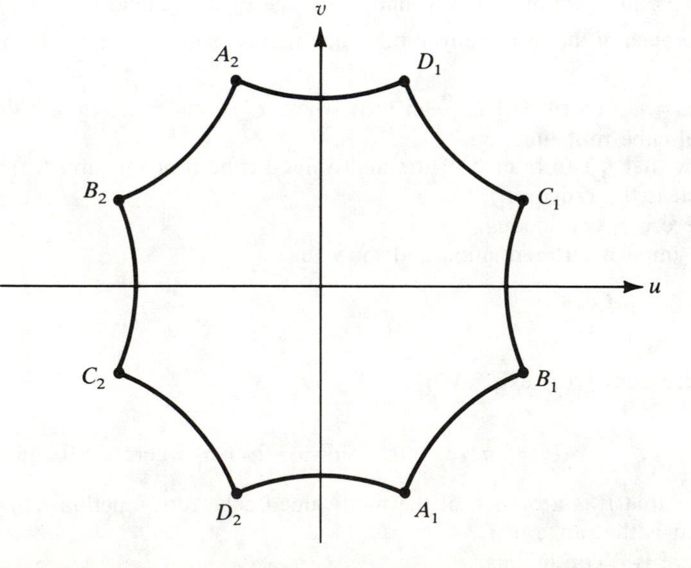

FIGURE 4.4 The points A_1, B_1, C_1, D_1, A_2, B_2, C_2, D_2, A_1 in the range of the mapping $w = f(z) = z^{1/2}$.

the sheets forming the surface. Proceed from d_2 to the other copy of the negative x axis along the sheet S_2, and then cross over to S_1 and down to a_1. Notice that we have wound around the origin $z = 0$ twice on our travel through the points $a_1, b_1, c_1, d_1, a_2, b_2, c_2, d_2, a_1$. If the image points are denoted A_1, $B_1, C_1, D_1, A_2, B_2, C_2, D_2, A_1$, then they have wound around the origin $w = 0$ once, as is shown in Figure 4.4.

EXERCISES FOR SECTION 4.1

1. Let $f_1(z)$ and $f_2(z)$ be the two branches of the "square root function" given by equations (1) and (2), respectively. Use the polar coordinate formulas in Section 2.3 to:
(a) Find the image of quadrant II, $x < 0$ and $y > 0$, under the mapping $w = f_1(z)$.
(b) Find the image of quadrant II, $x < 0$ and $y > 0$, under the mapping $w = f_2(z)$.
(c) Find the image of the right half-plane $\text{Re } z > 0$ under the mapping $w = f_1(z)$.
(d) Find the image of the right half-plane $\text{Re } z > 0$ under the mapping $w = f_2(z)$.

2. Let $f_1(z)$ be given by equation (1). Use formula (5) to:
(a) Find $f_1'(1)$. **(b)** Find $f_1'(i)$.
(c) Is f_1 differentiable at -1? Why?

3. Let $f_2(z)$ be given by equation (2). Use formula (5) to:
(a) Find $f_2'(1)$. **(b)** Find $f_2'(i)$.
(c) Is f_2 differentiable at -1? Why?

4. Let $\alpha = 0$ in equation (3), and find the range of the function $w = f(z)$.

5. Let $\alpha = 2\pi$ in equation (3), and find the range of the function $w = f(z)$.

6. Find a branch of the "square root function" that is continuous along the negative x axis.

7. Let $f_1(z) = r^{1/3}\cos(\theta/3) + ir^{1/3}\sin(\theta/3)$ where $r > 0$ and $-\pi < \theta \leqq \pi$ denote the principal cube root function.
(a) Show that f_1 is a branch of the multivalued cube root function $f(z) = z^{1/3}$.
(b) What is the range of f_1?
(c) Where is f_1 continuous?
(d) Use implicit differentiation and show that

$$f_1'(z) = \frac{1}{3}\frac{f_1(z)}{z}$$

(e) Where does $f_1'(z)$ exist? Why?
(f) Find $f_1'(i)$.

8. Let $f_2(z) = r^{1/3}\cos((\theta + 2\pi)/3) + ir^{1/3}\sin((\theta + 2\pi)/3)$ where $r > 0$ and $-\pi < \theta \leqq \pi$.
(a) Show that f_2 is a branch of the multivalued cube root function $f(z) = z^{1/3}$.
(b) What is the range of f_2?
(c) Where is f_2 continuous?
(d) What is the branch point associated with f?

9. Find a branch of the multivalued cube root function that is different from those in Exercises 7 and 8. State the domain and range of the branch you find.

10. Let $f(z) = z^{1/n}$ denote the multivalued "nth root function" where n is a positive integer.

(a) Show that f is in general an n-valued function.

(b) Write down the principal nth root function.

(c) Write down a branch of the multivalued "nth root function" that is different than the one in part (b).

(d) Use implicit differentiation to show that if $f_0(z)$ is a branch of the nth root function, then

$$f_0'(z) = \frac{1}{n} \frac{f_0(z)}{z}$$

holds in some suitably chosen domain.

11. Describe a Riemann surface for the domain of definition of the multivalued function $w = f(z) = z^{1/3}$.

12. Describe a Riemann surface for the domain of definition of the multivalued function $w = f(z) = z^{1/4}$.

13. Discuss how Riemann surfaces should be used for both the domain of definition and the range to help describe the behavior of the multivalued function $w = f(z) = z^{2/3}$.

4.2 The Exponential Function

One of the most important analytic functions is the exponential function. We wish to define the complex exponential function $f(z) = e^z$ so that it is analytic and coincides with the the real exponential when $z = x + 0i$, that is, $f(x + 0i) = e^x$. Recall that the real exponential is the solution to the differential equation

$$f'(x) = f(x) \qquad \text{with } f(0) = 1.$$

We must show how to construct $f(z)$ so that

$$(1) \qquad f \text{ is entire}, \qquad f'(z) = f(z), \qquad \text{and} \qquad f(0) = 1.$$

Starting with $f(z) = u(x, y) + iv(x, y)$ and $f'(z) = f(z)$, Theorem 3.3 tells us that $u_x = u$ and $v_x = v$. Let us separate the variables of u. The equation $u_x(x, y) - u(x, y) = 0$ leads to

$$[u_x(x, y) - u(x, y)]e^{-x} = 0,$$

and it is easily seen that

$$(2) \qquad \frac{\partial}{\partial x}[u(x, y)e^{-x}] = 0.$$

Partial integration of (2) with respect to x gives $u(x, y)e^{-x} = p(y)$ where $p(y)$ is a function of y alone. Therefore we must have

$$(3) \qquad u(x, y) = p(y)e^x.$$

Similarly, there exists function $q(y)$, of y alone, so that

$$v(x, y) = q(y)e^x. \tag{4}$$

The partial derivatives of u and v are

$$\begin{aligned} u_x(x, y) &= p(y)e^x, & v_x(x, y) &= q(y)e^x, \\ u_y(x, y) &= p'(y)e^x, & v_y(x, y) &= q'(y)e^x. \end{aligned} \tag{5}$$

The Cauchy-Riemann equations can be used with (5) to obtain $p(y)e^x = q'(y)e^x$ and $p'(y)e^x = -q(y)e^x$, which yield

$$p(y) = q'(y) \qquad \text{and} \qquad p'(y) = -q(y). \tag{6}$$

The initial condition $f(0) = 1$ implies that $u(0, 0) = 1$ and $v(0, 0) = 0$; hence (3) and (4) yield $p(0) = 1$ and $q(0) = 0$. When equations (6) are differentiated, we get $p''(y) = -q'(y)$ and $q''(y) = p'(y)$, which can be used with (6) to obtain the following two initial value problems:

$$\begin{aligned} p'' + p &= 0 \qquad \text{with } p(0) = 1 \text{ and } p'(0) = 0. \\ q'' + q &= 0 \qquad \text{with } q(0) = 0 \text{ and } q'(0) = 1. \end{aligned} \tag{7}$$

The solutions to (7) are easily found to be $p(y) = \cos y$ and $q(y) = \sin y$. Therefore the complex exponential function is $f(z) = e^x \cos y + ie^x \sin y$, where y is taken in radians, and it reduces to e^x when z is real, that is, $f(x + 0i) = e^x$.

Definition 4.1 (Exponential Function) *The complex exponential function is*

$$e^z = e^x \cos y + ie^x \sin y = e^x(\cos y + i \sin y). \tag{8}$$

When z is the pure imaginary number $i\theta$, then identity (8) becomes

$$e^{i\theta} = \cos \theta + i \sin \theta. \tag{9}$$

This is Euler's formula and was introduced in Section 1.4. The use of the symbol $e^{i\theta}$ given there is now seen to be consistent with our definition of the complex exponential. Another notation, $\exp z$, is used to denote the exponential function, and we write

$$\exp z = e^x \cos y + ie^x \sin y. \tag{10}$$

The notation in (10) is better than that in (8) for some situations. For example,

$$\exp(1/5) = 1.22140275816017 \tag{11}$$

is the positive fifth root of $e = 2.71828182845904$.

The notation $e^{1/5}$ could be interpreted as any one of the five fifth roots of the number e (see Section 1.5):

$$e^{1/5} = 1.22140275816017\left(\cos \frac{2\pi k}{5} + i \sin \frac{2\pi k}{5}\right), \tag{12}$$

for $k = 0, 1, \ldots, 4$.

To prevent confusion, we often use the notation (10) instead of notation (8) for the exponential. In this text

(13) $f(z) = \exp z = e^z$

will always be used to denote the single-valued exponential function given in equations (8) and (10).

A surprising fact is that the exponential function is periodic. Since the defining equation (8) involves the periodic functions $\cos y$ and $\sin y$, we must conclude that any point $w = \exp z$ in range of $\exp z$ is actually the image of an infinite number of points. If two points in the z plane have the same real part and their imaginary parts differ by an integral multiple of 2π, then they are mapped onto the same point (see Figure 4.5). Therefore the complex exponential function is periodic with period 2π; that is,

(14) $e^{z+2\pi i} = e^z$ holds for all values of z.

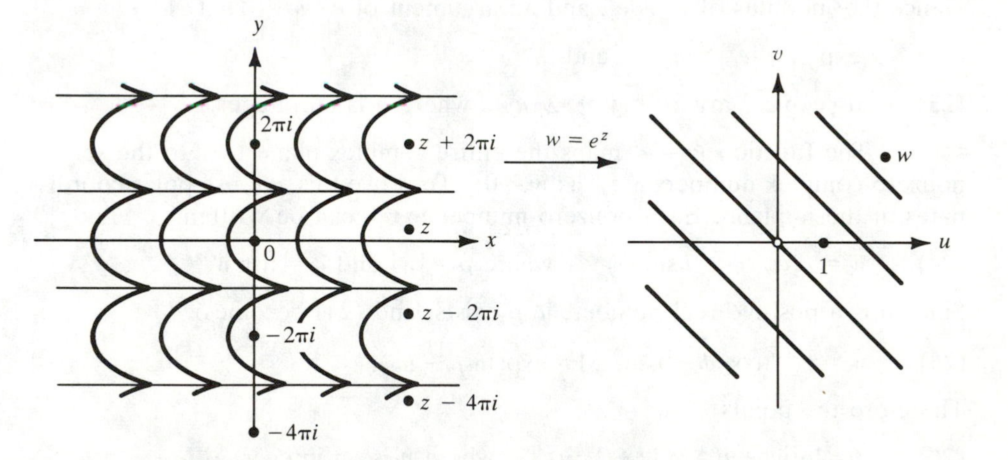

FIGURE 4.5 The many-to-one mapping $w = \exp z$.

Using (13) and (10) and $e^0 = 1$, we find that

(15) $e^z = 1$ for $z = 2n\pi i$ where n is an integer.

Another consequence is that the equation

(16) $\exp(z_1) = \exp(z_2)$

holds if and only if $z_2 = z_1 + 2k\pi i$ where k is some integer.

The additive property of the exponential function,

(17) $\exp(z_1 + z_2) = \exp(z_1)\exp(z_2)$

is valid for all complex values z_1 and z_2. To prove this, we write

(18) $\exp z_1 = e^{x_1}e^{iy_1}$ and $\exp z_2 = e^{x_2}e^{iy_2}$.

From our previous work with polar coordinates, identity (17) in Section 1.4 states that

$$e^{iy_1}e^{iy_2} = e^{i(y_1+y_2)}.$$

This together with equations (17) and (18) gives

$$\exp(z_1 + z_2) = e^{x_1+x_2}e^{i(y_1+y_2)} = e^{x_1}e^{x_2}e^{iy_1}e^{iy_2} \tag{19}$$
$$= e^{x_1}e^{iy_1}e^{x_2}e^{iy_2} = (\exp z_1)(\exp z_2).$$

Let us investigate the modulus and argument of $\exp z$. Definition 4.1 of the exponential function $w = e^z$ can be used with polar coordinates $w = \rho(\cos\phi + i\sin\phi)$ in the w plane to obtain

$$\rho(\cos\phi + i\sin\phi) = e^z = e^x(\cos y + i\sin y), \tag{20}$$

and it is easy to see in (20) that we must have

$$\rho = e^x \quad \text{and} \quad \phi = y. \tag{21}$$

Hence the modulus of e^z is e^x, and an argument of e^z is y. That is,

$$|\exp z| = |e^z| = e^x \quad \text{and} \tag{22}$$

$$\arg \exp z = \arg e^z = y + 2\pi n \quad \text{where } n \text{ is an integer.} \tag{23}$$

The function $w = e^z$ maps the entire complex plane C onto the set of nonzero complex numbers $S = \{w: w \neq 0\}$. To prove this, we use polar coordinates in the w plane. Each nonzero number $w \neq 0$ can be written

$$w = \rho(\cos\phi + i\sin\phi) \quad \text{where } \rho = |w| \text{ and } \phi = \arg w. \tag{24}$$

Since ρ is a positive real number, $\ln\rho$ exists, and (24) becomes

$$w = e^{\ln\rho}(\cos\phi + i\sin\phi) = \exp(\ln\rho + i\phi). \tag{25}$$

Therefore the points

$$z = \ln|w| + i(\operatorname{Arg} w + 2\pi n) \quad \text{where } n \text{ is an integer} \tag{26}$$

are mapped onto w by $w = \exp z$.

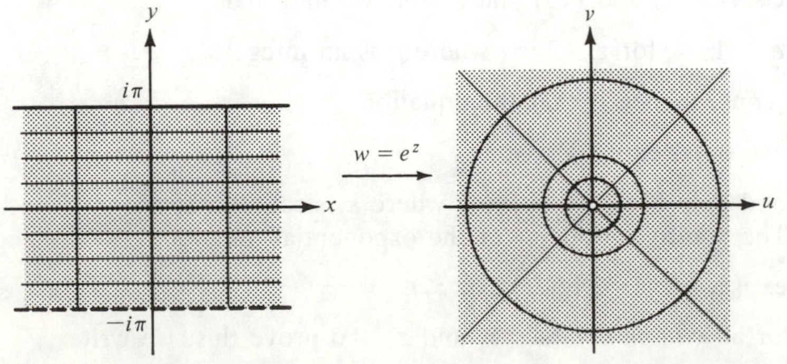

FIGURE 4.6 The fundamental period strip for the mapping $w = \exp z$.

Identity (26) shows that the range of f contains all the points in the w plane except the point $w = 0$. If we restrict the domain of definition of f to be the horizontal strip $-\pi < y \leqq \pi$, then the mapping $w = \exp z$ is a one-to-one mapping of this strip onto the range, since there is only one value $z = \ln|w| + i \operatorname{Arg} w$ that is mapped onto w. This strip is called the *fundamental period strip* for the mapping $w = \exp z$ and is illustrated in Figure 4.6.

EXAMPLE 4.3 Let us find all the values of z for which the equation

(27) $\exp z = \sqrt{3} - i = w$

holds true.

Solution Using the polar form $\sqrt{3} - i = 2e^{-i\pi/6}$, we see that

$\rho = |w| = |\sqrt{3} - i| = 2$ and

$$\phi = \arg w = \frac{-\pi}{6} + 2\pi n \qquad \text{where } n = 0, \pm 1, \pm 2, \ldots .$$

So that the solutions of (27) are given by

(28) $$z = \ln 2 + i\left(\frac{-\pi}{6} + 2\pi n\right) \qquad \text{where } n \text{ is an integer.}$$

EXERCISES FOR SECTION 4.2

1. Show that $\exp(z + i\pi) = \exp(z - i\pi)$ holds for all z.
2. Find the value of $e^z = u + iv$ for the following values of z.
   (a) $\dfrac{-i\pi}{3}$ **(b)** $\dfrac{1}{2} - \dfrac{i\pi}{4}$ **(c)** $-4 + i5$
   (d) $-1 + \dfrac{i3\pi}{2}$ **(e)** $1 + \dfrac{i5\pi}{4}$ **(f)** $\dfrac{\pi}{3} - 2i$
3. Find all values of z for which the following equations hold.
   (a) $e^z = -4$ **(b)** $e^z = 2 + 2i$
   (c) $e^z = \sqrt{3} - i$ **(d)** $e^z = -1 + i\sqrt{3}$
4. Express $\exp(z^2)$ and $\exp(1/z)$ in Cartesian form $u(x, y) + iv(x, y)$.
5. Show that:
   (a) $\exp(\overline{z}) = \overline{\exp z}$ holds for all z and that
   (b) $\exp(\overline{z})$ is nowhere analytic.
6. Show that $|e^{-z}| < 1$ if and only if $\operatorname{Re} z > 0$.
7. Show that:
   (a) $\displaystyle\lim_{z \to 0} \frac{e^z - 1}{z} = 1$ **(b)** $\displaystyle\lim_{z \to i\pi} \frac{e^z + 1}{z - i\pi} = -1$
8. Show that $f(z) = ze^z$ is analytic for all z by showing that its real and imaginary parts satisfy the Cauchy-Riemann equations.

9. Find the derivatives of the following:
(a) e^{iz} **(b)** $z^4 \exp(z^3)$
(c) $e^{(a+ib)z}$ **(d)** $\exp(1/z)$

10. Let n be a positive integer. Show that:
(a) $(\exp z)^n = \exp(nz)$ **(b)** $\dfrac{1}{(\exp z)^n} = \exp(-nz)$

11. Show that the image of the horizontal ray $x > 0$, $y = \pi/3$ under the mapping $w = \exp z$ is a ray.

12. Show that the image of the vertical line segment $x = 2$, $y = t$ where $\pi/6 < t < 7\pi/6$ under the mapping $w = e^z$ is half of a circle.

13. Show that $\exp(z_1 - z_2) = (\exp(z_1))/(\exp(z_2))$ holds for all z_1, z_2.

14. Use the fact that $\exp(z^2)$ is analytic to show that $e^{x^2-y^2} \sin 2xy$ is a harmonic function.

15. Show that the image of the line $x = t$, $y = 2\pi + t$ where $-\infty < t < \infty$ under the mapping $w = \exp z$ is a spiral.

16. Show that the image of the first quadrant $x > 0$, $y > 0$ under the mapping $w = \exp z$ is the region $|w| > 1$.

17. Let α be a real constant. Show that the mapping $w = \exp z$ maps the horizontal strip $\alpha < y \leqq \alpha + 2\pi$ one-to-one and onto the range $|w| > 0$.

18. Show that the function f given by condition (3) is the unique function with properties (1) and (2). *Hint*: Assume that g has properties (1) and (2). Let $F(z) = (g(z))/(\exp z)$, and show that $F'(z) \equiv 0$. Use the conditions $\exp(0) = 1$ and $g(0) = 1$, and show that $g(z) \equiv \exp z$.

4.3 Trigonometric and Hyperbolic Functions

The exponential function plays an important role in the development of the trigonometric and hyperbolic functions. Euler's formula can be written as $e^{ix} = \cos x + i \sin x$, and an immediate consequence is the formula $e^{-ix} = \cos x - i \sin x$. By adding and subtracting these equations we obtain the following formulas for $\cos x$ and $\sin x$, respectively:

$$\cos x = \tfrac{1}{2}(e^{ix} + e^{-ix}) \quad \text{and} \tag{1}$$

$$\sin x = \frac{1}{2i}(e^{ix} - e^{-ix}). \tag{2}$$

The natural extensions of (1) and (2) to functions of the complex variable z are given by the equations

$$\cos z = \tfrac{1}{2}(e^{iz} + e^{-iz}) \quad \text{and} \tag{3}$$

$$\sin z = \frac{1}{2i}(e^{iz} - e^{-iz}), \tag{4}$$

which serve as definitions for the complex sine and cosine functions.

The composite functions $e^{iz} = \exp(iz)$ and $e^{-iz} = \exp(-iz)$ are entire functions. Since identities (3) and (4) involve linear combinations of entire functions, it follows that both $\cos z$ and $\sin z$ are entire functions. An easy calculation reveals that their derivatives are given by the familiar rules

$$(5) \qquad \frac{d}{dz}\cos z = -\sin z \qquad \text{and} \qquad \frac{d}{dz}\sin z = \cos z.$$

The other trigonometric functions are defined in terms of the sine and cosine functions and are given by

$$(6) \qquad \tan z = \frac{\sin z}{\cos z} \qquad \text{and} \qquad \cot z = \frac{\cos z}{\sin z};$$

$$(7) \qquad \sec z = \frac{1}{\cos z} \qquad \text{and} \qquad \csc z = \frac{1}{\sin z}.$$

If we square both sides of identities (3) and (4) and add the resulting terms, then we obtain the trigonometric identity

$$(8) \qquad \cos^2 z + \sin^2 z = 1,$$

which holds for all complex values z. The details are straightforward and are left as an exercise. The rules for differentiating a quotient can now be used in (6) together with identity (8) to establish the differentiation rules

$$(9) \qquad \frac{d}{dz}\tan z = \sec^2 z \qquad \text{and} \qquad \frac{d}{dz}\cot z = -\csc^2 z.$$

The rule for differentiation of powers will help establish

$$(10) \qquad \frac{d}{dz}\sec z = \sec z \tan z \qquad \text{and} \qquad \frac{d}{dz}\csc z = -\csc z \cot z.$$

We will find it useful to have formulas to compute $\cos z$ and $\sin z$ that are given in the $u + iv$ form of Section 2.1. The applications in Chapters 8 and 9 will use these formulas. We would be wise to focus our attention on this problem. We start with $\cos z$, and use identity (3) to write

$$(11) \qquad \cos z = \tfrac{1}{2}(e^{-y+ix} + e^{y-ix}).$$

Then Euler's formula is used in (11) to obtain

$$(12) \qquad \cos z = \tfrac{1}{2}[e^{-y}(\cos x + i\sin x) + e^{y}(\cos x - i\sin x)]$$

$$= \frac{e^{y} + e^{-y}}{2}\cos x - i\,\frac{e^{y} - e^{-y}}{2}\sin x.$$

At this point we must observe that the real hyperbolic cosine and hyperbolic sine of the real variable y are given by

$$(13) \qquad \cosh y = \frac{e^{y} + e^{-y}}{2} \qquad \text{and} \qquad \sinh y = \frac{e^{y} - e^{-y}}{2}.$$

The identities in (13) are now used in (12), and the result is

$$(14) \qquad \cos z = \cos x \cosh y - i \sin x \sinh y.$$

A similar derivation leads to the identity

$$(15)\qquad \sin z = \sin x \cosh y + i \cos x \sinh y.$$

If we set $z = x + i0$ in identities (14) and (15) and use the facts that $\cosh 0 = 1$ and $\sinh 0 = 0$, then we see that

$$(16)\qquad \cos(x + i0) = \cos x \qquad \text{and} \qquad \sin(x + i0) = \sin x,$$

which shows that $\cos z$ and $\sin z$ are extensions of $\cos x$ and $\sin x$ to the complex case.

Identities (14) and (15) can be used to investigate the periodic character of the trigonometric functions, and we have

$$(17)\qquad \cos(z + 2\pi) = \cos z \qquad \text{and} \qquad \sin(z + 2\pi) = \sin z$$

and

$$(18)\qquad \cos(z + \pi) = -\cos z \qquad \text{and} \qquad \sin(z + \pi) = -\sin z.$$

The identities in (18) can be used to show that

$$(19)\qquad \tan(z + \pi) = \tan z \qquad \text{and} \qquad \cot(z + \pi) = \cot z.$$

A solution of the equation $f(z) = 0$ is called a *zero* of the given function f. The zeros of the sine and cosine function are real, and we find that

$$(20)\qquad \sin z = 0 \qquad \text{if and only if } z = n\pi$$

where $n = 0, \pm 1, \pm 2, \ldots$, and

$$(21)\qquad \cos z = 0 \qquad \text{if and only if } z = (n + \tfrac{1}{2})\pi \text{ where } n = 0, \pm 1, \pm 2, \ldots.$$

EXAMPLE 4.4 Let us verify identity (21). We start with equation (14) and write

$$(22)\qquad 0 = \cos x \cosh y - i \sin x \sinh y.$$

Equating the real and imaginary parts of equation (22), we obtain

$$(23)\qquad 0 = \cos x \cosh y \qquad \text{and} \qquad 0 = \sin x \sinh y.$$

Since the real-valued function $\cosh y$ is never zero, the first equation in (23) implies that $0 = \cos x$, from which we obtain $x = (n + \frac{1}{2})\pi$ for $n = 0, \pm 1, \pm 2, \ldots$. By using these values for x the second equation in (23) becomes

$$0 = \sin((n + \tfrac{1}{2})\pi) \sinh y = (-1)^n \sinh y.$$

Hence $y = 0$, and the zeros of $\cos z$ are the values $z = (n + \frac{1}{2})\pi$ where $n = 0, \pm 1, \pm 2, \ldots$.

The standard trigonometric identities are valid for complex variables:

$$(24)\qquad \sin(z_1 + z_2) = \sin z_1 \cos z_2 + \cos z_1 \sin z_2 \qquad \text{and}$$

$$(25)\qquad \cos(z_1 + z_2) = \cos z_1 \cos z_2 - \sin z_1 \sin z_2.$$

When $z_1 = z_2$, identities (24) and (25) become

$$\text{(26)} \qquad \sin 2z = 2 \sin z \cos z \qquad \text{and} \qquad \cos 2z = \cos^2 z - \sin^2 z.$$

Other useful identities are

$$\text{(27)} \qquad \sin(-z) = -\sin z \qquad \text{and} \qquad \cos(-z) = \cos z;$$

$$\text{(28)} \qquad \sin\left(\frac{\pi}{2} + z\right) = \sin\left(\frac{\pi}{2} - z\right) \qquad \text{and} \qquad \sin\left(\frac{\pi}{2} - z\right) = \cos z.$$

EXAMPLE 4.5 Let us show how identity (25) is proven in the complex case. We start with the definitions (3) and (4) and the right side of (25). Then we write

$$\cos z_1 \cos z_2 = \tfrac{1}{4}[e^{i(z_1+z_2)} + e^{i(z_1-z_2)} + e^{i(z_2-z_1)} + e^{-i(z_1+z_2)}] \quad \text{and}$$
$$-\sin z_1 \sin z_2 = \tfrac{1}{4}[e^{i(z_1+z_2)} - e^{i(z_1-z_2)} - e^{i(z_2-z_1)} + e^{-i(z_1+z_2)}].$$

When these expressions are added, we obtain

$$\cos z_1 \cos z_2 - \sin z_1 \sin z_2 = \tfrac{1}{2}[e^{i(z_1+z_2)} + e^{-i(z_1+z_2)}] = \cos(z_1 + z_2),$$

and identity (25) is established.

Identities involving moduli of cosine and sine are also important. If we start with identity (15) and compute the square of the modulus of $\sin z$, the result is

$$\begin{aligned} |\sin z|^2 &= |\sin x \cosh y + i \cos x \sinh y|^2 \\ &= \sin^2 x \cosh^2 y + \cos^2 x \sinh^2 y \\ &= \sin^2 x(\cosh^2 y - \sinh^2 y) + \sinh^2 y(\cos^2 x + \sin^2 x). \end{aligned}$$

Using the hyperbolic identity $\cosh^2 y - \sinh^2 y = 1$ yields

$$\text{(29)} \qquad |\sin z|^2 = \sin^2 x + \sinh^2 y.$$

A similar derivation shows that

$$\text{(30)} \qquad |\cos z|^2 = \cos^2 x + \sinh^2 y.$$

If we set $z = x_0 + iy$ in equation (29) and let $y \to \infty$, then the result is

$$\lim_{y\to\infty} |\sin(x_0 + iy)|^2 = \sin^2 x_0 + \lim_{y\to\infty} \sinh^2 y = \infty.$$

This shows that $\sin z$ is not a bounded function, and it is also evident that $\cos z$ is not a bounded function. This is one of the important differences between the real and complex cases of the functions sine and cosine.

From the periodic character of the trigonometric functions it is apparent that any point $w = \cos z$ in the range of $\cos z$ is actually the image of an infinite number of points.

EXAMPLE 4.6 Let us find all the values of z for which the equation $\cos z = \cosh 2$ holds true. Starting with (14), we write

$$\cos x \cosh y - i \sin x \sinh y = \cosh 2. \tag{31}$$

Equating the real and imaginary parts in equation (31) results in

$$\cos x \cosh y = \cosh 2 \quad \text{and} \quad \sin x \sinh y = 0. \tag{32}$$

The second equation in (32) implies either that $x = n\pi$ where n is an integer or that $y = 0$. Using the latter choice $y = 0$ and the first equation in (32) leads to the impossible situation $\cos x = (\cosh 2/\cosh 0) = \cosh 2 > 1$. Therefore $x = n\pi$ where n is an integer. Since $\cosh y \geqq 1$ for all values of y, we see that the term $\cos x$ in the first equation in (32) must also be positive. For this reason we eliminate the odd values of n and see that $x = 2\pi k$ where k is an integer.

We now solve the equation $\cosh y \cos 2\pi k = \cosh y = \cosh 2$ and use the fact that $\cosh y$ is an even function to conclude that $y = \pm 2$. Therefore the solutions to the equation $\cos z = \cosh 2$ are $z = 2\pi k \pm 2i$ where k is an integer.

The hyperbolic cosine and hyperbolic sine of a complex variable are defined by the equations

$$\cosh z = \tfrac{1}{2}(e^z + e^{-z}) \quad \text{and} \tag{33}$$

$$\sinh z = \tfrac{1}{2}(e^z - e^{-z}). \tag{34}$$

The other hyperbolic functions are given by the formulas

$$\tanh z = \frac{\sinh z}{\cosh z} \quad \text{and} \quad \coth z = \frac{\cosh z}{\sinh z}; \tag{35}$$

$$\operatorname{sech} z = \frac{1}{\cosh z} \quad \text{and} \quad \operatorname{csch} z = \frac{1}{\sinh z}. \tag{36}$$

The derivatives of the hyperbolic functions follow the same rules as in calculus:

$$\frac{d}{dz} \cosh z = \sinh z \quad \text{and} \quad \frac{d}{dz} \sinh z = \cosh z; \tag{37}$$

$$\frac{d}{dz} \tanh z = \operatorname{sech}^2 z \quad \text{and} \quad \frac{d}{dz} \coth z = -\operatorname{csch}^2 z; \tag{38}$$

$$\frac{d}{dz} \operatorname{sech} z = -\operatorname{sech} z \tanh z \quad \text{and} \quad \frac{d}{dz} \operatorname{csch} z = -\operatorname{csch} z \coth z. \tag{39}$$

The hyperbolic cosine and hyperbolic sine can be expressed as

$$\cosh z = \cosh x \cos y + i \sinh x \sin y \quad \text{and} \tag{40}$$

$$\sinh z = \sinh x \cos y + i \cosh x \sin y. \tag{41}$$

The trigonometric and hyperbolic functions are all defined in terms of the exponential function, and they can easily be shown to be related by the following identities:

$$\cosh(iz) = \cos z \quad \text{and} \quad \sinh(iz) = i \sin z; \tag{42}$$

$$\sin(iz) = i \sinh z \quad \text{and} \quad \cos(iz) = \cosh z. \tag{43}$$

Some of the identities involving the hyperbolic functions are

(44) $\cosh^2 z - \sinh^2 z = 1,$

(45) $\sinh(z_1 + z_2) = \sinh z_1 \cosh z_2 + \cosh z_1 \sinh z_2,$

(46) $\cosh(z_1 + z_2) = \cosh z_1 \cosh z_2 + \sinh z_1 \sinh z_2,$

(47) $\cosh(z + 2\pi i) = \cosh z,$

(48) $\sinh(z + 2\pi i) = \sinh z,$

(49) $\cosh(-z) = \cosh z,$ and

(50) $\sinh(-z) = -\sinh z.$

In the theory of electric circuits it is shown that the voltage drop E_R across a resistance R obeys Ohm's Law:

(51) $E_R = IR,$

where I is the current flowing through the resistor. It is also known that the current and voltage drop across an inductor L obey the equation

(52) $E_L = L \dfrac{dI}{dt}.$

The current and voltage drop across a capacitor C are related by

(53) $E_C = \dfrac{1}{C} \displaystyle\int_{t_0}^{t} I(\tau)\, d\tau.$

The voltages E_L, E_R, and E_C and the impressed voltage $E(t)$ in Figure 4.7 satisfy the equation

(54) $E_L + E_R + E_C = E(t).$

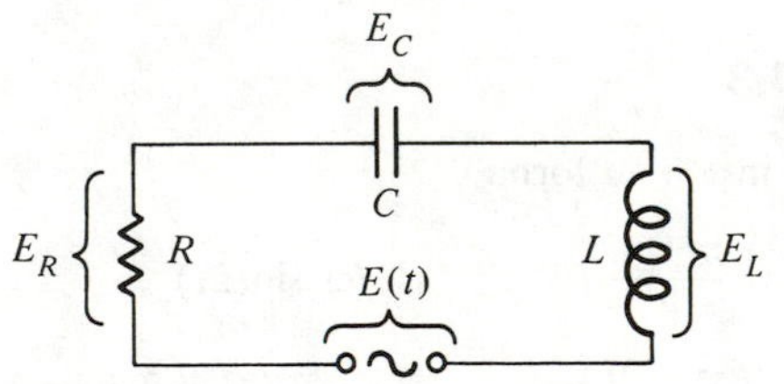

FIGURE 4.7 An LRC-circuit.

Suppose that the current $I(t)$ in the circuit is given by

(55) $I(t) = I_0 \sin \omega t.$

Using equations (51), (52), and (55), we obtain

(56) $E_R = RI_0 \sin \omega t$ and

(57) $E_L = \omega L I_0 \cos \omega t,$

and we can set $t_0 = \pi/2$ in (53) to obtain

(58) $$E_C = -\frac{1}{\omega C} I_0 \cos \omega t.$$

If we write (55) as a complex current

(59) $$I^* = I_0 e^{i\omega t}$$

and use the understanding that the actual physical current I is the imaginary part of I^*, then equations (56)–(58) can be written

(60) $$E_R^* = RI_0 e^{i\omega t} = RI^*,$$

(61) $$E_L^* = i\omega L I_0 e^{i\omega t} = i\omega L I^*, \qquad \text{and}$$

(62) $$E_C^* = \frac{1}{i\omega C} I_0 e^{i\omega t} = \frac{1}{i\omega C} I^*.$$

Substituting (60)–(62) into (54) results in

(63) $$E^* = E_R^* + E_L^* + E_C^* = \left[R + i\left(\omega L - \frac{1}{\omega C}\right)\right] I^*,$$

and the complex quantity Z defined by

(64) $$Z = R + i\left(\omega L - \frac{1}{\omega C}\right)$$

is called the *complex impedance*. Using equation (64), we can write

(65) $$E^* = ZI^*,$$

which is the complex extension of Ohm's Law.

EXERCISES FOR SECTION 4.3

1. Express the following quantities in $u + iv$ form:

(a) $\cos(1 + i)$ **(b)** $\sin\left(\frac{\pi + 4i}{4}\right)$ **(c)** $\sin(2i)$

(d) $\cos(-2 + i)$ **(e)** $\tan\left(\frac{\pi + 2i}{4}\right)$ **(f)** $\tan\left(\frac{\pi + i}{2}\right)$

2. Verify the differentiation formulas given in (5); that is:

(a) $\frac{d}{dz} \cos z = -\sin z$ **(b)** $\frac{d}{dz} \sin z = \cos z$

Use *two methods*. First use the Cauchy-Riemann equations, and then use the identities (3) and (4) and the chain rule fo differentiation of e^{iz} and e^{-iz}.

3. Show that:

(a) $\sin(\pi - z) = \sin z$ **(b)** $\sin\left(\frac{\pi}{2} - z\right) = \cos z$

4. Establish the identity $e^{i\pi} + 1 = 0$. It has been observed that this relation involves the five most important numbers in mathematics.
5. Find the derivatives of the following:
   (a) $\sin(1/z)$ **(b)** $z \tan z$ **(c)** $\sec(z^2)$ **(d)** $z \csc^2 z$
6. Establish identity (15).
7. Show that:
   (a) $\sin^2 z + \cos^2 z = 1$ **(b)** $e^{iz} = \cos z + i \sin z$
8. Show that:
   (a) $\sin \bar{z} = \overline{\sin z}$ holds for all z and that **(b)** $\sin \bar{z}$ is nowhere analytic.
9. Show that:
   (a) $\lim_{z \to 0} \frac{\cos z - 1}{z} = 0$ and that
   (b) $\lim_{y \to +\infty} \tan(x_0 + iy) = i$, where x_0 is any fixed real number.
10. Find all values of z for which the following equations hold:
    (a) $\sin z = \cosh 4$ **(b)** $\cos z = 2$ **(c)** $\sin z = i \sinh 1$
11. Show that the zeros of $\sin z$ are $z = n\pi$ where n is an integer.
12. Establish identity (24). **13.** Establish identity (30).
14. Establish the following relation: $|\sinh y| \leqq |\sin z| \leqq \cosh y$.
15. Use the result of Exercise 14 to help establish the inequality $|\cos z|^2 + |\sin z|^2 \geqq 1$, and show that equality holds if and only if z is a real number.
16. Show that the mapping $w = \sin z$ maps the y axis one-to-one and onto the v axis.
17. Use the fact that $\sin(iz)$ is analytic to show that $\sinh x \cos y$ is a harmonic function.
18. Show that the transformation $w = \sin z$ maps the ray $x = \pi/2$, $y > 0$ one-to-one and onto the ray $u > 1$, $v = 0$.
19. Express the following quantities in $u + iv$ form.
    (a) $\sinh(1 + i\pi)$ **(b)** $\cosh\left(\frac{i\pi}{2}\right)$ **(c)** $\cosh\left(\frac{4 - i\pi}{4}\right)$
20. Establish identity (44).
21. Show that:
    (a) $\sinh(z + i\pi) = -\sinh z$ **(b)** $\tanh(z + i\pi) = \tanh z$
22. Find all values of z for which the following equations hold:
    (a) $\sinh z = i/2$ **(b)** $\cosh z = 1$
23. Find the derivatives of the following:
    (a) $z \sinh z$ **(b)** $\cosh(z^2)$ **(c)** $z \tanh z$
24. Show that:
    (a) $\sin(iz) = i \sinh z$ **(b)** $\cosh(iz) = \cos z$
25. Establish identity (40).
26. Show that:
    (a) $\cosh \bar{z} = \overline{\cosh z}$ and that **(b)** $\cosh \bar{z}$ is nowhere analytic.
27. Establish identity (46).
28. Find the complex impedance Z if $R = 10$, $L = 10$, $C = 0.05$, and $\omega = 2$.
29. Find the complex impedance Z if $R = 15$, $L = 10$, $C = 0.05$, and $\omega = 4$.

4.4 Branches of the Logarithm Function

If w is a nonzero complex number, then, as we saw in equation (24) of Section 4.2, the equation $w = \exp z$ has infinitely many solutions. Since the function $\exp z$ is a many-to-one function, its inverse is necessarily multivalued.

We define $\log z$ as the inverse of the exponential function; that is,

$$(1) \qquad \log z = w \qquad \text{if and only if } z = \exp w.$$

The function $\log z$ is a multivalued function, and its values are given by the formula

$$(2) \qquad \log z = \ln|z| + i \arg z \qquad (z \neq 0)$$

where $\arg z$ is an argument of z and $\ln|z|$ denotes the natural logarithm of the positive real number $|z|$. Using identity (6) in Section 1.4, we write

$$(3) \qquad \log z = \ln|z| + i(\operatorname{Arg} z + 2\pi n) \qquad \text{where } n \text{ is an integer.}$$

We call any one of the values given in equations (2) or (3) a logarithm of z. We notice that the values of $\log z$ have the same real part and that their imaginary parts differ by an integral multiple of 2π.

The *principal value* of $\log z$ is denoted by $\operatorname{Log} z$ and is determined from equation (3) by choosing $n = 0$. We write

$$(4) \qquad \operatorname{Log} z = \ln|z| + i \operatorname{Arg} z \qquad (z \neq 0)$$

where $-\pi < \operatorname{Arg} z \leqq \pi$. The function $w = \operatorname{Log} z$ is single-valued. Its domain of definition is the set of all nonzero complex numbers, and its range is the horizontal strip $-\pi < \operatorname{Im} w \leqq \pi$ in the w plane. The function $\operatorname{Log} z$ is discontinuous at the point $z = 0$ and at each point along the negative x axis. This happens because $\operatorname{Arg} z$ is discontinuous at these points. To show this fact, we let $r_0 e^{\pm i\pi} = z_0$ denote a negative real number. We then compute the limit of $\operatorname{Log} z$ as z approaches z_0 through the upper half-plane $\operatorname{Im} z > 0$ and the limit of $\operatorname{Log} z$ as z approaches z_0 through the lower half-plane $\operatorname{Im} z < 0$. In polar coordinates these limits are given by

$$\lim_{(r,\theta)\to(r_0,\pi)} \operatorname{Log}(re^{i\theta}) = \lim_{(r,\theta)\to(r_0,\pi)} (\ln r + i\theta) = \ln r_0 + i\pi \qquad \text{and}$$

$$\lim_{(r,\theta)\to(r_0,-\pi)} \operatorname{Log}(re^{i\theta}) = \lim_{(r,\theta)\to(r_0,-\pi)} (\ln r + i\theta) = \ln r_0 - i\pi.$$

Since the two limits are distinct, the function $\operatorname{Log} z$ is discontinuous at z_0; hence the negative x axis is a branch cut.

In view of definitions (1) and (4), it follows that

$$(5) \qquad \exp(\operatorname{Log} z) = z \qquad \text{for all } z \neq 0.$$

Therefore the mapping $w = \operatorname{Log} z$ is a one-to-one mapping of the domain $|z| > 0$ onto the horizontal strip $-\pi < \operatorname{Im} w \leqq \pi$. The mapping $w = \operatorname{Log} z$ and its branch cut are illustrated in Figure 4.8.

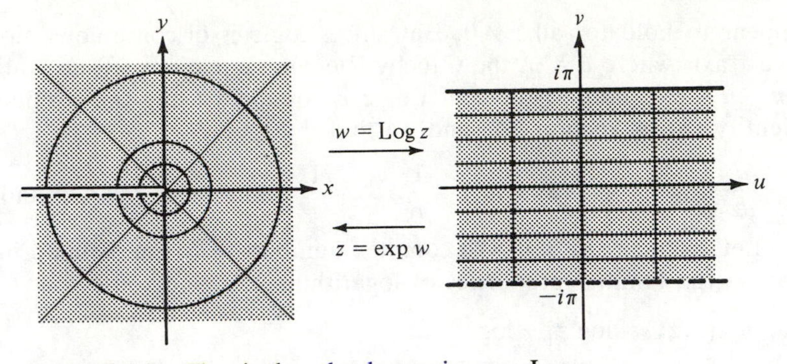

FIGURE 4.8 The single valued mapping $w = \text{Log } z$.

EXAMPLE 4.7 Calculation reveals that

$$\log(1+i) = \ln|1+i| + i \arg(1+i) = \ln\sqrt{2} + i\left(\frac{\pi}{4} + 2\pi n\right)$$

where n is an integer, and the principal value is

$$\text{Log}(1+i) = \ln\sqrt{2} + i\,\frac{\pi}{4}.$$

The complex logarithm can be used to find the logarithm of negative real numbers.

EXAMPLE 4.8 Calculation reveals that

$$\log(-e) = \ln|-e| + i \arg(-e) = 1 + i(\pi + 2\pi n)$$

where n is an integer, and the principal value is

$$\text{Log}(-e) = 1 + i\pi.$$

When $z = x + i0$ where x is a positive real number, the principal value of the complex logarithm of z is

(6) $\quad \text{Log}(x + i0) = \ln x \quad$ where $x > 0$.

Hence Log z is an extension of the real function $\ln x$ to the complex case. To find the derivative of Log z, we use polar coordinates $z = re^{i\theta}$, and formula (4) becomes

(7) $\quad \text{Log } z = \ln r + i\theta \quad$ where $-\pi < \theta \leqq \pi$ and $r > 0$.

In (7) we have $u(r, \theta) = \ln r$ and $v(r, \theta) = \theta$. The polar form of the Cauchy-Riemann equations are

(8) $\quad u_r = \frac{1}{r} v_\theta = \frac{1}{r} \quad$ and $\quad v_r = \frac{-1}{r} u_\theta = 0$

and appear to hold for all $z \neq 0$. But since $\text{Log}\, z$ is discontinuous along the negative x axis where $\theta = \pi$, the Cauchy-Riemann equations (8) are valid only for $-\pi < \theta < \pi$. The derivative of $\text{Log}\, z$ is found by using the results of (8) and identity (5) of Section 3.2, and we find that

$$(9) \qquad \frac{d}{dz} \text{Log}\, z = e^{-i\theta}(u_r + iv_r) = \frac{1}{r} e^{-i\theta} = \frac{1}{z} \quad \text{where } -\pi < \theta < \pi \text{ and } r > 0.$$

Let z_1 and z_2 be nonzero complex numbers. The multivalued function $\log z$ obeys the familiar properties of logarithms:

$$(10) \qquad \log(z_1 z_2) = \log z_1 + \log z_2,$$

$$(11) \qquad \log(z_1/z_2) = \log z_1 - \log z_2, \quad \text{and}$$

$$(12) \qquad \log(1/z) = -\log z.$$

Identity (10) is easy to establish. Using identity (7) in Section 1.4 concerning the argument of a product, we write

$$\begin{aligned} \log(z_1 z_2) &= \ln|z_1||z_2| + i \arg(z_1 z_2) \\ &= \ln|z_1| + \ln|z_2| + i \arg z_1 + i \arg z_2 \\ &= (\ln|z_1| + i \arg z_1) + (\ln|z_2| + i \arg z_2) = \log z_1 + \log z_2. \end{aligned}$$

Identities (11) and (12) are easy to verify and are left as exercises.

It should be noted that identities (10)–(12) do not in general hold true when $\log z$ is replaced everywhere by $\text{Log}\, z$. For example, if we make the specific choices $z_1 = -\sqrt{3} + i$ and $z_2 = -1 + i\sqrt{3}$, then their product is $z_1 z_2 = -4i$. Computing the principal value of the logarithms, we find that

$$\text{Log}(z_1 z_2) = \text{Log}(-4i) = \ln 4 - \frac{i\pi}{2}.$$

The sum of the logarithms is given by

$$\begin{aligned} \text{Log}\, z_1 + \text{Log}\, z_2 &= \text{Log}(-\sqrt{3} + i) + \text{Log}(-1 + i\sqrt{3}) \\ &= \ln 2 + \frac{i5\pi}{6} + \ln 2 + \frac{i2\pi}{3} = \ln 4 + \frac{i3\pi}{2}. \end{aligned}$$

and identity (10) does not hold for these principal values of the logarithm.

We can construct many different branches of the multivalued logarithm function. Let α denote a fixed real number, and choose the value of $\theta = \arg z$ that lies in the range $\alpha < \theta \leqq \alpha + 2\pi$. Then the function

$$(13) \qquad f(z) = \ln r + i\theta \qquad \text{where } r > 0 \text{ and } \alpha < \theta \leqq \alpha + 2\pi$$

is a single-valued branch of the logarithm function. The branch cut for f is the ray $\theta = \alpha$, and each point along this ray is a point of discontinuity of f. Since $\exp(f(z)) = z$, we conclude that the mapping $w = f(z)$ is a one-to-one mapping of the domain $|z| > 0$ onto the horizontal strip $\alpha < \text{Im}\, w \leqq \alpha + 2\pi$. If $\alpha < c < d < \alpha + 2\pi$, then the function $w = f(z)$ maps the set $D = \{re^{i\theta} : a < r < b,$

$c < \theta < d\}$ one-to-one and onto the rectangle $R = \{u + iv: \ln a < u < \ln b,\ c < v < d\}$. The mapping $w = f(z)$, its branch cut $\theta = \alpha$, and the set D and its image R are shown in Figure 4.9.

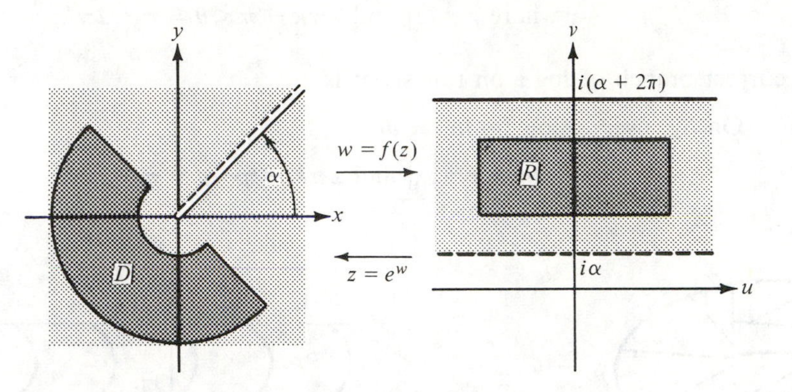

FIGURE 4.9 The branch $w = f(z)$ of the logarithm.

Any branch of the logarithm function is discontinuous along its branch cut. In particular, $f(z) = \text{Log } z$ is discontinuous at $z = 0$ and at points on the negative x axis. Hence the derivative of $\text{Log } z$ does not exist at $z = 0$ and at points on the negative x axis. However, if we focus our attention on the multivalued function $w = \log z$, then implicit differentiation will permit us to use the formula $dw/dz = 1/z$. This can be justified by starting with the second formula in (1) and differentiating implicitly with respect to z to get

$$1 = \exp(w)\,\frac{dw}{dz}. \tag{14}$$

We can substitute $\exp w = z$ in (14) and obtain

$$\frac{dw}{dz} = \frac{1}{\exp(w)} = \frac{1}{z}.$$

Therefore we have shown that

$$\frac{d}{dz}\log z = \frac{1}{z} \tag{15}$$

holds for all $z \neq 0$.

It is appropriate to consider the Riemann surface for the multivalued function $w = \log z$. This requires infinitely many copies of the z plane cut along the negative x axis, which we will label S_k for $k = \ldots, -n, \ldots, -1, 0, 1, \ldots, n, \ldots$. Now stack these cut planes directly upon each other so that the corresponding points have the same position. Join the sheet S_k to S_{k+1} as follows. For each integer k the edge of the sheet S_k in the upper half-plane is joined to the edge of the sheet S_{k+1} in the lower half-plane. The Riemann surface for the domain of $\log z$ looks like a spiral staircase that extends upward on the sheets $S_1, S_2, \ldots$ and downward on the sheets $S_{-1}, S_{-2}, \ldots$ as shown in Figure 4.10. Polar coordinates are used for z on each sheet:

(16) On S_k, use $z = r(\cos\theta + i\sin\theta)$

where $r = |z|$ and $2\pi k - \pi < \theta \leqq \pi + 2\pi k$.

The correct branch of $\log z$ on this sheet is

(17) On S_k, use $\log z = \ln r + i\theta$

where $r = |z|$ and $2\pi k - \pi < \theta \leqq \pi + 2\pi k$.

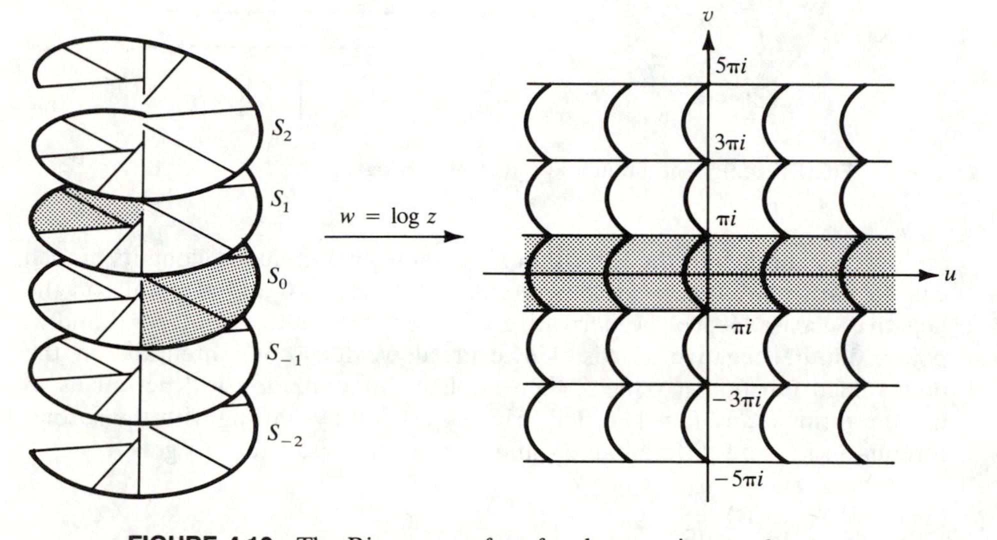

FIGURE 4.10 The Riemann surface for the mapping $w = \log z$.

EXERCISES FOR SECTION 4.4

1. Find the principal value $\text{Log } z = u + iv$ for the following:
(a) $\text{Log}(ie^2)$ **(b)** $\text{Log}(\sqrt{3} - i)$
(c) $\text{Log}(i\sqrt{2} - \sqrt{2})$ **(d)** $\text{Log}((1+i)^4)$

2. Find *all* the values of $\log z$ for the following:
(a) $\log(-3)$ **(b)** $\log(4i)$
(c) $\log(8)$ **(d)** $\log(-\sqrt{3} - i)$

3. Find *all* the values of z for which the following equations hold:
(a) $\text{Log } z = 1 - \dfrac{i\pi}{4}$ **(b)** $\text{Log}(z-1) = \dfrac{i\pi}{2}$
(c) $\exp z = -ie$ **(d)** $\exp(z+1) = i$

4. Use properties of $\arg z$ in Section 1.4 to establish identity (11).

5. Use properties of $\arg z$ in Section 1.4 to establish identity (12).

6. Show that $\log z = \operatorname{Log} z + i2\pi n$ where n is an integer.

7. Let $w = \log(f(z))$. Use implicit differentiation to find dw/dz at points where $f(z)$ is analytic and nonzero.

8. Use implicit differentiation to find dw/dz for the following:
(a) $w = \log(z^2 - z + 2)$ **(b)** $w = z \log z$

9. Show that $f(z) = \operatorname{Log}(iz)$ is analytic everywhere except at points on the ray $y \geqq 0$, $x = 0$. Find $f'(z)$.

10. Show that $f(z) = (\operatorname{Log}(z+5))/(z^2 + 3z + 2)$ is analytic everywhere except at the points -1, -2 and at points on the ray $x \leqq -5$, $y = 0$.

11. Show that if $\operatorname{Re} z > 0$, then $\operatorname{Log} z = \frac{1}{2}\ln(x^2 + y^2) + i \arctan(y/x)$, where the principal branch of the real function $\arctan t$ is used; that is, $-\pi/2 < \arctan t < \pi/2$.

12. Show that:
(a) $\ln(x^2 + y^2)$ **(b)** $\arctan(y/x)$
are harmonic functions in the right half-plane $\operatorname{Re} z > 0$.

13. Show that $z^n = \exp(n \log z)$ where n is an integer.

14. **(a)** Show that $\operatorname{Log}(z_1 z_2) = \operatorname{Log} z_1 + \operatorname{Log} z_2$ holds true provided that $\operatorname{Re} z_1 > 0$ and $\operatorname{Re} z_2 > 0$.
(b) Does $\operatorname{Log}((-1+i)^2) = 2\operatorname{Log}(-1+i)$?

15. **(a)** Is it always true that $\operatorname{Log}(1/z) = -\operatorname{Log} z$? Why?
(b) Is it always true that $\dfrac{d}{dz}\operatorname{Log} z = \dfrac{1}{z}$? Why?

16. Construct branches of $\log(z+2)$ that are analytic at all points in the plane except at points on the following rays:
(a) $x \geqq -2$, $y = 0$ **(b)** $x = -2$, $y \geqq 0$ **(c)** $x = -2$, $y \leqq 0$

17. Construct a branch of $\log(z+4)$ that is analytic at the point $z = -5$ and takes on the value $7\pi i$ there.

18. Using the polar coordinate notation $z = re^{i\theta}$, discuss the possible interpretations of the function $f(z) = \ln r + i\theta$.

19. Show that the mapping $w = \operatorname{Log} z$ maps the ray $r > 0$, $\theta = \pi/3$ one-to-one and onto the horizontal line $v = \pi/3$.

20. Show that the mapping $w = \operatorname{Log} z$ maps the semicircle $r = 2$, $-\pi/2 \leqq \theta \leqq \pi/2$ one-to-one and onto the vertical line segment $u = \ln 2$, $-\pi/2 \leqq v \leqq \pi/2$.

21. Find specific values of z_1 and z_2 so that $\operatorname{Log}(z_1/z_2) \neq \operatorname{Log} z_1 - \operatorname{Log} z_2$.

22. Show that $\log(\exp z) = z + i2\pi n$ where n is an integer.

4.5 Complex Exponents

We now define what is meant by a complex number raised to a complex power. Let c be a complex number, and let $z \neq 0$. Then z^c is defined by the equation

(1) $$z^c = \exp(c \log z).$$

Since $\log z$ is a multivalued function, the function z^c will in general be multivalued. The function f given by

$$(2) \qquad f(z) = \exp(c \operatorname{Log} z)$$

is called the *principal branch* of the multivalued function z^c. The principal branch of z^c is obtained from equation (1) by using the principal value of the logarithm.

EXAMPLE 4.9 The values of $(-1)^i$ are given by

$$(-1)^i = \exp(i \log(-1)) = \exp(i(1+2n)\pi i) = e^{-(1+2n)\pi}$$

where n is an integer,

and the principal value of $(-1)^i$ is given by

$$(-1)^i = e^{-\pi}.$$

We now consider various cases of formula (1):

Case (*i*). Let us fix $c = k$, where k is an integer, and $z = re^{i\theta}$, then

$$k \log z = k \ln r + ik(\theta + 2\pi n) \qquad \text{where } n \text{ is an integer.}$$

Since the complex exponential function has period $2\pi i$, formula (1) becomes

$$(3) \qquad z^k = \exp(k \ln r + ik\theta) = r^k(\cos k\theta + i \sin k\theta),$$

which is the single-valued kth power of z that was studied in Section 1.5.

Case (*ii*). If $c = 1/k$, where k is an integer, and $z = re^{i\theta}$, then

$$\frac{1}{k} \log z = \frac{1}{k} \ln r + \frac{i(\theta + 2\pi n)}{k} \qquad \text{where } n \text{ is an integer.}$$

Since the complex exponential function has period $2\pi i$, formula (1) becomes

$$(4) \qquad z^{1/k} = r^{1/k}\left(\cos \frac{\theta + 2\pi n}{k} + i \sin \frac{\theta + 2\pi n}{k}\right),$$

and there are k distinct values corresponding to $n = 0, 1, \ldots, k-1$. Therefore the fractional power $z^{1/k}$ is the multivalued k root function.

Case (*iii*). If j and k are positive integers that have no common factors and $c = j/k$, then formula (1) becomes

$$(5) \qquad z^{j/k} = r^{j/k}\left(\cos \frac{j(\theta + 2\pi n)}{k} + i \sin \frac{j(\theta + 2\pi n)}{k}\right),$$

and there are k distinct values corresponding to $n = 0, 1, \ldots, k-1$.

Case (*iv*). If c is not an integer or a rational number, then there are infinitely many values for z^c.

EXAMPLE 4.10 Calculation reveals that

$$2^{\sqrt{2}} = \exp(\sqrt{2}\log 2) = \exp(\sqrt{2}(\ln 2 + i2\pi n))$$
$$= e^{\sqrt{2}\ln 2}(\cos 2\sqrt{2}\pi n + i \sin 2\sqrt{2}\pi n)$$

where n is an integer, and the principal value of $2^{\sqrt{2}}$ is

$$2^{\sqrt{2}} = e^{\sqrt{2}\ln 2} = 2.665144142\ldots.$$

Some of the rules for exponents carry over from the real case. If c and d are complex numbers and $z \neq 0$, then

$$(6)\qquad z^{-c} = \frac{1}{z^c};$$

$$(7)\qquad z^c z^d = z^{c+d};$$

$$(8)\qquad \frac{z^c}{z^d} = z^{c-d};$$

$$(9)\qquad (z^c)^n = z^{cn} \quad \text{where } n \text{ is an integer.}$$

Identity (9) does not hold if n is replaced with an arbitrary complex value.

EXAMPLE 4.11

$$(i^2)^i = \exp(i\log(-1)) = e^{-(1+2n)\pi} \quad \text{where } n \text{ is an integer and}$$
$$(i)^{2i} = \exp(2i\log(i)) = e^{-(1+4n)\pi} \quad \text{where } n \text{ is an integer.}$$

Since these sets of solutions are not equal, identity (9) does not hold.

The derivative of $f(z) = \exp(c \operatorname{Log} z)$ can be found by using the chain rule, and we see that

$$(10)\qquad f'(z) = \frac{c}{z}\exp(c \operatorname{Log} z).$$

If we use $z^c = \exp(c \operatorname{Log} z)$, then equation (10) can be written in the familiar form we learn in calculus. That is,

$$(11)\qquad \frac{d}{dz} z^c = \frac{c}{z} z^c,$$

which holds true when z^c is the principal value, when z lies in the domain $r > 0$, $-\pi < \theta < \pi$, and when c is a complex number.

We can use definition (1) to define the exponential function with base b where $b \neq 0$ is a complex number, and we write

$$(12)\qquad b^z = \exp(z \log b).$$

If a value of $\log b$ is specified, then b^z in equation (12) can be made single-valued, and the rules of differentiation can be used to show that the resulting branch of b^z is an entire function. The derivative of b^z is then given by the familiar rule

$$\frac{d}{dz} b^z = b^z \log b. \tag{13}$$

EXERCISES FOR SECTION 4.5

1. Find the principal value of the following:
   (a) 4^i **(b)** $(1+i)^{\pi i}$ **(c)** $(-1)^{1/\pi}$ **(d)** $(1+i\sqrt{3})^{i/2}$
2. Find *all* values of the following:
   (a) i^i **(b)** $(-1)^{\sqrt{2}}$ **(c)** $(i)^{2/\pi}$ **(d)** $(1+i)^{2-i}$
3. Show that if $z \neq 0$, then z^0 has a unique value.
4. Find *all* values of $(-1)^{3/4}$ and $(i)^{2/3}$.
5. Use polar coordinates $z = re^{i\theta}$, and show that the principal branch of z^i is given by the equation

   $$z^i = e^{-\theta}(\cos(\ln r) + i \sin(\ln r)) \qquad \text{where } r > 0 \text{ and } -\pi < \theta \leqq \pi.$$

6. Let α be a real number. Show that the principal branch of z^α is given by the equation

   $$z^\alpha = r^\alpha \cos \alpha\theta + ir^\alpha \sin \alpha\theta \qquad \text{where } -\pi < \theta \leqq \pi.$$

   Find $(d/dz)z^\alpha$.
7. Establish identity (13); that is, $(d/dz)b^z = b^z \log b$.
8. Let $z_n = (1+i)^n$ for $n = 1, 2, \ldots$. Show that the sequence $\{z_n\}$ is a solution to the linear difference equation

   $$z_n = 2z_{n-1} - 2z_{n-2} \qquad \text{for } n = 3, 4, \ldots.$$

   Hint: Show that the equation holds true when the values z_n, z_{n-1}, z_{n-2} are substituted.
9. Verify identity (6).
10. Verify identity (7).
11. Verify identity (8).
12. Verify identity (9).
13. Is 1 raised to any power always equal to 1? Why?
14. Construct an example which shows that the principal value of $(z_1 z_2)^{1/3}$ need not be equal to the product of the principal values of $z_1^{1/3}$ and $z_2^{1/3}$.

4.6 Inverse Trigonometric and Hyperbolic Functions

The trigonometric and hyperbolic functions were defined in Section 4.3 in terms of the exponential function. When we solve for their inverses, we will obtain formulas that involve the logarithm. Since the trigonometric and hyperbolic functions are all periodic, they are many-to-one. Hence their

re necessarily multivalued. The formulas for the inverse trigonomet-
ions are given by

$$\arcsin z = -i \log[iz + (1 - z^2)^{1/2}],$$

$$\arccos z = -i \log[z + i(1 - z^2)^{1/2}], \quad \text{and}$$

$$(3) \qquad \arctan z = \frac{i}{2} \log\left[\frac{i + z}{i - z}\right].$$

The derivatives of the functions in formulas (1)–(3) can be found by implicit differentiation and are given by the formulas:

$$(4) \qquad \frac{d}{dz} \arcsin z = \frac{1}{(1 - z^2)^{1/2}},$$

$$(5) \qquad \frac{d}{dz} \arccos z = \frac{-1}{(1 - z^2)^{1/2}}, \quad \text{and}$$

$$(6) \qquad \frac{d}{dz} \arctan z = \frac{1}{1 + z^2}.$$

We shall establish equations (1) and (4) and leave the others as exercises. Starting with $w = \arcsin z$, we write

$$z = \sin w = \frac{1}{2i}(e^{iw} - e^{-iw}),$$

which also can be written as

$$(7) \qquad e^{iw} - 2iz - e^{-iw} = 0.$$

If each term in (7) is multiplied by e^{iw}, the result is

$$(8) \qquad (e^{iw})^2 - 2ize^{iw} - 1 = 0,$$

which is a quadratic equation in terms of e^{iw}. Using the quadratic equation to solve for e^{iw} in (8), we obtain

$$(9) \qquad e^{iw} = \frac{2iz + (4 - 4z^2)^{1/2}}{2} = iz + (1 - z^2)^{1/2},$$

where the square root is a multivalued function. Taking the logarithm of both sides of (9) leads to the desired equation

$$w = \arcsin z = -i \log[iz + (1 - z^2)^{1/2}],$$

where the multivalued logarithm is used. To construct a specific branch of $\arcsin z$, we must first select a branch of the square root and then select a branch of the logarithm.

The derivative of $w = \arcsin z$ is found by starting with the equation $\sin w = z$ and using implicit differentiation to obtain

$$\frac{dw}{dz} = \frac{1}{\cos w}.$$

The trigonometric identity $\cos w = (1 - \sin^2 w)^{1/2}$, together with $\sin w = z$, c be used to conclude that

$$\frac{dw}{dz} = \frac{d}{dz} \arcsin z = \frac{1}{(1 - z^2)^{1/2}},$$

and formula (4) is established. It is important to note that the form of equations (1) and (4) have been chosen to be compatible. Once a branch of the square root has been chosen in (1), it must be used in (4) to obtain the corresponding value of the derivative.

EXAMPLE 4.12 The values of $\arcsin \sqrt{2}$ are given by

$$(10) \qquad \arcsin \sqrt{2} = -i \log[i\sqrt{2} + (1 - (\sqrt{2})^2)^{1/2}] = -i \log[i\sqrt{2} \pm i].$$

Using straightforward techniques, we simplify (10) and obtain

$$\begin{aligned} \arcsin \sqrt{2} &= -i \log[(\sqrt{2} \pm 1)i] \\ &= -i\left[\ln(\sqrt{2} \pm 1) + i\left(\frac{\pi}{2} + 2\pi n\right)\right] \\ &= \frac{\pi}{2} + 2\pi n - i \ln(\sqrt{2} \pm 1) \quad \text{where } n \text{ is an integer.} \end{aligned}$$

If we observe that

$$\ln(\sqrt{2} - 1) = \ln \frac{(\sqrt{2} - 1)(\sqrt{2} + 1)}{\sqrt{2} + 1} = \ln \frac{1}{\sqrt{2} + 1} = -\ln(\sqrt{2} + 1),$$

then we can write

$$\arcsin \sqrt{2} = \frac{\pi}{2} + 2\pi n \pm i \ln(\sqrt{2} + 1) \qquad \text{where } n \text{ is an integer.}$$

EXAMPLE 4.13 Suppose that we make specific choices in equation (10). We select $+i$ as the value of the square root $(1 - (\sqrt{2})^2)^{1/2}$ and use the principal value of the logarithm. The result will be

$$f(\sqrt{2}) = \arcsin \sqrt{2} = -i \operatorname{Log}(i\sqrt{2} + i) = \frac{\pi}{2} - i \ln(\sqrt{2} + 1),$$

and the corresponding value of the derivative is given by

$$f'(\sqrt{2}) = \frac{1}{(1 - (\sqrt{2})^2)^{1/2}} = \frac{1}{i} = -i.$$

The inverse hyperbolic functions are given by the equations:

$$(11) \qquad \operatorname{arcsinh} z = \log[z + (z^2 + 1)^{1/2}],$$

$$(12) \qquad \operatorname{arccosh} z = \log[z + (z^2 - 1)^{1/2}], \quad \text{and}$$

$$(13) \qquad \operatorname{arctanh} z = \frac{1}{2} \log\left[\frac{1 + z}{1 - z}\right].$$

The trigonometric identity $\cos w = (1 - \sin^2 w)^{1/2}$, together with $\sin w = z$, can be used to conclude that

$$\frac{dw}{dz} = \frac{d}{dz} \arcsin z = \frac{1}{(1 - z^2)^{1/2}},$$

and formula (4) is established. It is important to note that the form of equations (1) and (4) have been chosen to be compatible. Once a branch of the square root has been chosen in (1), it must be used in (4) to obtain the corresponding value of the derivative.

EXAMPLE 4.12 The values of $\arcsin \sqrt{2}$ are given by

$$\text{(10)} \qquad \arcsin \sqrt{2} = -i \log[i\sqrt{2} + (1 - (\sqrt{2})^2)^{1/2}] = -i \log[i\sqrt{2} \pm i].$$

Using straightforward techniques, we simplify (10) and obtain

$$\begin{aligned} \arcsin \sqrt{2} &= -i \log[(\sqrt{2} \pm 1)i] \\ &= -i\left[\ln(\sqrt{2} \pm 1) + i\left(\frac{\pi}{2} + 2\pi n\right)\right] \\ &= \frac{\pi}{2} + 2\pi n - i \ln(\sqrt{2} \pm 1) \qquad \text{where } n \text{ is an integer.} \end{aligned}$$

If we observe that

$$\ln(\sqrt{2} - 1) = \ln \frac{(\sqrt{2} - 1)(\sqrt{2} + 1)}{\sqrt{2} + 1} = \ln \frac{1}{\sqrt{2} + 1} = -\ln(\sqrt{2} + 1),$$

then we can write

$$\arcsin \sqrt{2} = \frac{\pi}{2} + 2\pi n \pm i \ln(\sqrt{2} + 1) \qquad \text{where } n \text{ is an integer.}$$

EXAMPLE 4.13 Suppose that we make specific choices in equation (10). We select $+i$ as the value of the square root $(1 - (\sqrt{2})^2)^{1/2}$ and use the principal value of the logarithm. The result will be

$$f(\sqrt{2}) = \arcsin \sqrt{2} = -i \operatorname{Log}(i\sqrt{2} + i) = \frac{\pi}{2} - i \ln(\sqrt{2} + 1),$$

and the corresponding value of the derivative is given by

$$f'(\sqrt{2}) = \frac{1}{(1 - (\sqrt{2})^2)^{1/2}} = \frac{1}{i} = -i.$$

The inverse hyperbolic functions are given by the equations:

$$\text{(11)} \qquad \operatorname{arcsinh} z = \log[z + (z^2 + 1)^{1/2}],$$

$$\text{(12)} \qquad \operatorname{arccosh} z = \log[z + (z^2 - 1)^{1/2}], \quad \text{and}$$

$$\text{(13)} \qquad \operatorname{arctanh} z = \frac{1}{2} \log\left[\frac{1 + z}{1 - z}\right].$$

inverses are necessarily multivalued. The formulas for the inverse trigonometric functions are given by

$$(1)\qquad \arcsin z = -i \log[iz + (1 - z^2)^{1/2}],$$

$$(2)\qquad \arccos z = -i \log[z + i(1 - z^2)^{1/2}], \qquad \text{and}$$

$$(3)\qquad \arctan z = \frac{i}{2} \log\left[\frac{i + z}{i - z}\right].$$

The derivatives of the functions in formulas (1)–(3) can be found by implicit differentiation and are given by the formulas:

$$(4)\qquad \frac{d}{dz} \arcsin z = \frac{1}{(1 - z^2)^{1/2}},$$

$$(5)\qquad \frac{d}{dz} \arccos z = \frac{-1}{(1 - z^2)^{1/2}}, \qquad \text{and}$$

$$(6)\qquad \frac{d}{dz} \arctan z = \frac{1}{1 + z^2}.$$

We shall establish equations (1) and (4) and leave the others as exercises. Starting with $w = \arcsin z$, we write

$$z = \sin w = \frac{1}{2i} (e^{iw} - e^{-iw}),$$

which also can be written as

$$(7)\qquad e^{iw} - 2iz - e^{-iw} = 0.$$

If each term in (7) is multiplied by e^{iw}, the result is

$$(8)\qquad (e^{iw})^2 - 2ize^{iw} - 1 = 0,$$

which is a quadratic equation in terms of e^{iw}. Using the quadratic equation to solve for e^{iw} in (8), we obtain

$$(9)\qquad e^{iw} = \frac{2iz + (4 - 4z^2)^{1/2}}{2} = iz + (1 - z^2)^{1/2},$$

where the square root is a multivalued function. Taking the logarithm of both sides of (9) leads to the desired equation

$$w = \arcsin z = -i \log[iz + (1 - z^2)^{1/2}],$$

where the multivalued logarithm is used. To construct a specific branch of $\arcsin z$, we must first select a branch of the square root and then select a branch of the logarithm.

The derivative of $w = \arcsin z$ is found by starting with the equation $\sin w = z$ and using implicit differentiation to obtain

$$\frac{dw}{dz} = \frac{1}{\cos w}.$$

The derivatives of the inverse hyperbolic functions are given by

$$(14)\qquad \frac{d}{dz}\operatorname{arcsinh} z = \frac{1}{(z^2+1)^{1/2}},$$

$$(15)\qquad \frac{d}{dz}\operatorname{arccosh} z = \frac{1}{(z^2-1)^{1/2}}, \qquad \text{and}$$

$$(16)\qquad \frac{d}{dz}\operatorname{arctanh} z = \frac{1}{1-z^2}.$$

To establish identity (13), we start with $w = \operatorname{arctanh} z$ and obtain

$$z = \tanh w = \frac{e^w - e^{-w}}{e^w + e^{-w}} = \frac{e^{2w}-1}{e^{2w}+1},$$

which can be solved for e^{2w} to yield $e^{2w} = (1+z)/(1-z)$. After taking the logarithms of both sides, we obtain the result

$$w = \operatorname{arctanh} z = \frac{1}{2}\log\left[\frac{1+z}{1-z}\right],$$

and identity (13) is established.

EXAMPLE 4.14 Calculation reveals that

$$\operatorname{arctanh}(1+2i) = \frac{1}{2}\log\frac{1+1+2i}{1-1-2i} = \frac{1}{2}\log(-1+i)$$
$$= \frac{1}{4}\ln 2 + i\left(\frac{3}{8}+n\right)\pi \qquad \text{where } n \text{ is an integer.}$$

EXERCISES FOR SECTION 4.6

1. Find *all* values of the following:
(a) $\arcsin \frac{5}{4}$ **(b)** $\arccos \frac{5}{3}$ **(c)** $\arcsin 3$
(d) $\arccos 3i$ **(e)** $\arctan 2i$ **(f)** $\arctan i$

2. Find *all* values of the following:
(a) $\operatorname{arcsinh} i$ **(b)** $\operatorname{arcsinh} \frac{3}{4}$ **(c)** $\operatorname{arccosh} i$
(d) $\operatorname{arccosh} \frac{1}{2}$ **(e)** $\operatorname{arctanh} i$ **(f)** $\operatorname{arctanh} i\sqrt{3}$

3. Establish equations (2) and (5).

4. Establish equations (3) and (6).

5. Establish the identity $\arcsin z + \arccos z = (\pi/2) + 2\pi n$ where n is an integer.

6. Establish equation (16).

7. Establish equations (11) and (14).

8. Establish equations (12) and (15).

5 Complex Integration

5.1 Complex Integrals

In Chapter 3 we saw how the derivative of a complex function is defined. We now turn our attention to the problem of integrating complex functions. We will find that integrals of analytic functions are well behaved and that many properties from calculus carry over to the complex case. To introduce the integral of a complex function, we start by defining what is meant by the integral of a complex-valued function of a real variable. Let

$$f(t) = u(t) + iv(t) \qquad \text{for } a \leqq t \leqq b,$$

where $u(t)$ and $v(t)$ are real-valued functions of the real variable t. If u and v are continuous functions on the interval, then from calculus we know that u and v are integrable functions of t. Therefore we make the following definition for the definite integral of f:

$$(1) \qquad \int_a^b f(t)\,dt = \int_a^b u(t)\,dt + i\int_a^b v(t)\,dt.$$

Definition (1) is implemented by finding the antiderivatives of u and v and evaluating the definite integrals on the right side of equation (1). That is, if $U'(t) = u(t)$ and $V'(t) = v(t)$, then we write

$$(2) \qquad \int_a^b f(t)\,dt = U(b) - U(a) + i[V(b) - V(a)].$$

EXAMPLE 5.1 Let us show that

$$(3) \qquad \int_0^1 (t-i)^3\,dt = \frac{-5}{4}.$$

Since the complex integral is defined in terms of real integrals, we write the integrand in (3) in terms of its real and imaginary parts: $f(t) = (t-i)^3 = t^3 - 3t + i(-3t^2 + 1)$. Here we see that u and v are given by $u(t) = t^3 - 3t$ and

$v(t) = -3t^2 + 1$. The integrals of u and v are easy to compute, and we find that

$$\int_0^1 (t^3 - 3t)\, dt = \frac{-5}{4} \qquad \text{and} \qquad \int_0^1 (-3t^2 + 1)\, dt = 0.$$

Hence definition (1) can be used to conclude that

$$\int_0^1 (t - i)^3\, dt = \int_0^1 u(t)\, dt + i \int_0^1 v(t)\, dt = \frac{-5}{4}.$$

Our knowledge about the elementary functions can be used to find their integrals.

EXAMPLE 5.2 Let us show that

$$\int_0^{\pi/2} \exp(t + it)\, dt = \frac{1}{2}(e^{\pi/2} - 1) + \frac{i}{2}(e^{\pi/2} + 1).$$

Using the method suggested by (1) and (2), we obtain

$$\int_0^{\pi/2} \exp(t + it)\, dt = \int_0^{\pi/2} e^t \cos t\, dt + i \int_0^{\pi/2} e^t \sin t\, dt.$$

The integrals can be found in a table of integrals, and we have

$$\int_0^{\pi/2} \exp(t + it)\, dt = \frac{1}{2} e^t(\cos t + \sin t) + \frac{i}{2} e^t(\sin t - \cos t)\Big|_{t=0}^{t=\pi/2}$$
$$= \frac{1}{2}(e^{\pi/2} - 1) + \frac{i}{2}(e^{\pi/2} + 1).$$

Complex integrals have properties that are similar to those of real integrals. Let $f(t) = u(t) + iv(t)$ and $g(t) = p(t) + iq(t)$ be continuous on the interval $a \leqq t \leqq b$. Then the integral of their sum is the sum of their integrals; so we can write

$$(4) \qquad \int_a^b [f(t) + g(t)]\, dt = \int_a^b f(t)\, dt + \int_a^b g(t)\, dt.$$

It is sometimes convenient to divide the interval $a \leqq t \leqq b$ into two parts $a \leqq t \leqq c$ and $c \leqq t \leqq b$ and integrate $f(t)$ over these intervals. Hence we obtain the formula

$$(5) \qquad \int_a^b f(t)\, dt = \int_a^c f(t)\, dt + \int_c^b f(t)\, dt.$$

Constant multiples are dealt with in the same manner as in calculus. If $c + id$ denotes a complex constant, then

$$(6) \qquad \int_a^b (c + id) f(t)\, dt = (c + id) \int_a^b f(t)\, dt.$$

If the limits of integration are reversed, then

$$(7) \qquad \int_b^a f(t)\,dt = -\int_a^b f(t)\,dt.$$

Let us emphasize that we are dealing with complex integrals. We write the integral of the product as follows:

$$(8) \qquad \int_a^b f(t)g(t)\,dt = \int_a^b [u(t)p(t) - v(t)q(t)]\,dt + i\int_a^b [u(t)q(t) + v(t)p(t)]\,dt.$$

EXAMPLE 5.3 Let us prove equation (6). We start by writing

$$(c + id)f(t) = cu(t) - dv(t) + i[cv(t) + du(t)].$$

Using definition (1), the left side of (6) can be written as

$$(9) \qquad c\int_a^b u(t)\,dt - d\int_a^b v(t)\,dt + ic\int_a^b v(t)\,dt + id\int_a^b u(t)\,dt.$$

which is easily seen to be equivalent to the product

$$(10) \qquad [c + id]\left[\int_a^b u(t)\,dt + i\int_a^b v(t)\,dt\right].$$

It is worthwhile to point out the similarity between (2) and its counterpart in calculus. If U and V are differentiable on $a < t < b$ and $F(t) = U(t) + iV(t)$, then $F'(t)$ is defined to be

$$F'(t) = U'(t) + iV'(t),$$

and equation (2) takes on the familiar form

$$(11) \qquad \int_a^b f(t)\,dt = F(b) - F(a) \qquad \text{where } F'(t) = f(t).$$

This can be viewed as an extension of the Fundamental Theorem of Calculus. In Section 5.5 we will see how the extension is made to the case of analytic functions of a complex variable. In particular, we have the following important case of (11):

$$(12) \qquad \int_a^b f'(t)\,dt = f(b) - f(a).$$

EXAMPLE 5.4 Let us use (11) to show that $\int_0^\pi \exp(it)\,dt = 2i$.

Solution If we let $F(t) = -i\exp(it) = \sin t - i\cos t$ and $f(t) = \exp(it) = \cos t + i\sin t$, then notice that $F'(t) = f(t)$, and from (11) we obtain

$$\int_0^\pi \exp(it)\,dt = \int_0^\pi f(t)\,dt = F(\pi) - F(0) = -ie^{i\pi} + ie^0 = 2i.$$

EXERCISES FOR SECTION 5.1

For Exercises 1–4, use equations (1) and (2) to find the following definite integrals.

1. $\int_0^1 (3t - i)^2\, dt$ **2.** $\int_0^1 (t + 2i)^3\, dt$ **3.** $\int_0^{\pi/2} \cosh(it)\, dt$ **4.** $\int_0^2 \frac{t}{t+i}\, dt$

5. Find $\int_0^{\pi/4} t \exp(it)\, dt$.

6. Let m and n be integers. Show that

$$\int_0^{2\pi} e^{imt} e^{-int}\, dt = \begin{cases} 0 & \text{when } m \neq n, \\ 2\pi & \text{when } m = n. \end{cases}$$

7. Show that $\int_0^\infty e^{-zt}\, dt = 1/z$ provided that $\text{Re } z > 0$.

8. Let $f(t) = u(t) + iv(t)$ where u and v are differentiable. Show that $\int_a^b f(t)f'(t)\, dt = \frac{1}{2}[f(b)]^2 - \frac{1}{2}[f(a)]^2$.

9. Establish identity (4).

10. Establish identity (5).

11. Establish identity (7).

12. Establish identity (8).

5.2 Contours and Contour Integrals

In Section 1.6 we introduced the concept of a curve C in the plane, and we used the parametric notation

$$(1) \qquad C\colon z(t) = x(t) + iy(t) \qquad \text{for } a \leqq t \leqq b$$

where $x(t)$ and $y(t)$ are continuous functions. We now want to place a few more restrictions on the type of curve that we will be studying. The following discussion will lead to the concept of a contour, which is a type of curve that is adequate for the study of integration.

Recall that C is said to be *simple* if it does not cross itself, which is expressed by requiring that $z(t_1) \neq z(t_2)$ whenever $t_1 \neq t_2$. A curve C with the property that $z(b) = z(a)$ is said to be a *closed curve*. If $z(b) = z(a)$ is the only point of intersection, then we say that C is a *simple closed curve*. As the parameter t increases from the value a to the value b, the point $z(t)$ starts at the *initial point* $z(a)$, moves along the curve C, and ends up at the *terminal point* $z(b)$. If C is simple, then $z(t)$ moves continuously from $z(a)$ to $z(b)$ as t increases, and the curve is given an *orientation*, which we indicate by drawing arrows along the curve. Figure 5.1 illustrates how the terms "simple" and "closed" can be used to describe a curve.

The complex-valued function $z(t)$ in (1) is said to be *differentiable* if both $x(t)$ and $y(t)$ are differentiable for $a \leqq t \leqq b$. Here the one-sided derivatives* of $x(t)$ and $y(t)$ are required to exist at the endpoints of the

* The derivative on the right $x'(a^+)$ and on the left $x'(b^-)$ are defined by the following limits:

$$x'(a^+) = \lim_{t \to a^+} \frac{x(t) - x(a)}{t - a} \qquad \text{and} \qquad x'(b^-) = \lim_{t \to b^-} \frac{x(t) - x(b)}{t - b}.$$

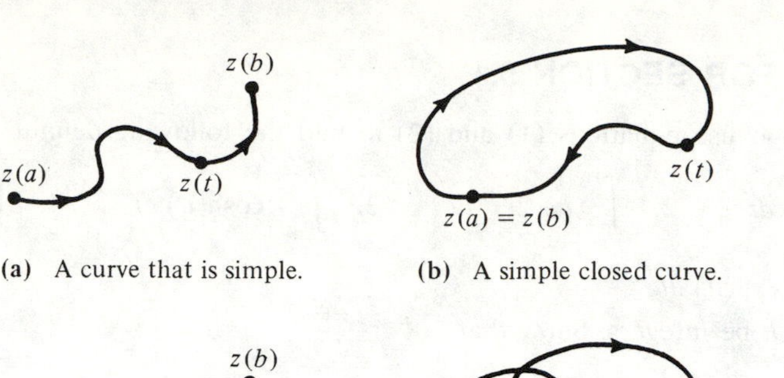

(c) A curve that is not simple, and not closed.
(d) A closed curve that is not simple.

FIGURE 5.1 The terms "simple" and "closed" used to describe curves.

interval. The derivative $z'(t)$ with respect to t is defined by the equation

$$z'(t) = x'(t) + iy'(t) \qquad \text{for } a \leqq t \leqq b. \tag{2}$$

The curve C defined by (1) is said to be *smooth* if $z'(t)$, given by (2), is continuous and nonzero on the interval. If C is a smooth curve, then C has a nonzero tangent vector at each point $z(t)$, which is given by the vector $z'(t)$. If $x'(t_0) = 0$, then the tangent vector $z'(t_0) = iy'(t_0)$ is vertical. If $x'(t_0) \neq 0$, then the slope dy/dx of the tangent line to C at the point $z(t_0)$ is given by $y'(t_0)/x'(t_0)$. Hence the angle of inclination $\theta(t)$ of the tangent vector $z'(t)$ is defined for all values of t and is the continuous function given by

$$\theta(t) = \arg(z'(t)) = \arg(x'(t) + iy'(t)).$$

Therefore a smooth curve has a continuously turning tangent vector. A smooth curve has no corners or cusps. Figure 5.2 illustrates this concept.

If C is a smooth curve, then ds, the differential of arc length, is given by

$$ds = \sqrt{(x'(t))^2 + (y'(t))^2}\, dt = |z'(t)|\, dt. \tag{3}$$

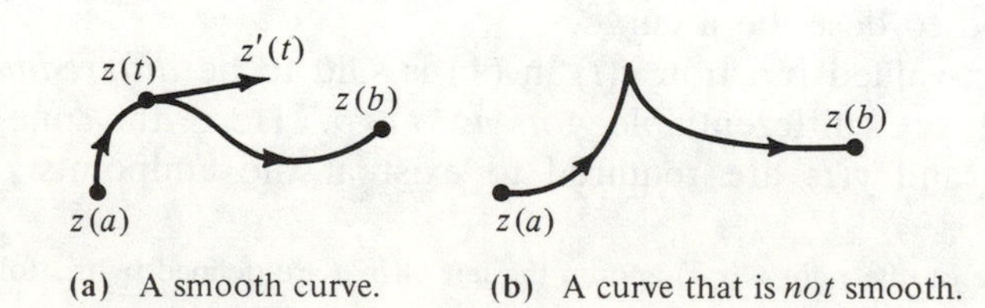

(a) A smooth curve. (b) A curve that is *not* smooth.

FIGURE 5.2 The term "smooth" used to describe curves.

Since $x'(t)$ and $y'(t)$ are continuous functions, then so is the function $\sqrt{(x'(t))^2 + (y'(t))^2}$, and the length L of the curve C is given by the definite integral

(4) $$L = \int_a^b \sqrt{(x'(t))^2 + (y'(t))^2}\, dt = \int_a^b |z'(t)|\, dt.$$

Now consider C to be a curve with parameterization

$$C\colon z_1(t) = x(t) + iy(t) \qquad \text{for } a \leqq t \leqq b.$$

The *opposite curve* $-C$ traces out the same set of points in the plane but in the reverse order, and it has the parameterization

$$-C\colon z_2(t) = x(-t) + iy(-t) \qquad \text{for } -b \leqq t \leqq -a.$$

Since $z_2(t) = z_1(-t)$, it is easy to see that $-C$ is merely C traversed in the opposite sense. This is illustrated in Figure 5.3.

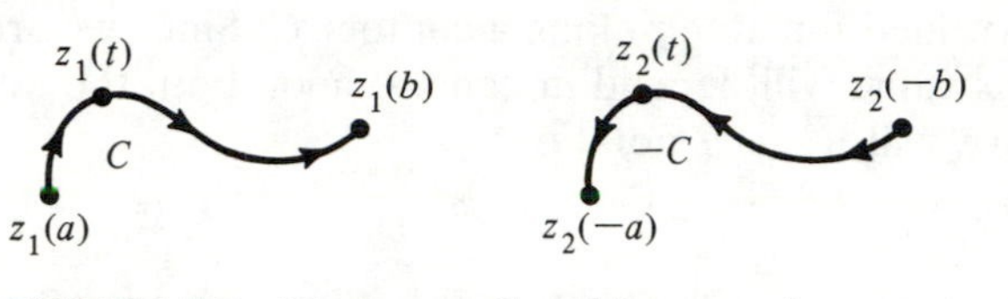

FIGURE 5.3 The curve C and its opposite curve $-C$.

A curve C that is constructed by joining finitely many smooth curves end to end is called a *contour*. Let $C_1, C_2, \ldots, C_n$ denote n smooth curves such that the terminal point of C_k coincides with the initial point of C_{k+1} for $k = 1, 2, \ldots, n-1$. Then the contour C is expressed by the equation

(5) $$C = C_1 + C_2 + \cdots + C_n.$$

A synonym for contour is *path*.

EXAMPLE 5.5 Let us find a parameterization of the polygonal path C from $-1+i$ to $3-i$, which is shown in Figure 5.4.

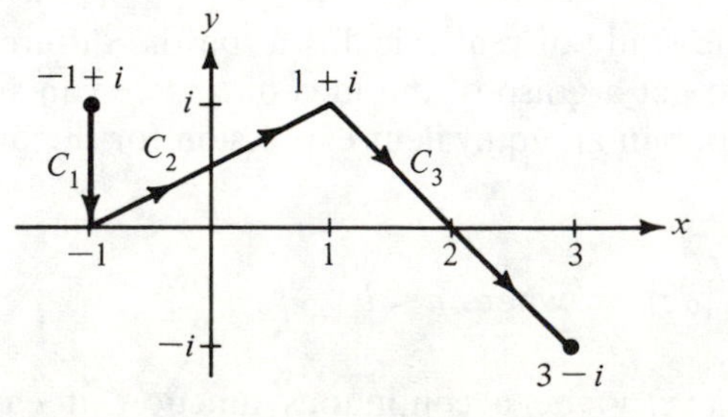

FIGURE 5.4 The polygonal path $C = C_1 + C_2 + C_3$ from $-1+i$ to $3-i$.

Solution The contour is conveniently expressed as three smooth curves $C = C_1 + C_2 + C_3$. A formula for the straight line segment joining two points was given by equation (2) in Section 1.6. If we set $z_0 = -1 + i$ and $z_1 = -1$, then the segment C_1 joining z_0 to z_1 is given by

$$C_1\colon z_1(t) = z_0 + t(z_1 - z_0) = [-1 + i] + t[-1 - (-1 + i)],$$

which can be simplified to obtain

$$C_1\colon z_1(t) = -1 + i(1 - t) \qquad \text{for } 0 \leqq t \leqq 1.$$

In a similar fashion the segments C_2 and C_3 are given by

$$C_2\colon z_2(t) = (-1 + 2t) + it \qquad \text{for } 0 \leqq t \leqq 1 \qquad \text{and}$$

$$C_3\colon z_3(t) = (1 + 2t) + i(1 - 2t) \qquad \text{for } 0 \leqq t \leqq 1.$$

We are now ready to define what is meant by a contour integral or line integral of a complex-valued function f along a contour C. Since we are dealing with a line integral, its value will depend in general upon both the integrand f and the contour C. We will use the notation

$$(6) \qquad \int_C f(z)\,dz$$

where the contour C is called the *path of integration*. The line integral can be rigorously defined as the limit of Riemann sums, and the reader is referred to more advanced texts for this development. We choose to give a development based on the method of evaluating complex integrals in Section 5.1, together with property (2) of a contour.

To start with, let us write dz/dt instead of $z'(t)$ in equation (2), and then multiply both sides of this equation by dt. The result is

$$(7) \qquad dz = z'(t)\,dt = [x'(t) + iy'(t)]\,dt,$$

which is an expression for the *complex differential* dz. If we write $dz = dx + i\,dy$, then the real and imaginary parts of (7) are given by

$$(8) \qquad dx = x'(t)\,dt \qquad \text{and} \qquad dy = y'(t)\,dt,$$

which are recognized from the study of real calculus to be the differentials of $x(t)$ and $y(t)$, respectively. To get a grasp of the idea of dz, we can substitute dz given by (7) into (4) and obtain an equivalent expression for the arc length L of a contour C. We write

$$(9) \qquad L = \int_a^b |z'(t)|\,dt = \int_C |dz| \qquad \text{where } a < b.$$

Let $f(z) = u(x, y) + iv(x, y)$ be a continuous function at each point $z = (x, y) = (x(t), y(t))$ of a contour C with parameterization given by equations (1) and (2). We are able to make the following important definition.

Definition 5.1 (Contour Integral) *The contour integral of $f(z)$ along C is defined to be*

$$(10) \qquad \int_C f(z)\,dz = \int_a^b f(z(t))z'(t)\,dt.$$

Since definition (5.1) involves the integral of the complex function $f(z(t))z'(t)$ of the real variable t, we can use identity (8) of Section 5.1 to express (10) in the form

$$(11) \qquad \int_C f(z)\,dz = \int_a^b (ux' - vy')\,dt + i\int_a^b (vx' + uy')\,dt$$

where u, v, x, y, x', y' are all functions of t; that is,

$$u = u(x(t), y(t)), \qquad v = v(x(t), y(t)), \qquad x' = x'(t), \qquad y' = y'(t).$$

If differentials are used, then (10) can be written in terms of line integrals of the real-valued functions u and v, and we write

$$(12) \qquad \int_C f(z)\,dz = \int_C u\,dx - v\,dy + i\int_C v\,dx + u\,dy,$$

which is easy to remember if we recall that $f(z)\,dz = (u + iv)(dx + i\,dy)$. Expressions (10)–(12) all state the same mathematical concept. The reader is encouraged to use the expression that is easiest for the reader to grasp.

EXAMPLE 5.6 Let us show that

$$\int_C \exp z\,dz = -1 + ie^2$$

where C is the line segment from 0 to $2 + (i\pi/2)$. The contour C can be parameterized as follows:

$$C\colon z(t) = t + \frac{i\pi}{4}t \qquad \text{and} \qquad dz = \left(1 + \frac{i\pi}{4}\right)dt \qquad \text{for } 0 \leqq t \leqq 2.$$

Along the contour,

$$f(z(t)) = \exp\left(t + \frac{i\pi}{4}t\right) = e^t\cos\left(\frac{\pi}{4}t\right) + ie^t\sin\left(\frac{\pi}{4}t\right).$$

Computing the value of the integral given in equation (11), we use the facts that $x'(t) = 1$ and $y'(t) = \pi/4$ and obtain

$$\int_0^2 \left[e^t\cos\left(\frac{\pi}{4}t\right) - \frac{\pi}{4}e^t\sin\left(\frac{\pi}{4}t\right)\right]dt$$
$$+\, i\int_0^2 \left[e^t\sin\left(\frac{\pi}{4}t\right) + \frac{\pi}{4}e^t\cos\left(\frac{\pi}{4}t\right)\right]dt.$$

A table of integrals can be used to find the integrals, and we obtain

$$\int_C \exp z \, dz = e^t \exp\left(\frac{\pi}{4} t\right) + ie^t \sin\left(\frac{\pi}{4} t\right)\Big|_{t=0}^{t=2}$$

$$= e^2 \cos \frac{\pi}{2} + ie^2 \sin \frac{\pi}{2} - 1 = -1 + ie^2.$$

EXAMPLE 5.7 Let us show that

$$\int_{C_1} z \, dz = \int_{C_2} z \, dz = 4 + 2i$$

where C_1 is the line segment from $-1 - i$ to $3 + i$ and C_2 is the portion of the parabola $x = y^2 + 2y$ joining $-1 - i$ to $3 + i$ as indicated in Figure 5.5.

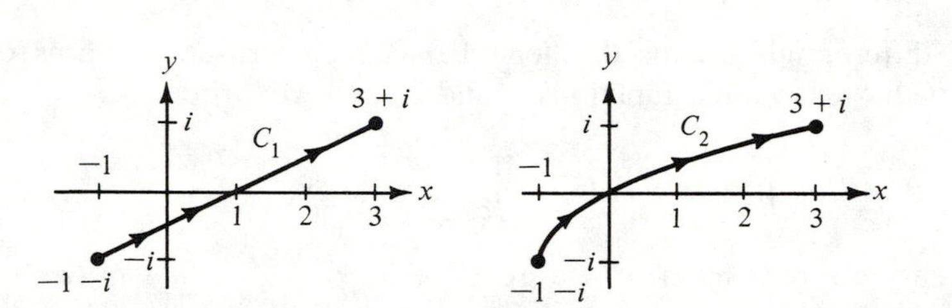

(a) The line segment. **(b)** A portion of a parabola.

FIGURE 5.5 The two contours C_1 and C_2 joining $-1 - i$ to $3 + i$.

The line segment joining $(-1, -1)$ to $(3, 1)$ is given by the slope intercept formula $y = \frac{1}{2}x - \frac{1}{2}$, which can be written as $x = 2y + 1$. It is convenient to choose the parameterization $y = t$ and $x = 2t + 1$. Then the segment C_1 can be given by

$$C_1\colon z(t) = 2t + 1 + it \quad \text{and} \quad dz = (2 + i)\, dt \quad \text{for } -1 \leqq t \leqq 1.$$

Along C_1, we have $f(z(t)) = 2t + 1 + it$. Computing the value of the integral in (10), we obtain

$$\int_{C_1} z \, dz = \int_{-1}^{1} (2t + 1 + it)(2 + i)\, dt,$$

which can be evaluated by using straightforward techniques to obtain

$$\int_{C_1} z \, dz = \int_{-1}^{1} (3t + 2)\, dt + i \int_{-1}^{1} (4t + 1)\, dt = 4 + 2i.$$

Similarly, for the portion of the parabola $x = y^2 + 2y$ joining $(-1, -1)$ to $(3, 1)$, it is convenient to choose the parameterization $y = t$ and $x = t^2 + 2t$.

Then C_2 can be given by

$$C_2\colon z(t) = t^2 + 2t + it \quad \text{and} \quad dz = (2t + 2 + i)\,dt \quad \text{for } -1 \leqq t \leqq 1.$$

Along C_2 we have $f(z(t)) = t^2 + 2t + it$. Computing the value of the integral in (10), we obtain

$$\int_{C_2} z\,dz = \int_{-1}^{1} (t^2 + 2t + it)(2t + 2 + i)\,dt$$
$$= \int_{-1}^{1} (2t^3 + 6t^2 + 3t)\,dt + i\int_{-1}^{1} (3t^2 + 4t)\,dt = 4 + 2i.$$

In this example the value of the two integrals is the same. This does not hold in general, as is shown in the next example.

EXAMPLE 5.8 Let us show that

$$\int_{C_1} \bar{z}\,dz = -\pi i \quad \text{and} \quad \int_{C_2} \bar{z}\,dz = -4i$$

where C_1 is the semicircular path from -1 to 1 and C_2 is the polygonal path from -1 to 1, respectively, that are shown in Figure 5.6.

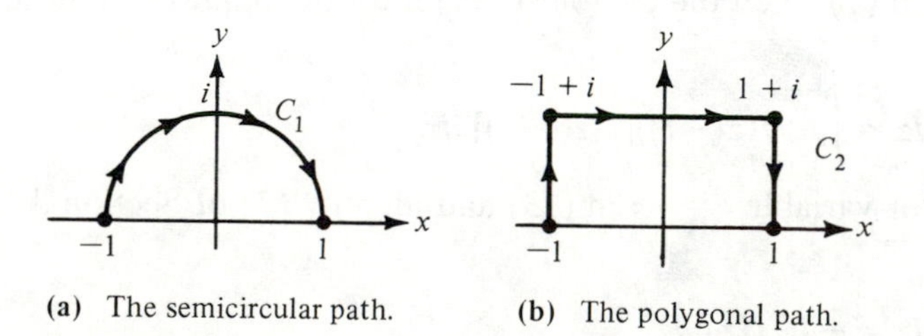

FIGURE 5.6 The two contours C_1 and C_2 joining -1 to 1.

Solution The semicircle C_1 can be parameterized by

$$C_1\colon z(t) = -\cos t + i\sin t \quad \text{and} \quad dz = (\sin t + i\cos t)\,dt \quad \text{for } 0 \leqq t \leqq \pi.$$

Along C_1 we have $f(z(t)) = -\cos t - i\sin t$. Computing the value of the integral in (10), we obtain

$$\int_{C_1} \bar{z}\,dz = \int_0^{\pi} (-\cos t - i\sin t)(\sin t + i\cos t)\,dt$$
$$= -i\int_0^{\pi} (\cos^2 t + \sin^2 t)\,dt = -\pi i.$$

The polygonal path C_2 must be parameterized in three parts, one for each line segment:

$$z_1(t) = -1 + it, \qquad dz_1 = i\,dt, \qquad f(z_1(t)) = -1 - it,$$
$$z_2(t) = -1 + 2t + i, \qquad dz_2 = 2\,dt, \qquad f(z_2(t)) = -1 + 2t - i,$$
$$z_3(t) = 1 + i(1-t), \qquad dz_3 = -i\,dt, \qquad f(z_3(t)) = 1 - i(1-t),$$

where all of the parameters t are to be taken on the interval $0 \leqq t \leqq 1$. The value of the integral in (10) is obtained by adding the three integrals along the above three segments, and the result is

$$\int_0^1 (-1 - it)i\,dt + \int_0^1 (-1 + 2t - i)2\,dt + \int_0^1 (1 - i(1-t))(-i)\,dt.$$

A straightforward calculation now shows that

$$\int_{C_2} \bar{z}\,dz = \int_0^1 (6t-3)\,dt + i\int_0^1 (-4)\,dt = -4i.$$

We remark that the value of the contour integral along C_1 is *not* the same as the value of the contour integral along C_2, although both integrals have the same initial and terminal points.

Contour integrals have properties that are similar to those of integrals of a complex function of a real variable, which were studied in Section 5.1. If C is given by equation (1), then the contour integral for the opposite contour $-C$ is given by

$$(13) \qquad \int_{-C} f(z)\,dz = \int_{-b}^{-a} f(z(-\tau))[-z'(-\tau)]\,d\tau.$$

Using the change of variable $t = -\tau$ in (13) and identity (7) of Section 5.1, we obtain

$$(14) \qquad \int_{-C} f(z)\,dz = -\int_C f(z)\,dz.$$

If two functions f and g can be integrated over the same path of integration C, then their sum can be integrated over C, and we have the familiar result

$$(15) \qquad \int_C [f(z) + g(z)]\,dz = \int_C f(z)\,dz + \int_C g(z)\,dz.$$

Constant multiples are dealt with in the same manner as in identity (6) in Section 5.1:

$$(16) \qquad \int_C (c + id)f(z)\,dz = (c + id)\int_C f(z)\,dz.$$

If two contours C_1 and C_2 are placed end to end so that the terminal point of C_1 coincides with the initial point of C_2, then the contour $C = C_1 + C_2$ is a *continuation* of C_1, and we have the property

$$(17) \qquad \int_{C_1 + C_2} f(z)\,dz = \int_{C_1} f(z)\,dz + \int_{C_2} f(z)\,dz.$$

If the contour C has two parameterizations

$$C\colon z_1(t) = x_1(t) + iy_1(t) \qquad \text{for } a \leqq t \leqq b \qquad \text{and}$$

$$C\colon z_2(\tau) = x_2(\tau) + iy_2(\tau) \qquad \text{for } \alpha \leqq \tau \leqq \beta$$

and there exists a differentiable function ϕ such that

$$\alpha = \phi(a), \qquad \beta = \phi(b), \qquad \text{and} \qquad \phi'(t) > 0 \qquad \text{for } a < t < b, \tag{18}$$

then we say that $z_2(\tau)$ is a *reparameterization* of the contour C. If f is continuous on C, then we have

$$\int_a^b f(z_1(t))z_1'(t)\,dt = \int_\alpha^\beta f(z_2(\tau))z_2'(\tau)\,d\tau. \tag{19}$$

Identity (19) shows that the value of a contour integral is invariant under a change in the parametric representation of its contour if the reparameterization satisfies equations (18).

There are a few important inequalities relating to complex integrals, which we now state.

Lemma 5.1 (Integral Triangle Inequality) *If $f(t) = u(t) + iv(t)$ is a continuous function of the real parameter t, then*

$$\left|\int_a^b f(t)\,dt\right| \leqq \int_a^b |f(t)|\,dt. \tag{20}$$

Proof Write the value of the integral in polar form:

$$r_0 e^{i\theta_0} = \int_a^b f(t)\,dt \qquad \text{and} \qquad r_0 = \int_a^b e^{-i\theta_0} f(t)\,dt. \tag{21}$$

Taking the real part of the second integral in (21), we write

$$r_0 = \int_a^b \operatorname{Re}(e^{-i\theta_0} f(t))\,dt.$$

Using equation (2) of Section 1.3, we obtain the relation

$$\operatorname{Re}(e^{-i\theta_0} f(t)) \leqq |e^{-i\theta_0} f(t)| \leqq |f(t)|.$$

The left and right sides can be used as integrands, and then familiar results from calculus can be used to obtain

$$r_0 = \int_a^b \operatorname{Re}(e^{-i\theta_0} f(t))\,dt \leqq \int_a^b |f(t)|\,dt.$$

Since

$$r_0 = \left|\int_a^b f(t)\,dt\right|,$$

we have established inequality (20).

Lemma 5.2 (ML Inequality) *If $f(z) = u(x, y) + iv(x, y)$ is continuous on the contour C, then*

$$\left|\int_C f(z)\,dz\right| \leqq ML \tag{22}$$

where L is the length of the contour C and M is an upper bound for the modulus $|f(z)|$ on C.

Proof When inequality (20) is used with Definition 5.1, we get

$$\left|\int_C f(z)\,dz\right| \leqq \left|\int_a^b f(z(t))z'(t)\,dt\right| \leqq \int_a^b |f(z(t))z'(t)|\,dt. \tag{23}$$

Let M be the positive real constant such that

$$|f(z)| \leqq M \qquad \text{for all } z \text{ on } C.$$

Then (9) and (23) imply that

$$\left|\int_C f(z)\,dz\right| \leqq \int_a^b M|z'(t)|\,dt = ML.$$

Therefore inequality (22) is proven.

EXAMPLE 5.9 Let us use inequality (23) to show that

$$\left|\int_C \frac{1}{z^2+1}\,dz\right| \leqq \frac{1}{2\sqrt{5}}$$

where C is the straight line segment from 2 to $2 + i$. Here $|z^2 + 1| = |z - i| \times |z + i|$, and the terms $|z - i|$ and $|z + i|$ represent the distance from the point z to the points i and $-i$, respectively. We refer to Figure 5.7 and use a geometric argument to see that

$$|z - i| \geqq 2 \qquad \text{and} \qquad |z + i| \geqq \sqrt{5} \qquad \text{for } z \text{ on } C.$$

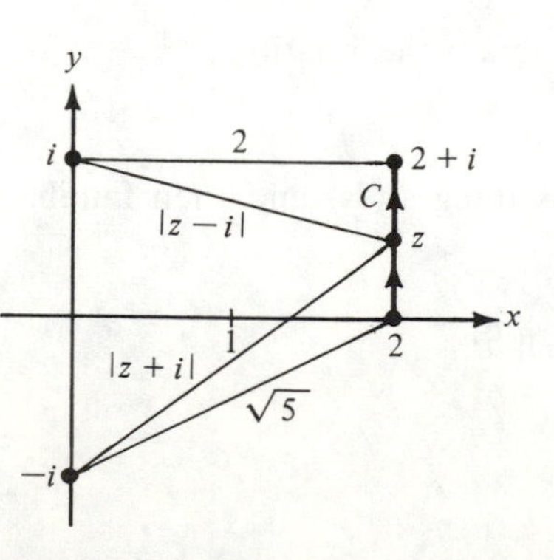

FIGURE 5.7 The distances $|z - i|$ and $|z + i|$ for z on C.

Here we have

$$|f(z)| = \frac{1}{|z-i||z+i|} \leqq \frac{1}{2\sqrt{5}} = M$$

and $L = 1$, so (23) implies that

$$\left| \int_C \frac{1}{z^2+1}\, dz \right| \leqq ML = \frac{1}{2\sqrt{5}}.$$

EXERCISES FOR SECTION 5.2

1. Sketch the following curves.
   (a) $z(t) = t^2 - 1 + i(t+4)$ for $1 \leqq t \leqq 3$.
   (b) $z(t) = \sin t + i \cos 2t$ for $-\pi/2 \leqq t \leqq \pi/2$.
   (c) $z(t) = 5 \cos t - i3 \sin t$ for $\pi/2 \leqq t \leqq 2\pi$.
2. Give a parameterization of the contour $C = C_1 + C_2$ indicated in Figure 5.8.
3. Give a parameterization of the contour $C = C_1 + C_2 + C_3$ indicated in Figure 5.9.

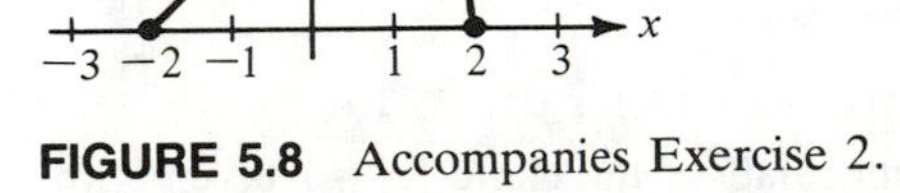

FIGURE 5.8 Accompanies Exercise 2.

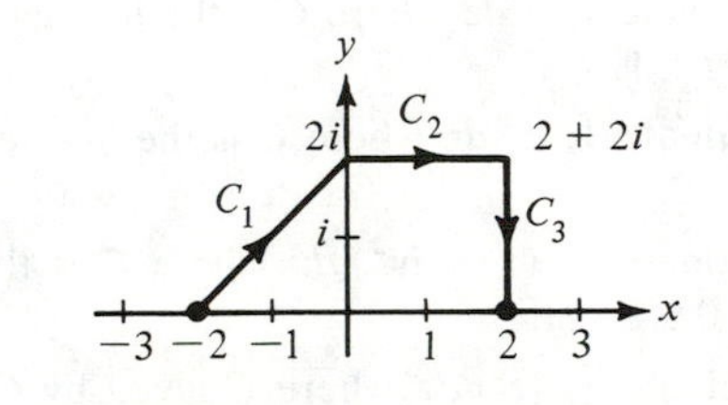

FIGURE 5.9 Accompanies Exercise 3.

4. Evaluate $\int_C y\, dz$ for $-i$ to i along the following contours as shown in Figures 5.10(a) and 5.10(b).
   (a) The polygonal path C with vertices $-i$, $-1-i$, -1, i.
   (b) The contour C that is the left half of the circle $|z| = 1$.

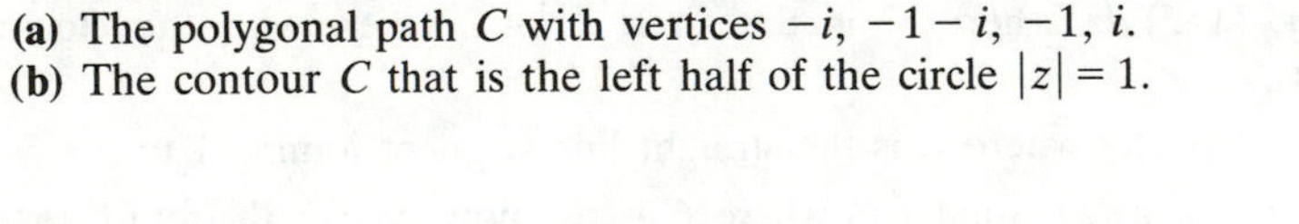

FIGURE 5.10 Accompanies Exercise 4.

5. Evaluate $\int_C x\,dz$ from -4 to 4 along the following contours as shown in Figures 5.11(a) and 5.11(b).
(a) The polygonal path C with vertices -4, $-4+4i$, $4+4i$, 4.
(b) The contour C that is the upper half of the circle $|z| = 4$.

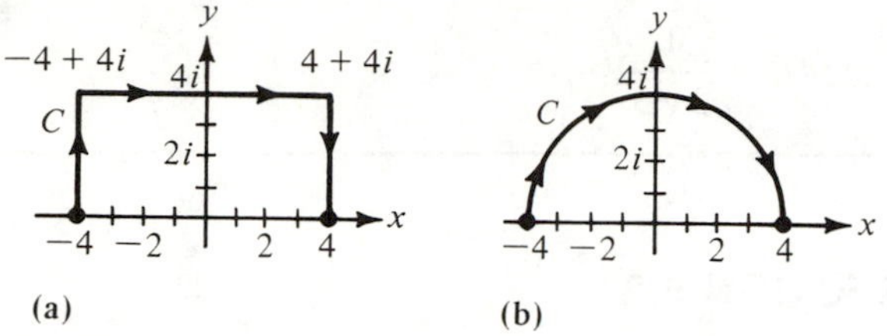

FIGURE 5.11 Accompanies Exercise 5.

6. Evaluate $\int_C z\,dz$ where C is the circle $|z| = 4$ taken with the counterclockwise orientation. *Hint*: Let C: $z(t) = 4\cos t + i4\sin t$ for $0 \leqq t \leqq 2\pi$.

7. Evaluate $\int_C \bar{z}\,dz$ where C is the circle $|z| = 4$ taken with the counterclockwise orientation.

8. Evaluate $\int_C (z+1)\,dz$ where C given by C: $z(t) = \cos t + i\sin t$ for $0 \leqq t \leqq \pi/2$.

9. Evaluate $\int_C z\,dz$ where C is the line segment from i to 1 and $z(t) = t + (1-t)i$ for $0 \leqq t \leqq 1$.

10. Evaluate $\int_C z^2\,dz$ where C is the line segment from 1 to $1+i$ and $z(t) = 1 + it$ for $0 \leqq t \leqq 1$.

11. Evaluate $\int_C (x^2 - iy^2)\,dz$ where C is the upper semicircle C: $z(t) = \cos t + i\sin t$ for $0 \leqq t \leqq \pi$.

12. Evaluate $\int_C |z^2|\,dz$ where C given by C: $z(t) = t + it^2$ for $0 \leqq t \leqq 1$.

13. Evaluate $\int_C |z-1|^2\,dz$ where C is the upper half of the circle $|z| = 1$ taken with the counterclockwise orientation.

14. Evaluate $\int_C (1/z)\,dz$ where C is the circle $|z| = 2$ taken with the clockwise orientation. *Hint*: C: $z(t) = 2\cos t - i2\sin t$ for $0 \leqq t \leqq 2\pi$.

15. Evaluate $\int_C (1/\bar{z})\,dz$ where C is the circle $|z| = 2$ taken with the clockwise orientation.

16. Evaluate $\int_C \exp z\,dz$ where C is the straight line segment joining 1 to $1 + i\pi$.

17. Show that $\int_C \cos z\,dz = \sin(1+i)$ where C is the polygonal path from 0 to $1+i$ that consists of the line segments from 0 to 1 and 1 to $1+i$.

18. Show that $\int_C \exp z\,dz = \exp(1+i) - 1$ where C is the straight line segment joining 0 to $1+i$.

19. Evaluate $\int_C \bar{z}\exp z\,dz$ where C is the square with vertices 0, 1, $1+i$, and i taken with the counterclockwise orientation.

20. Let $z(t) = x(t) + iy(t)$ for $a \leqq t \leqq b$ be a smooth curve. Give a meaning for each of the following expressions.

(a) $z'(t)$ **(b)** $|z'(t)|\,dt$ **(c)** $\int_a^b z'(t)\,dt$ **(d)** $\int_a^b |z'(t)|\,dt$

21. Let f be a continuous function on the circle $|z - z_0| = R$. Let the circle C have the parameterization $C: z(\theta) = z_0 + \mathrm{Re}^{i\theta}$ for $0 \leqq \theta \leqq 2\pi$. Show that

$$\int_C f(z)\, dz = iR \int_0^{2\pi} f(z_0 + \mathrm{Re}^{i\theta}) e^{i\theta}\, d\theta.$$

22. Use the results of Exercise 21 to show that

(a) $\displaystyle\int_C \frac{1}{z - z_0}\, dz = 2\pi i$ and

(b) $\displaystyle\int_C \frac{1}{(z - z_0)^n}\, dz = 0$, where $n \neq 1$ is an integer,

where the contour C is the circle $|z - z_0| = R$ taken with the counterclockwise orientation.

5.3 The Cauchy-Goursat Theorem

The Cauchy-Goursat Theorem states that within certain domains the integral of an analytic function over a simple closed contour is zero. An extension of this theorem will allow us to replace integrals over certain complicated contours with integrals over contours that are easy to evaluate. We will show how to use the technique of partial fractions together with the Cauchy-Goursat Theorem to evaluate certain integrals. In Section 5.4 we will see that the Cauchy-Goursat Theorem implies that an analytic function has an antiderivative. To start with, we need to introduce a few new concepts.

Associated with each simple closed contour C are two disjoint domains, each of which has C as its boundary. The contour C divides the plane into two domains. One domain is bounded and is called the *interior* of C, and the other domain is unbounded and is called the *exterior* of C. Figure 5.12 illustrates this concept. This result is known as the Jordan Curve Theorem. It seems to be an obvious fact, yet the proof is quite complicated and can be found in advanced textbooks on topology.

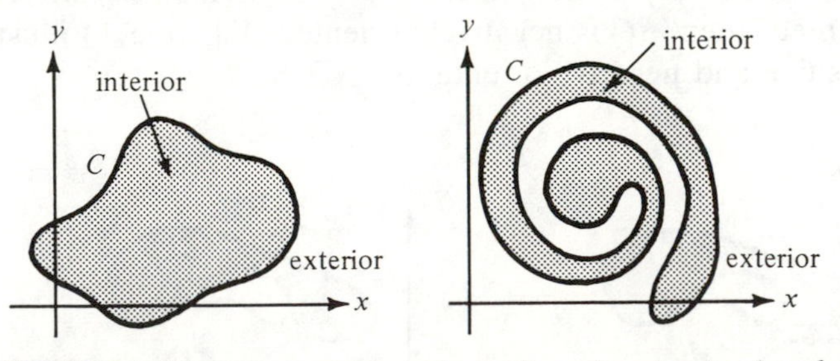

FIGURE 5.12 The interior and exterior of a simple closed contour.

In Section 1.6 we saw that a domain D is an open connected set. In particular, if z_1 and z_2 are any pair of points in D, then they can be joined by a curve that lies entirely in D. A domain D is said to be *simply connected* if it has the property that any simple closed contour C contained in D has its interior contained in D. In other words, there are no "holes" in a simply connected domain. A domain that is not simply connected is said to be a *multiply connected domain*. Figure 5.13 illustrates the use of the terms "simply connected" and "multiply connected."

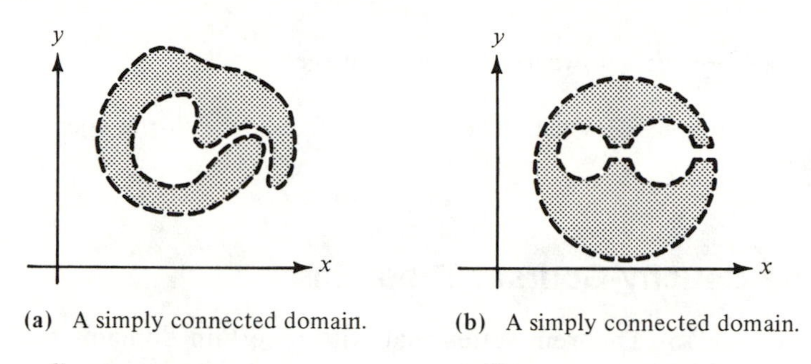

(a) A simply connected domain. (b) A simply connected domain.

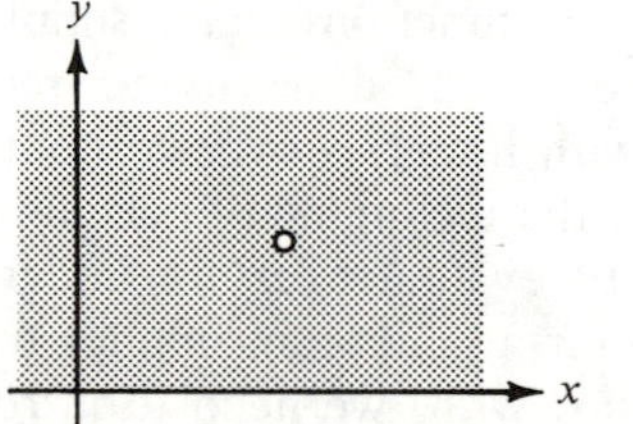

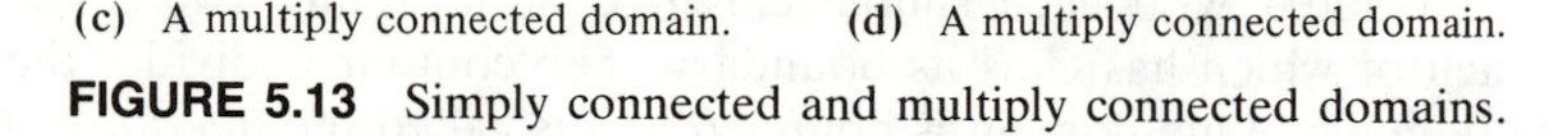

(c) A multiply connected domain. (d) A multiply connected domain.

FIGURE 5.13 Simply connected and multiply connected domains.

Let the simple closed contour C have the parameterization $C\colon z(t) = x(t) + iy(t)$ for $a \leqq t \leqq b$. If C is parameterized so that the interior of C is kept on the left as $z(t)$ moves around C, then we say that C is oriented in the *positive* (counterclockwise) sense; otherwise, C is oriented *negatively*. If C is positively oriented, then $-C$ is negatively oriented. Figure 5.14 illustrates the concept of positive and negative orientation.

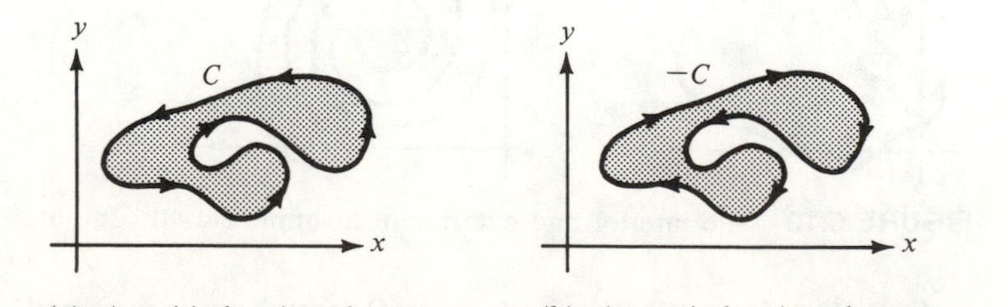

(a) A positively oriented contour. (b) A negatively oriented contour.

FIGURE 5.14 Simple closed contours that are positively and negatively oriented.

An important result from the calculus of real variables is known as Green's Theorem and is concerned with the line integral of real-valued functions.

Theorem 5.1 (Green's Theorem) *Let C be a simple closed contour with positive orientation, and let R be the domain that forms the interior of C. If P and Q are continuous and have continuous partial derivatives P_x, P_y, Q_x, and Q_y at all points on C and R, then*

$$(1) \qquad \int_C P(x, y)\,dx + Q(x, y)\,dy = \iint_R [Q_x(x, y) - P_y(x, y)]\,dx\,dy.$$

Proof for a Standard Region* If R is a standard region, then there exist functions $y = g_1(x)$ and $y = g_2(x)$ for $a \leqq x \leqq b$ whose graphs form the lower and upper portions of C, respectively, as indicated in Figure 5.15. Since C is to be given the positive (counterclockwise) orientation, these functions can be used to express C as the sum of two contours C_1 and C_2 where

$$C_1\colon z_1(t) = t + ig_1(t) \qquad \text{for } a \leqq t \leqq b \qquad \text{and}$$

$$C_2\colon z_2(t) = -t + ig_2(-t) \qquad \text{for } -b \leqq t \leqq -a.$$

We now use the functions $g_1(x)$ and $g_2(x)$ to express the double integral of $-P_y(x, y)$ over R as an iterated integral, first with respect to y and second with respect to x, as follows:

$$(2) \qquad -\iint_R P_y(x, y)\,dx\,dy = -\int_a^b \left[\int_{g_1(x)}^{g_2(x)} P_y(x, y)\,dy\right] dx.$$

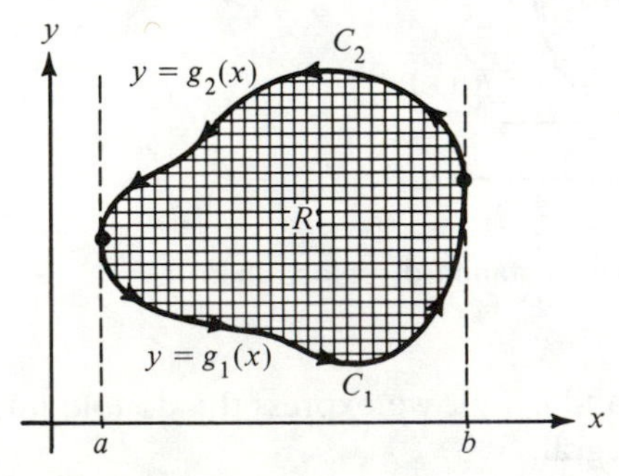

FIGURE 5.15 Integration over a standard region where $C = C_1 + C_2$.

* A standard region is bounded by a contour C, which can be expressed in the two forms $C = C_1 + C_2$ and $C = C_3 + C_4$ that are used in the proof.

Computing the first iterated integral on the right side of equation (2), we obtain

$$(3)\qquad -\iint_R P_y(x, y)\,dx\,dy = \int_a^b P(x, g_1(x))\,dx - \int_a^b P(x, g_2(x))\,dx.$$

In the second integral on the right side of equation (3) we can use the change of variable $x = -t$ and manipulate the integral to obtain

$$(4)\qquad -\iint_R P_y(x, y)\,dx\,dy = \int_a^b P(x, g_1(x))\,dx + \int_{-b}^{-a} P(-t, g_2(-t))(-1)\,dt.$$

When the two integrals on the right side of equation (4) are interpreted as contour integrals along C_1 and C_2, respectively, we see that

$$(5)\qquad -\iint_R P_y(x, y)\,dx\,dy = \int_{C_1} P(x, y)\,dx + \int_{C_2} P(x, y)\,dx = \int_C P(x, y)\,dx.$$

To complete the proof, we rely on the fact that for a standard region, there exist functions $x = h_1(y)$ and $x = h_2(y)$ for $c \leqq y \leqq d$ whose graphs form the left and right portions of C, respectively, as indicated in Figure 5.16. Since C has the positive orientation, it can be expressed as the sum of two contours C_3 and C_4 where

$$C_3\colon z_3(t) = h_1(-t) - it \qquad \text{for } -d \leqq t \leqq -c \qquad \text{and}$$

$$C_4\colon z_4(t) = h_2(t) + it \qquad \text{for } c \leqq t \leqq d.$$

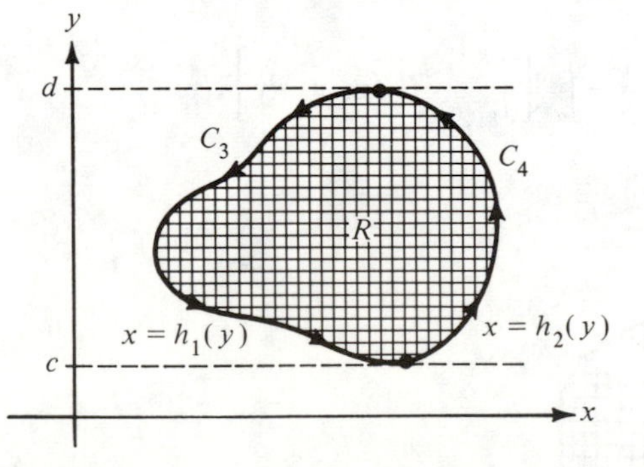

FIGURE 5.16 Integration over a standard region where $C = C_3 + C_4$.

Using the functions $h_1(y)$ and $h_2(y)$, we express the double integral of $Q_x(x, y)$ over R as an iterated integral:

$$(6)\qquad \iint_R Q_x(x, y)\,dx\,dy = \int_c^d \left[\int_{h_1(y)}^{h_2(y)} Q_x(x, y)\,dx\right] dy.$$

A similar derivation will show that (6) is equivalent to

$$(7) \qquad \iint_R Q_x(x, y)\, dx\, dy = \int_C Q(x, y)\, dy.$$

When (5) and (7) are added, the result is (1), and the proof is complete.

We are now ready to state our main result in this section.

Theorem 5.2 (Cauchy-Goursat Theorem) *Let f be analytic in a simply connected domain D. If C is a simple closed contour that lies in D, then*

$$(8) \qquad \int_C f(z)\, dz = 0.$$

Proof A proof that does not require the continuity of $f'(z)$ was devised by Edouard Goursat (1858–1936) in 1883 and is given in Appendix A at the end of this chapter. If we add the additional hypothesis that the derivative $f'(z)$ is also continuous, the proof is more intuitive. It was Augustin Cauchy (1789–1857) who first proved this theorem under the hypothesis that $f'(z)$ is continuous. His proof, which we will now state, used Green's Theorem.

Proof Using Green's Theorem We assume that C is oriented in the positive sense and use (12) in Section 5.2 to write

$$(9) \qquad \int_C f(z)\, dz = \int_C u\, dx - v\, dy + i \int_C v\, dx + u\, dy.$$

If we use Green's Theorem on the real part of the right side of equation (9) with $P = u$ and $Q = -v$, then we obtain

$$(10) \qquad \int_C u\, dx - v\, dy = \iint_R [-v_x - u_y]\, dx\, dy$$

where R is the region that is the interior of C. If we use Green's Theorem on the imaginary part, the result will be

$$(11) \qquad \int_C v\, dx + u\, dy = \iint_R [u_x - v_y]\, dx\, dy.$$

The Cauchy-Riemann equations $u_x = v_y$ and $u_y = -v_x$ can be used in (10) and (11) to see that the value of (9) is given by

$$\int_C f(z)\, dz = \iint_R [0]\, dx\, dy + i \iint_R [0]\, dx\, dy = 0,$$

and the proof is complete.

EXAMPLE 5.10 Let us recall that $\exp z$, $\cos z$, and z^n where n is a positive integer are all entire functions and have continuous derivatives. The Cauchy-Goursat Theorem implies that for any simple closed contour we have

$$\int_C \exp z\, dz = 0, \qquad \int_C \cos z\, dz = 0, \qquad \int_C z^n\, dz = 0.$$

EXAMPLE 5.11 If C is a simple closed contour such that the origin does not lie interior to C, then there is a simply connected domain D that contains C in which $f(z) = 1/z^n$ is analytic, as is indicated in Figure 5.17. The Cauchy-Goursat Theorem implies that

$$\int_C \frac{1}{z^n}\, dz = 0 \qquad \text{provided that the origin does not lie interior to } C.$$

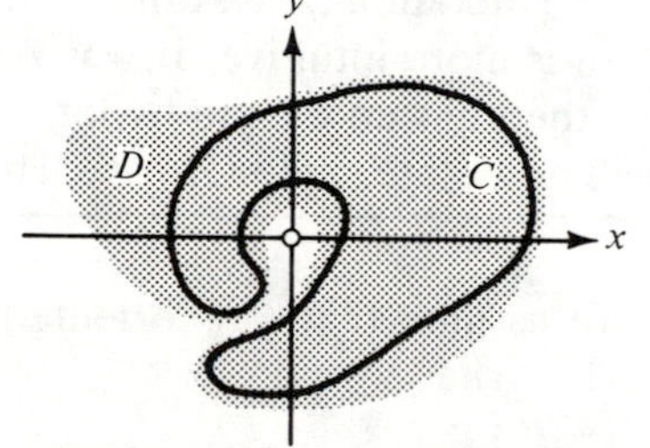

FIGURE 5.17 A simple connected domain D containing the simple closed contour C that does not contain the origin.

It is desirable to be able to replace integrals over certain complicated contours with integrals that are easy to evaluate. If C_1 is a simple closed contour that can be continuously deformed into another simple closed contour C_2 without passing through a point where f is not analytic, then the value of the contour integral of f over C_1 is the same as the value of the integral of f over C_2. To be precise, we state the following result.

Theorem 5.3 (Deformation of Contour) *Let C_1 and C_2 be two simple closed positively oriented contours such that C_1 lies interior to C_2. If f is analytic in a domain D that contains both C_1 and C_2 and the region between them, as shown in Figure 5.18, then*

$$\int_{C_1} f(z)\, dz = \int_{C_2} f(z)\, dz.$$

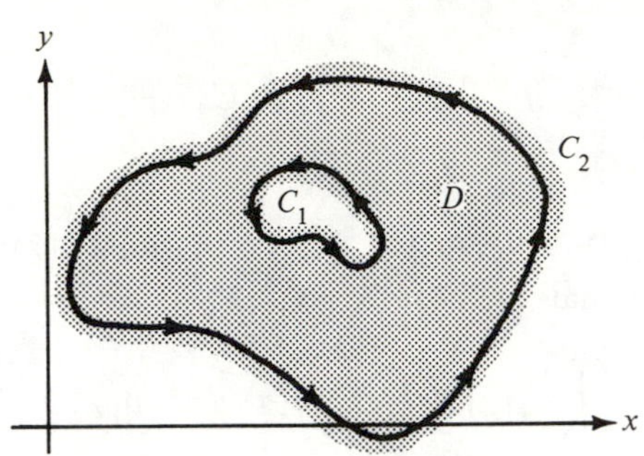

FIGURE 5.18 The domain D that contains the simple closed contours C_1 and C_2 and the region between them.

Proof Assume that both C_1 and C_2 have positive (counterclockwise) orientation. We construct two disjoint contours or *cuts* L_1 and L_2 that join C_1 to C_2. Hence the contour C_1 will be cut into two contours C_1^* and C_1^{**}, and the contour C_2 will be cut into C_2^* and C_2^{**}. We now form two new contours:

$$K_1 = -C_1^* + L_1 + C_2^* - L_2 \quad \text{and} \quad K_2 = -C_1^{**} + L_2 + C_2^{**} - L_1,$$

which are shown in Figure 5.19. The function f will be analytic on a simply connected domain D_1 that contains K_1, and f will be analytic on the simply connected domain D_2 that contains K_2, as is illustrated in Figure 5.19.

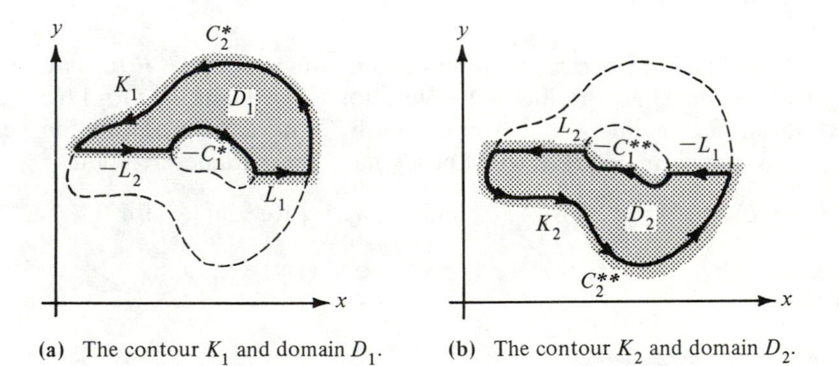

(a) The contour K_1 and domain D_1. (b) The contour K_2 and domain D_2.

FIGURE 5.19 The cuts L_1 and L_2 and the contours K_1 and K_2 used to prove the Deformation Theorem.

The Cauchy-Goursat Theorem can be applied to the contours K_1 and K_2, and the result is

$$\text{(12)} \qquad \int_{K_1} f(z)\,dz = 0 \quad \text{and} \quad \int_{K_2} f(z)\,dz = 0.$$

Adding contours, we observe that

$$(13)\qquad K_1 + K_2 = -C_1^* + L_1 + C_2^* - L_2 - C_1^{**} + L_2 + C_2^{**} - L_1$$
$$= C_2^* + C_2^{**} - C_1^* - C_1^{**} = C_2 - C_1.$$

We can use identities (14) and (17) of Section 5.2 and equations (12) and (13) given in this section to conclude that

$$\int_{C_2} f(z)\,dz - \int_{C_1} f(z)\,dz = \int_{K_1} f(z)\,dz + \int_{K_2} f(z)\,dz = 0,$$

which completes the proof of Theorem 5.3.

We now state an important result that is proven by the Deformation Theorem. This result will occur several times in the theory to be developed and is an important tool for computations.

EXAMPLE 5.12 Let z_0 denote a fixed complex value. If C is a simple closed contour with positive orientation such that z_0 lies interior to C, then

$$(14)\qquad \int_C \frac{dz}{z - z_0} = 2\pi i \quad \text{and} \quad \int_C \frac{dz}{(z - z_0)^n} = 0 \quad \text{where } n \neq 1 \text{ is an integer.}$$

Solution Since z_0 lies interior to C, we can choose R so that the circle C_R will center z_0 and radius R lies interior to C. Hence $f(z) = 1/(z - z_0)^n$ is analytic in a domain D that contains both C and C_R and the region between them, as shown in Figure 5.20. Let C_R have the parameterization

$$C_R\colon z(\theta) = z_0 + Re^{i\theta} \quad \text{and} \quad dz = i\,Re^{i\theta}\,d\theta \quad \text{for } 0 \leqq \theta \leqq 2\pi.$$

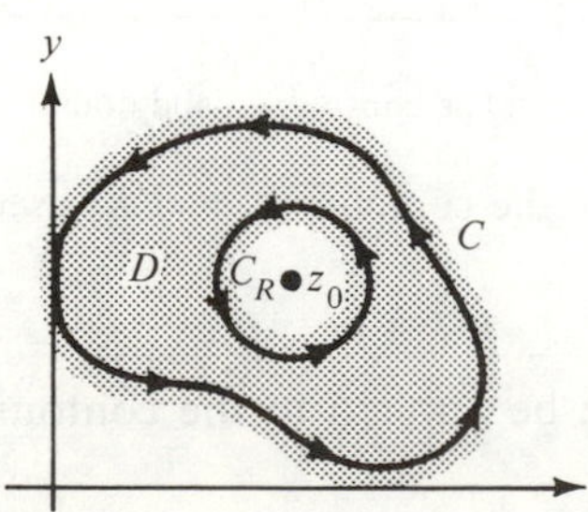

FIGURE 5.20 The domain D that contains both C and C_R.

The Deformation Theorem implies that the integral of f over C_R has the same value as the integral of f over C, and we obtain

$$\int_C \frac{dz}{z - z_0} = \int_{C_R} \frac{dz}{z - z_0} = \int_0^{2\pi} \frac{i\,\mathrm{R}e^{i\theta}}{\mathrm{R}e^{i\theta}}\,d\theta = i\int_0^{2\pi} d\theta = 2\pi i$$

and

$$\int_C \frac{dz}{(z - z_0)^n} = \int_{C_R} \frac{dz}{(z - z_0)^n} = \int_0^{2\pi} \frac{i\,\mathrm{R}e^{i\theta}}{R^n e^{in\theta}}\,d\theta = iR^{1-n}\int_0^{2\pi} e^{i(1-n)\theta}\,d\theta$$

$$= \frac{R^{1-n}}{1-n} e^{i(1-n)\theta}\Big|_{\theta=0}^{\theta=2\pi} = \frac{R^{1-n}}{1-n} - \frac{R^{1-n}}{1-n} = 0.$$

The Deformation Theorem is an extension of the Cauchy-Goursat Theorem to a doubly connected domain in the following sense. Let D be a domain that contains C_1 and C_2 and the region between them as shown in Figure 5.18. Then the contour $C = C_2 - C_1$ is a parameterization of the boundary of the region R that lies between C_1 and C_2 so that the points of R lie to the left of C as a point $z(t)$ moves around C. Hence C is a positive orientation of the boundary of R, and Theorem 5.3 implies that

$$\int_C f(z)\,dz = 0.$$

We can extend Theorem 5.3 to multiply connected domains with more than one "hole." The proof, which is left for the reader, involves the introduction of several cuts and is similar to the proof of Theorem 5.3.

Theorem 5.4 (Extended Cauchy-Goursat Theorem) *Let C, C_1, $C_2, \ldots, C_n$ be simple closed positively oriented contours with the property that C_k lies interior to C for $k = 1, 2, \ldots, n$, and the set interior to C_k has no points in common with the set interior to C_j if $k \neq j$. Let f be analytic on a domain D that contains all the contours and the region between C and $C_1 + C_2 + \cdots + C_n$, which is shown in Figure 5.21. Then*

$$(15)\qquad \int_C f(z)\,dz = \sum_{k=1}^{n} \int_{C_k} f(z)\,dz.$$

EXAMPLE 5.13 If C is the circle $|z| = 2$ taken with positive orientation, then

$$(16)\qquad \int_C \frac{2z\,dz}{z^2 + 2} = 4\pi i.$$

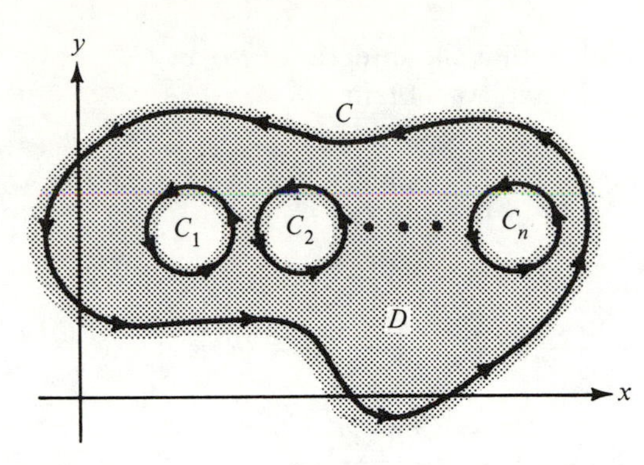

FIGURE 5.21 The multiply connected domain D and the contours C and C_1, $C_2, \ldots, C_n$ in the statement of the Extended Cauchy-Goursat Theorem.

Solution Using partial fractions, the integral in equation (16) can be written as

$$\int_C \frac{2z\,dz}{z^2+2} = \int_C \frac{dz}{z+i\sqrt{2}} + \int_C \frac{dz}{z-i\sqrt{2}}. \tag{17}$$

Since the points $z = \pm i\sqrt{2}$ lie interior to C, Example 5.12 implies that

$$\int_C \frac{dz}{z \pm i\sqrt{2}} = 2\pi i. \tag{18}$$

The results in (18) can be used in (17) to conclude that

$$\int_C \frac{2z\,dz}{z^2+2} = 2\pi i + 2\pi i = 4\pi i.$$

EXAMPLE 5.14 If C is the circle $|z - i| = 1$ taken with positive orientation, then

$$\int_C \frac{2z\,dz}{z^2+2} = 2\pi i. \tag{19}$$

Solution Using partial fractions, the integral in equation (19) can be written as

$$\int_C \frac{2z\,dz}{z^2+2} = \int_C \frac{dz}{z+i\sqrt{2}} + \int_C \frac{dz}{z-i\sqrt{2}}. \tag{20}$$

In this case, only the point $z = i\sqrt{2}$ lies interior to C, so the second integral on the right side of equation (20) has the value $2\pi i$. The function $f(z) = 1/(z + i\sqrt{2})$ is analytic on a simply connected domain that contains C. Hence by the Cauchy-Goursat Theorem the first integral on the right side of equation (20) is zero (see Figure 5.22). Therefore

$$\int_C \frac{2z\,dz}{z^2+2} = 0 + 2\pi i = 2\pi i.$$

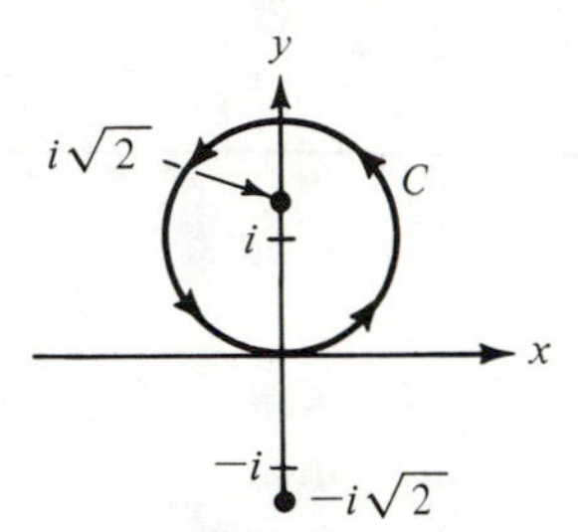

FIGURE 5.22 The circle $|z - i| = 1$ and the points $z = \pm i\sqrt{2}$.

EXAMPLE 5.15 Show that

$$\int_C \frac{z-2}{z^2-z}\,dz = -6\pi i$$

where C is the "figure-eight" contour shown in Figure 5.23(a).

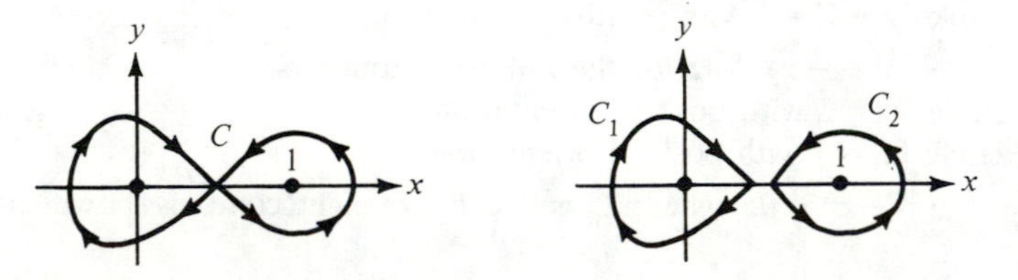

(a) The "figure eight" contour C. (b) The contours C_1 and C_2.

FIGURE 5.23 The contour $C = C_1 + C_2$.

Solution Partial fractions can be used to express the integral as

$$\int_C \frac{z-2}{z^2-z}\,dz = 2\int_C \frac{1}{z}\,dz - \int_C \frac{1}{z-1}\,dz. \tag{21}$$

Using the Cauchy-Goursat Theorem and property (14) of Section 5.2 together with Example 5.11, we compute the value of the first integral on the right side of equation (21):

$$2\int_C \frac{1}{z}\,dz = 2\int_{C_1} \frac{1}{z}\,dz + 2\int_{C_2} \frac{1}{z}\,dz \tag{22}$$

$$= -2\int_{-C_1} \frac{1}{z}\,dz + 0 = -2(2\pi i) = -4\pi i.$$

In a similar fashion we find that

$$(23) \qquad -\int_C \frac{dz}{z-1} = -\int_{C_1} \frac{dz}{z-1} - \int_{C_2} \frac{dz}{z-1} = 0 - 2\pi i = -2\pi i.$$

The results of equations (22) and (23) can be used in equation (21) to conclude that

$$\int_C \frac{z-2}{z^2-z}\, dz = -4\pi i - 2\pi i = -6\pi i.$$

EXERCISES FOR SECTION 5.3

1. Determine the domain of analyticity for the following functions, and conclude that $\int_C f(z)\, dz = 0$ where C is the circle $|z| = 1$ with positive orientation.

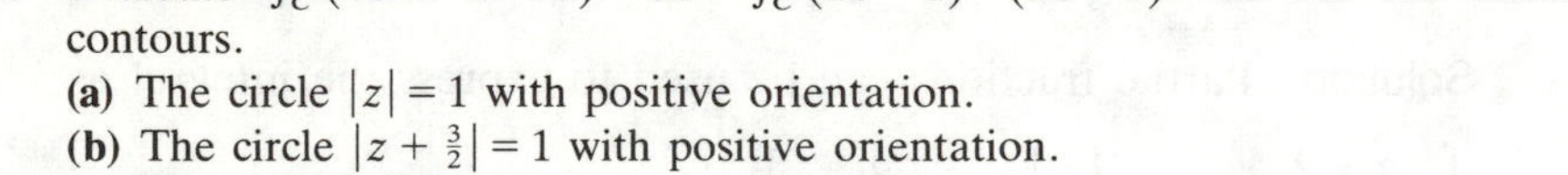

   (a) $f(z) = \dfrac{z}{z^2+2}$ **(b)** $f(z) = \dfrac{1}{z^2+2z+2}$

   (c) $f(z) = \tan z$ **(d)** $f(z) = \text{Log}(z+5)$

2. Show that $\int_C z^{-1}\, dz = 2\pi i$ where C is the square with vertices $1 \pm i$, $-1 \pm i$ with positive orientation.
3. Show that $\int_C (4z^2 - 4z + 5)^{-1}\, dz = 0$ where C is the unit circle $|z| = 1$ with positive orientation.
4. Find $\int_C (z^2 - z)^{-1}\, dz$ for the following contours.
   (a) The circle $|z-1| = 2$ with positive orientation.
   (b) The circle $|z-1| = \frac{1}{2}$ with positive orientation.
5. Find $\int_C (2z-1)(z^2-z)^{-1}\, dz$ for the following contours.
   (a) The circle $|z| = 2$ with positive orientation.
   (b) The circle $|z| = \frac{1}{2}$ with positive orientation.
6. Evaluate $\int_C (z^2 - z)^{-1}\, dz$ where C is the figure-eight contour shown in Figure 5.23(a).
7. Evaluate $\int_C (2z-1)(z^2-z)^{-1}\, dz$ where C is the figure-eight contour shown in Figure 5.23(a).
8. Evaluate $\int_C (4z^2 + 4z - 3)^{-1}\, dz = \int_C (2z-1)^{-1}(2z+3)^{-1}\, dz$ for the following contours.
   (a) The circle $|z| = 1$ with positive orientation.
   (b) The circle $|z + \frac{3}{2}| = 1$ with positive orientation.
   (c) The circle $|z| = 3$ with positive orientation.
9. Evaluate $\int_C (z^2 - 1)^{-1}\, dz$ for the contours given in Figure 5.24.

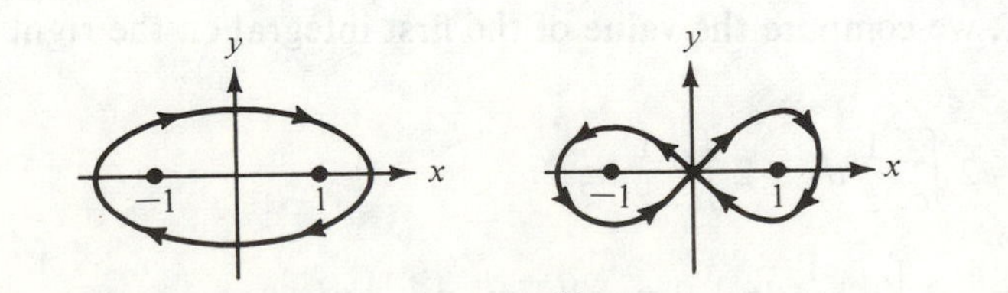

FIGURE 5.24 Accompanies Exercise 9.

10. Let C be the triangle with vertice 0, 1, and i with positive orientation. Parameterize C and show that

$$\int_C 1\,dz = 0 \qquad \text{and} \qquad \int_C z\,dz = 0.$$

11. Let the circle $|z| = 1$ be given the parameterization

$$C\colon z(t) = \cos t + i \sin t \qquad \text{for } -\pi \leqq t \leqq \pi.$$

Use the principal branch of the square root function:

$$z^{1/2} = r^{1/2} \cos \frac{\theta}{2} + i r^{1/2} \sin \frac{\theta}{2} \qquad \text{for } -\pi < \theta \leqq \pi$$

and find $\int_C z^{1/2}\,dz$.

12. Evaluate $\int_C |z|^2 \exp z\,dz$ where C is the unit circle $|z| = 1$ with positive orientation.

13. Let $f(z) = u(r, \theta) + iv(r, \theta)$ be analytic for all values of $z = re^{i\theta}$. Show that

$$\int_0^{2\pi} [u(r, \theta) \cos \theta - v(r, \theta) \sin \theta]\,d\theta = 0.$$

Hint: Integrate f around the circle $|z| = 1$.

14. Show by using Green's Theorem that the area enclosed by a simple closed contour C is $\frac{1}{2} \int_C x\,dy - y\,dx$.

5.4 The Fundamental Theorems of Integration

Let f be analytic in the simply connected domain D. The theorems in this section show that an antiderivative F can be constructed by contour integration. A consequence will be the fact that in a simply connected domain, the integral of an analytic function f along any contour joining z_1 to z_2 is the same, and its value is given by $F(z_2) - F(z_1)$. Hence we will be able to use the antiderivative formulas from calculus to compute the value of definite integrals.

Theorem 5.5 (Indefinite Integrals or Antiderivatives) *Let f be analytic in the simply connected domain D. If z_0 is a fixed value in D and if C is any contour in D with initial point z_0 and terminal point z, then the function given by*

$$F(z) = \int_C f(\xi)\,d\xi = \int_{z_0}^{z} f(\xi)\,d\xi \tag{1}$$

is analytic in D and

$$F'(z) = f(z). \tag{2}$$

Proof We first establish that the integral is independent of the path of integration. Hence we will need to keep track only of the endpoints, and we can use the notation

$$\int_C f(\xi)\,d\xi = \int_{z_0}^{z} f(\xi)\,d\xi.$$

Let C_1 and C_2 be two contours in D, both with the initial point z_0 and the terminal point z, as shown in Figure 5.25. Then $C = C_1 - C_2$ is a simple closed contour, and the Cauchy-Goursat Theorem implies that

$$\int_{C_1} f(\xi)\, d\xi - \int_{C_2} f(\xi)\, d\xi = \int_{C_1 - C_2} f(\xi)\, d\xi = 0.$$

Therefore the contour integral in equation (1) is independent of path. Here we have taken the liberty of drawing contours that intersect only at the endpoints. A slight modification of the foregoing proof will show that a finite number of other points of intersection are permitted.

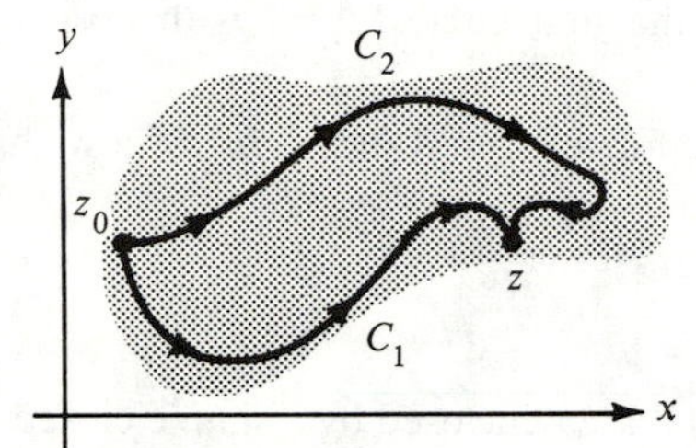

FIGURE 5.25 The contours C_1 and C_2 joining z_0 to z.

We now show that $F'(z) = f(z)$. Let z be held fixed, and let Δz be chosen small enough so that the point $z + \Delta z$ also lie in the domain D. Since z is held fixed, $f(z) = K$ where K is a constant, and equation (12) of Section 5.1 implies that

$$(3) \qquad \int_z^{z+\Delta z} f(z)\, d\xi = \int_z^{z+\Delta z} K\, d\xi = K\,\Delta z = f(z)\,\Delta z.$$

Using the additive property of contours and the definition of F given in equation (1), it follows that

$$(4) \qquad F(z + \Delta z) - F(z) = \int_{z_0}^{z+\Delta z} f(\xi)\, d\xi - \int_{z_0}^{z} f(\xi)\, d\xi$$

$$= \int_{C_2} f(\xi)\, d\xi - \int_{C_1} f(\xi)\, d\xi = \int_C f(\xi)\, d\xi$$

where the contour C is the straight line segment joining z to $z + \Delta z$ and C_1 and C_2 join z_0 to z and z_0 to $z + \Delta z$, respectively, as shown in Figure 5.26.

Since f is continuous at z, then if $\varepsilon > 0$, there is a $\delta > 0$ so that

$$(5) \qquad |f(\xi) - f(z)| < \varepsilon \qquad \text{whenever } |\xi - z| < \delta.$$

If we require that $|\Delta z| < \delta$, then using (3), (4), (5) and inequality (22) of Section 5.2, we obtain the following estimate:

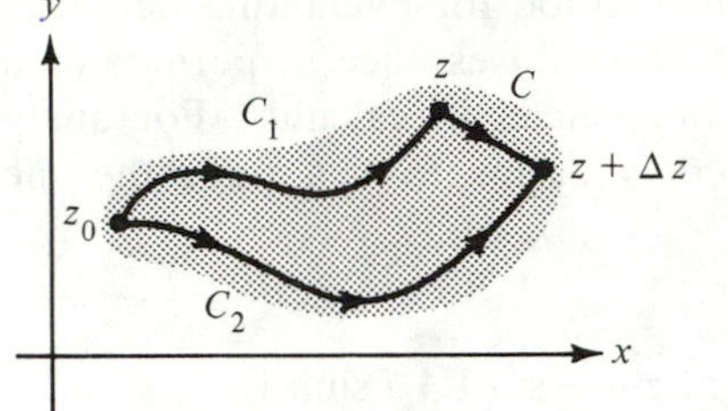

FIGURE 5.26 The contours C_1, C_2 and the line segment $C = -C_1 + C_2$.

$$(6)\qquad \left|\frac{F(z+\Delta z)-F(z)}{\Delta z} - f(z)\right| = \frac{1}{|\Delta z|}\left|\int_C f(\xi)\,d\xi - \int_C f(z)\,d\xi\right|$$

$$\leqq \frac{1}{|\Delta z|}\int_C |f(\xi)-f(z)||d\xi|$$

$$< \frac{1}{|\Delta z|}\,\varepsilon|\Delta z| = \varepsilon.$$

Consequently, the left side of (6) tends to 0 as $\Delta z \to 0$; that is, $F'(z) = f(z)$, and the theorem is proven.

It is important to notice that the line integral of an analytic function is independent of path. Let us recall Example 5.7. There we showed that

$$\int_{C_1} z\,dz = \int_{C_2} z\,dz = 4 + 2i$$

where C_1 and C_2 were contours joining $-1 - i$ to $3 + i$. Since the integrand $f(z) = z$ is an analytic function, Theorem 5.5 implies that the value of the two integrals is the same; hence one calculation would suffice.

If we set $z = z_1$ in Theorem 5.5, then we obtain the following familiar result for evaluating a definite integral of an analytic function.

Theorem 5.6 (Definite Integrals) *Let f be analytic in a simply connected domain D. If z_0 and z_1 are two points in D, then*

$$(7)\qquad \int_{z_0}^{z_1} f(z)\,dz = F(z_1) - F(z_0)$$

where F is an antiderivative of f.

Proof If F is chosen to be the function in equation (1), then equation (7) holds true. If G is any other antiderivative of f, then $H(z) = G(z) - F(z)$ is analytic, and $H'(z) = 0$ for all points z in D. Hence $H(z) = K$ where K is a constant, and $G(z) = F(z) + K$. Therefore $G(z_1) - G(z_0) = F(z_1) - F(z_0)$, and Theorem 5.6 is proven.

Theorem 5.6 is an important method for evaluating definite integrals when the integrand is an analytic function. In essence, it permits us to use all the rules of integration that were introduced in calculus. For analytic integrands, application of Theorem 5.6 is easier to use than the method of parameterization of a contour.

EXAMPLE 5.16 Show that $\int_1^i \cos z\, dz = -\sin 1 + i \sinh 1$.

Solution An antiderivative of $f(z) = \cos z$ is $F(z) = \sin z$. Hence

$$\int_1^i \cos z\, dz = \sin(i) - \sin(1) = -\sin 1 + i \sinh 1.$$

EXAMPLE 5.17 Show that $\int_{1-i}^{1+i} z^3\, dz = 0$.

Solution An antiderivative of $f(z) = z^3$ is $F(z) = \frac{1}{4}z^4$. Hence

$$\int_{1-i}^{1+i} z^3\, dz = \tfrac{1}{4}(1+i)^4 - \tfrac{1}{4}(1-i)^4 = -1 + 1 = 0.$$

EXAMPLE 5.18 Show that

$$\int_4^{8+6i} \frac{dz}{2z^{1/2}} = 1 + i$$

where $z^{1/2}$ is the principal branch of the square root function and the integral is to be taken along the line segment joining 4 to $8 + 6i$.

Solution Example 3.8 showed that if $F(z) = z^{1/2}$ then $F'(z) = 1/2z^{1/2}$ where the principal branch of the square root function is used in both the formulas for F and F'. Hence

$$\int_4^{8+6i} \frac{dz}{2z^{1/2}} = (8+6i)^{1/2} - 4^{1/2} = 3 + i - 2 = 1 + i.$$

EXAMPLE 5.19 Let $D = \{z = re^{i\theta} : r > 0 \text{ and } -\pi < \theta < \pi\}$ be the simply connected domain shown in Figure 5.27. Then $F(z) = \text{Log } z$ is analytic in D, and its derivative is $F'(z) = 1/z$. If C is a contour in D that joins the point z_1 to the point z_2, then Theorem 5.6 implies that

$$\int_{z_1}^{z_2} \frac{dz}{z} = \int_C \frac{dz}{z} = \text{Log } z_2 - \text{Log } z_1.$$

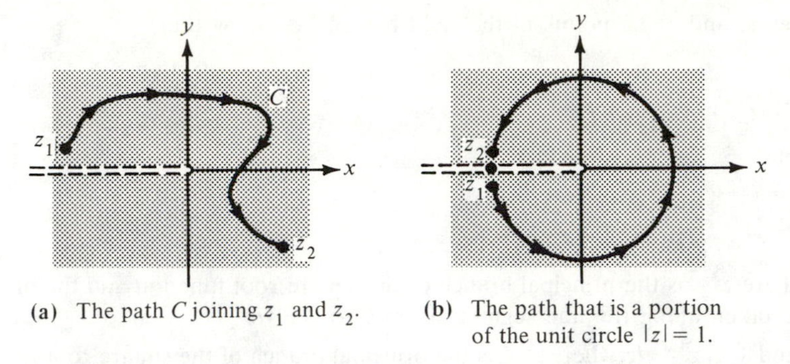

(a) The path C joining z_1 and z_2. (b) The path that is a portion of the unit circle $|z| = 1$.

FIGURE 5.27 The simply connected domain D in Examples 5.19 and 5.20.

EXAMPLE 5.20 As a consequence of Example 5.19, let us show that

$$\int_C \frac{dz}{z} = 2\pi i \qquad \text{where } C \text{ is the unit circle } |z| = 1$$

taken with positive orientation.

Solution If we let z_2 approach -1 through the upper half-plane and z_1 approaches -1 through the lower half-plane, then we can integrate around the portion of the circle shown in Figure 5.27(b) and take limits to obtain

$$\int_C \frac{dz}{z} = \lim_{\substack{z_1 \to -1 \\ z_2 \to -1}} \int_{z_1}^{z_2} \frac{dz}{z} = \lim_{\substack{z_2 \to -1 \\ \text{Im } z_2 > 0}} \text{Log } z_2 - \lim_{\substack{z_1 \to -1 \\ \text{Im } z_1 < 0}} \text{Log } z_1$$

$$= i\pi - (-i\pi) = 2\pi i.$$

EXERCISES FOR SECTION 5.4

For Exercises 1–14, use antiderivatives to find the value of the definite integral.

1. $\int_{1+i}^{2+i} z^2 \, dz$ **2.** $\int_1^i \frac{1+z}{z} \, dz$ (use Log z) **3.** $\int_2^{i\pi/2} \exp z \, dz$

4. $\int_i^{1+i} (z^2 + z^{-2}) \, dz$ **5.** $\int_{-i}^{1+i} \cos z \, dz$ **6.** $\int_0^{\pi - 2i} \sin \frac{z}{2} \, dz$

7. $\int_{-1-i\pi/2}^{2+\pi i} z \exp z \, dz$ **8.** $\int_{1-2i}^{1+2i} z \exp(z^2) \, dz$ **9.** $\int_0^i z \cos z \, dz$

10. $\int_0^i \sin^2 z \, dz$ **11.** $\int_1^{1+i} \text{Log } z \, dz$ **12.** $\int_2^{2+i} \frac{dz}{z^2 - z}$

13. $\int_2^{2+i} \frac{2z-1}{z^2 - z} \, dz$ **14.** $\int_2^{2+i} \frac{z-2}{z^2 - z} \, dz$

15. Show that $\int_{z_1}^{z_2} 1 \, dz = z_2 - z_1$ by parameterizing the line segment from z_1 to z_2.

16. Let z_1 and z_2 be points in the right half-plane. Show that

$$\int_{z_1}^{z_2} \frac{dz}{z^2} = \frac{1}{z_1} - \frac{1}{z_2}.$$

17. Find

$$\int_9^{3+4i} \frac{dz}{2z^{1/2}}$$

where $z^{1/2}$ is the principal branch of the square root function and the integral is to be taken along the line segment from 9 to $3+4i$.

18. Find $\int_{-2i}^{2i} z^{1/2}\, dz$ where $z^{1/2}$ is the principal branch of the square root function and the integral is to be taken along the right half of the circle $|z| = 2$.

19. Using the equation

$$\frac{1}{z^2+1} = \frac{i}{2}\frac{1}{z+i} - \frac{i}{2}\frac{1}{z-i},$$

show that if z lies in the right half-plane, then

$$\int_0^z \frac{d\xi}{\xi^2+1} = \arctan z = \frac{i}{2}\operatorname{Log}(z+i) - \frac{i}{2}\operatorname{Log}(z-i).$$

20. Let f' and g' be analytic for all z. Show that

$$\int_{z_1}^{z_2} f(z)g'(z)\, dz = f(z_2)g(z_2) - f(z_1)g(z_1) - \int_{z_1}^{z_2} f'(z)g(z)\, dz.$$

5.5 Integral Representations for Analytic Functions

We now present some major results in the theory of functions of a complex variable. The first result is known as Cauchy's Integral Formula and shows that the value of an analytic function f can be represented by a certain contour integral. The nth derivative, $f^{(n)}(z)$, will have a similar representation. In Chapter 6 we will show how the Cauchy Integral Formulae are used to prove Taylor's Theorem, and we will establish the power series representation for analytic functions. The Cauchy Integral Formulae will also be a convenient tool for evaluating certain contour integrals.

Theorem 5.7 (Cauchy's Integral Formula) *Let f be analytic in the simply connected domain D, and let C be a simple closed positively oriented contour that lies in D. If z_0 is a point that lies interior to C, then*

$$f(z_0) = \frac{1}{2\pi i}\int_C \frac{f(z)}{z-z_0}\, dz. \tag{1}$$

Proof Since f is continuous at z_0, if $\varepsilon > 0$ is given, there is a $\delta > 0$ such that

$$(2) \qquad |f(z) - f(z_0)| < \varepsilon \qquad \text{whenever } |z - z_0| < \delta.$$

Also the circle $C_0\colon |z - z_0| = \frac{1}{2}\delta$ lies interior to C as shown in Figure 5.28.

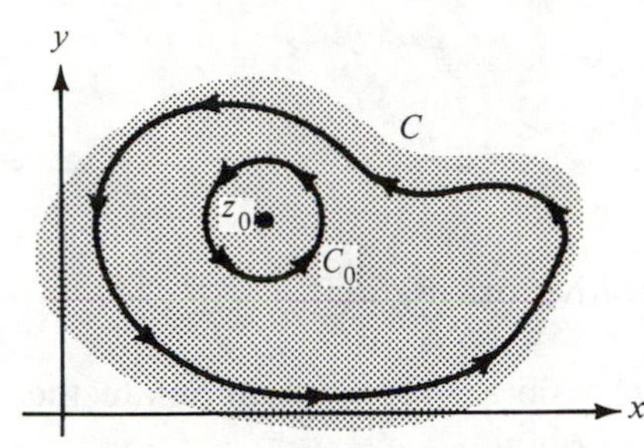

FIGURE 5.28 The contours C and C_0 in the proof of Cauchy's Integral Formula.

Since $f(z_0)$ is a fixed value, we can use the result of Example 5.12 to conclude that

$$(3) \qquad f(z_0) = \frac{f(z_0)}{2\pi i}\int_{C_0} \frac{dz}{z - z_0} = \frac{1}{2\pi i}\int_{C_0} \frac{f(z_0)}{z - z_0}\, dz.$$

Using the Deformation Theorem we see that

$$(4) \qquad \frac{1}{2\pi i}\int_C \frac{f(z)}{z - z_0}\, dz = \frac{1}{2\pi i}\int_{C_0} \frac{f(z)}{z - z_0}\, dz.$$

Using (2), (3), (4) and inequality (22) of Section 5.2, we obtain the following estimate:

$$(5) \qquad \left|\frac{1}{2\pi i}\int_C \frac{f(z)\, dz}{z - z_0} - f(z_0)\right| = \left|\frac{1}{2\pi i}\int_{C_0} \frac{f(z)\, dz}{z - z_0} - \frac{1}{2\pi i}\int_{C_0} \frac{f(z_0)\, dz}{z - z_0}\right|.$$

$$\leq \frac{1}{2\pi}\int_{C_0} \frac{|f(z) - f(z_0)|}{|z - z_0|}\, |dz|$$

$$\leq \frac{1}{2\pi}\, \frac{\varepsilon}{(1/2)\delta}\, \pi\delta = \varepsilon.$$

Since ε can be made arbitrarily small, the theorem is proven.

EXAMPLE 5.21 Show that

$$\int_C \frac{\exp z}{z - 1}\, dz = i2\pi e$$

where C is the circle $|z| = 2$ with positive orientation.

Solution Here we have $f(z) = \exp z$ and $f(1) = e$. The point $z_0 = 1$ lies interior to C, so Cauchy's Integral Formula implies that

$$e = f(1) = \frac{1}{2\pi i} \int_C \frac{\exp z}{z-1}\, dz,$$

and multiplication by $2\pi i$ will establish the desired result.

EXAMPLE 5.22 Show that

$$\int_C \frac{\sin z}{4z+\pi}\, dz = \frac{-\sqrt{2}\pi i}{4}$$

where C is the circle $|z| = 1$ with positive orientation.

Solution Here we have $f(z) = \sin z$. We can manipulate the integral and use Cauchy's Integral Formula to obtain

$$\int_C \frac{\sin z}{4z+\pi}\, dz = \frac{1}{4}\int_C \frac{\sin z}{z+(\pi/4)}\, dz = \frac{1}{4}\int_C \frac{f(z)}{z-(-\pi/4)}\, dz$$

$$= \frac{1}{4}(2\pi i) f\left(\frac{-\pi}{4}\right) = \frac{\pi i}{2}\sin\left(\frac{-\pi}{4}\right) = \frac{-\sqrt{2}\pi i}{4}.$$

We now state a general result that shows how differentiation under the integral sign can be accomplished. The proof can be found in some advanced texts. See, for instance, Rolf Nevanlinna and V. Paatero, *Introduction to Complex Analysis* (Reading, Massachusetts: Addison-Wesley Publishing Company, 1969), Section 9.7.

Theorem 5.8 (Leibniz's Rule) *Let D be a simply connected domain, and let I: $a \leqq t \leqq b$ be an interval of real numbers. Let $f(z, t)$ and its partial derivative $f_z(z, t)$ with respect to z be continuous functions for all z in D and all t in I. Then*

$$F(z) = \int_a^b f(z, t)\, dt \tag{6}$$

is analytic for z in D, and

$$F'(z) = \int_a^b f_z(z, t)\, dt.$$

We now show how Theorem 5.7 can be generalized to give an integral representation for the nth derivative, $f^{(n)}(z)$. Leibniz's Rule will be used in the proof, and we shall see that this method of proof will be a mnemonic device for remembering how the denominator is written.

Theorem 5.9 (Cauchy's Integral Formulae for Derivatives) *Let f be analytic in the simply connected domain D, and let C be a simple closed positively oriented contour that lies in D. If z is a point that lies interior to C, then*

$$f^{(n)}(z) = \frac{n!}{2\pi i}\int_C \frac{f(\xi)}{(\xi - z)^{n+1}}\,d\xi. \tag{7}$$

Proof We will establish the theorem for the case $n = 1$. We start by using the parameterization

$$C\colon \xi = \xi(t) \qquad \text{and} \qquad d\xi = \xi'(t)\,dt \qquad \text{for } a \leqq t \leqq b.$$

We use Theorem 5.7 and write

$$f(z) = \frac{1}{2\pi i}\int_C \frac{f(\xi)}{\xi - z}\,d\xi = \frac{1}{2\pi i}\int_a^b \frac{f(\xi(t))\xi'(t)\,dt}{\xi(t) - z}. \tag{8}$$

The integrand on the right side of (8) can be considered as a function $f(z, t)$ of the two variables z and t where

$$f(z, t) = \frac{f(\xi(t))\xi'(t)}{\xi(t) - z} \qquad \text{and} \qquad f_z(z, t) = \frac{f(\xi(t))\xi'(t)}{(\xi(t) - z)^2}. \tag{9}$$

Using (9) and Leibniz's Rule, we see that $f'(z)$ is given by

$$f'(z) = \frac{1}{2\pi i}\int_a^b \frac{f(\xi(t))\xi'(t)\,dt}{(\xi(t) - z)^2} = \frac{1}{2\pi i}\int_C \frac{f(\xi)\,d\xi}{(\xi - z)^2},$$

and the proof for the case $n = 1$ is complete. We can apply the same argument to the analytic function $f'(z)$ and show that its derivative $f''(z)$ has representation (7) with $n = 2$. The principle of mathematical induction will establish the theorem for any value of n.

EXAMPLE 5.23 Let z_0 denote a fixed complex value. If C is a simple closed positively oriented contour such that z_0 lies interior to C, then

$$\int_C \frac{dz}{z - z_0} = 2\pi i \qquad \text{and} \qquad \int_C \frac{dz}{(z - z_0)^{n+1}} = 0 \tag{10}$$

where $n \geqq 1$ is a positive integer.

Solution Here we have $f(z) = 1$ and the nth derivative is $f^{(n)}(z) = 0$. Theorem 5.7 implies that the value of the first integral in (10) is given by

$$\int_C \frac{dz}{z - z_0} = 2\pi i f(z_0) = 2\pi i,$$

and Theorem 5.9 can be used to conclude that

$$\int_C \frac{dz}{(z - z_0)^{n+1}} = \frac{2\pi i}{n!} f^{(n)}(z_0) = 0.$$

We remark that this is the same result that was proven earlier in Example 5.12. It should be obvious that the technique of using Theorems 5.7 and 5.9 is easier.

EXAMPLE 5.24 Show that

$$\int_C \frac{\exp(z^2)}{(z - i)^4}\, dz = \frac{-4\pi}{3e}$$

where C is the circle $|z| = 2$ with positive orientation.

Solution Here we have $f(z) = \exp(z^2)$, and a straightforward calculation shows that $f^{(3)}(z) = (12z + 8z^3)\exp(z^2)$. Using Cauchy's Integral Formulae with $n = 3$, we conclude that

$$\int_C \frac{\exp(z^2)}{(z - i)^4}\, dz = \frac{2\pi i}{3!} f^{(3)}(i) = \frac{2\pi i}{6}\frac{4i}{e} = \frac{-4\pi}{3e}.$$

EXAMPLE 5.25 Show that

$$\int_C \frac{\exp(i\pi z)\, dz}{2z^2 - 5z + 2} = \frac{2\pi}{3}$$

where C is the circle $|z| = 1$ with positive orientation.

Solution By factoring the denominator we obtain $2z^2 - 5z + 2 = (2z - 1)(z - 2)$. Only the root $z_0 = \frac{1}{2}$ lies interior to C. Now we set $f(z) = (\exp(i\pi z))/(z - 2)$ and use Theorem 5.7 to conclude that

$$\int_C \frac{\exp(i\pi z)\, dz}{2z^2 - 5z + 2} = \frac{1}{2}\int_C \frac{f(z)\, dz}{z - (1/2)} = \frac{1}{2}(2\pi i) f\left(\frac{1}{2}\right) = \pi i \frac{\exp(i\pi/2)}{(1/2) - 2}$$
$$= \frac{2\pi}{3}.$$

We now state two important corollaries to Theorem 5.9.

Corollary 5.1 *If f is analytic in the domain D, then all derivatives f', f'', . . . , $f^{(n)}$, . . . exist and are analytic in D.*

Proof For each point z_0 in D, there exists a closed disk $|z - z_0| \leqq R$ that is contained in D. The circle C: $|z - z_0| = R$ can be used in Theorem 5.9 to show that $f^{(n)}(z_0)$ exists for all n.

This result is interesting, since the definition of analytic function means that the derivative f' exists at all points in D. Here we find something more, that the derivatives of all orders exist!

Corollary 5.2 *If u is a harmonic function at each point (x, y) in the domain D, then all partial derivatives u_x, u_y, u_{xx}, u_{xy}, and u_{yy} exist and are harmonic functions.*

Proof For each point (x_0, y_0) in D there exists a closed disk $|z - z_0| \leqq R$ that is contained in D. A conjugate harmonic function v exists in this disk, so the function $f(z) = u + iv$ is an analytic function. We use the Cauchy-Riemann equations and see that $f'(z) = u_x + iv_x = v_y - iu_y$. Since f' is analytic, the functions u_x and u_y are harmonic. Again, we can use the Cauchy-Riemann equations to see that

$$f''(z) = u_{xx} + iv_{xx} = v_{yx} - iu_{yx} = -u_{yy} - iv_{yy}.$$

Since f'' is analytic, the functions u_{xx}, u_{xy}, and u_{yy} are harmonic.

EXERCISES FOR SECTION 5.5

For Exercises 1–15, assume that the contour C has positive orientation.

1. Find $\int_C (\exp z + \cos z)z^{-1}\, dz$ where C is the circle $|z| = 1$.

2. Find $\int_C (z+1)^{-1}(z-1)^{-1}\, dz$ where C is the circle $|z-1| = 1$.

3. Find $\int_C (z+1)^{-1}(z-1)^{-2}\, dz$ where C is the circle $|z-1| = 1$.

4. Find $\int_C (z^3-1)^{-1}\, dz$ where C is the circle $|z-1| = 1$.

5. Find $\int_C (z \cos z)^{-1}\, dz$ where C is the circle $|z| = 1$.

6. Find $\int_C z^{-4} \sin z\, dz$ where C is the circle $|z| = 1$.

7. Find $\int_C z^{-3} \sinh(z^2)\, dz$ where C is the circle $|z| = 1$.

8. Find $\int_C z^{-2} \sin z\, dz$ along the following contours:
(a) The circle $|z - (\pi/2)| = 1$. **(b)** The circle $|z - (\pi/4)| = 1$.

9. Find $\int_C z^{-n} \exp z\, dz$ where C is the circle $|z| = 1$ and n is a positive integer.

10. Find $\int_C z^{-2}(z^2 - 16)^{-1} \exp z\, dz$ along the following contours:
(a) The circle $|z| = 1$. **(b)** The circle $|z - 4| = 1$.

11. Find $\int_C (z^4 + 4)^{-1}\, dz$ where C is the circle $|z - 1 - i| = 1$.

12. Find $\int_C (z^2+1)^{-1}\, dz$ along the following contours:
(a) The circle $|z - i| = 1$. **(b)** The circle $|z + i| = 1$.

13. Find $\int_C (z^2+1)^{-1} \sin z\, dz$ along the following contours:
(a) The circle $|z - i| = 1$. **(b)** The circle $|z + i| = 1$.

14. Find $\int_C (z^2+1)^{-2}\, dz$ where C is the circle $|z - i| = 1$.

15. Find $\int_C z^{-1}(z-1)^{-1} \exp z\, dz$ along the following contours:
(a) The circle $|z| = 1/2$. **(b)** The circle $|z| = 2$.

For Exercises 16–19, assume that the contour C has positive orientation.

16. Let $P(z) = a_0 + a_1 z + a_2 z^2 + a_3 z^3$ be a cubic polynomial. Find $\int_C P(z) z^{-n}\, dz$ where C is the circle $|z| = 1$ and n is a positive integer.
17. Let f be analytic in the simply connected domain D, and let C be a simple closed contour in D. Suppose that z_0 lies exterior to C. Find $\int_C f(z)(z - z_0)^{-1}\, dz$.
18. Let z_1 and z_2 be two complex numbers that lie interior to the simple closed contour C. Show that $\int_C (z - z_1)^{-1}(z - z_2)^{-1}\, dz = 0$.
19. Let f be analytic in the simply connected domain D, and let z_1 and z_2 be two complex numbers that lie interior to the simple closed contour C that lies in D. Show that

$$\frac{f(z_2) - f(z_1)}{z_2 - z_1} = \frac{1}{2\pi i} \int_C \frac{f(z)\, dz}{(z - z_1)(z - z_2)}.$$

State what happens when $z_2 \to z_1$.

20. The *Legendre polynomial* $P_n(z)$ is defined by

$$P^n(z) = \frac{1}{2^n n!} \frac{d^n}{dz^n} [(z^2 - 1)^n].$$

Use Cauchy's Integral Formula to show that

$$P_n(z) = \frac{1}{2\pi i} \int_C \frac{(\xi^2 - 1)^n\, d\xi}{2^n(\xi - z)^{n+1}}$$

where z lies inside C.

5.6 The Theorems of Morera and Liouville and Some Applications

In this section we investigate some of the qualitative properties of analytic and harmonic functions. Our first result shows that the existence of an antiderivative for a continuous function is equivalent to the statement that the integral of f is independent of the path of integration. This result is stated in a form that will serve as a converse to the Cauchy-Goursat Theorem.

Theorem 5.10 (Morera's Theorem) *Let f be a continuous function in a simply connected domain D. If*

$$\int_C f(z)\, dz = 0$$

for every closed contour in D, then f is analytic in D.

Proof Select a point z_0 in D and define $F(z)$ by the following integral:

$$F(z) = \int_{z_0}^{z} f(\xi)\, d\xi.$$

The function $F(z)$ is uniquely defined because if C_1 and C_2 are two contours in

D, both with initial point z_0 and terminal point z, then $C = C_1 - C_2$ is a closed contour in D, and

$$0 = \int_C f(\xi)\, d\xi = \int_{C_1} f(\xi)\, d\xi - \int_{C_2} f(\xi)\, d\xi.$$

Since $f(z)$ is continuous, then if $\varepsilon > 0$, there exists a $\delta > 0$ such that $|\xi - z| < \delta$ implies that $|f(\xi) - f(z)| < \varepsilon$. Now we can use the identical steps to those in the proof of Theorem 5.5 to show that $F'(z) = f(z)$. Hence $F(z)$ is analytic on D, and Corollary 5.1 implies that $F'(z)$ and $F''(z)$ are also analytic. Therefore $f'(z) = F''(z)$ exists for all z in D, and we have proven that $f(z)$ is analytic on D.

Cauchy's Integral Formula shows how the value $f(z_0)$ can be represented by a certain contour integral. If we choose the contour of integration C to be a circle with center z_0, then we can show that the value $f(z_0)$ is the integral average of the values of $f(z)$ at points z on the circle C.

Theorem 5.11 (Gauss's Mean Value Theorem) *If f is analytic in a simply connected domain D that contains the circle $C: |z - z_0| = R$, then*

(1) $$f(z_0) = \frac{1}{2\pi} \int_0^{2\pi} f(z_0 + Re^{i\theta})\, d\theta.$$

Proof The circle C can be given the parameterization

(2) $$C: z(\theta) = z_0 + Re^{i\theta} \quad \text{and} \quad dz = i\,Re^{i\theta}\, d\theta \quad \text{for } 0 \leqq \theta \leqq 2\pi.$$

We can use the parameterization (2) and Cauchy's Integral Formula to obtain

$$f(z_0) = \frac{1}{2\pi i} \int_0^{2\pi} \frac{f(z_0 + Re^{i\theta})\, i\, Re^{i\theta}\, d\theta}{Re^{i\theta}} = \frac{1}{2\pi} \int_0^{2\pi} f(z_0 + Re^{i\theta})\, d\theta,$$

and Theorem 5.11 is proven.

We now prove an important result concerning the modulus of an analytic function.

Theorem 5.12 (Maximum Modulus Principle) *Let f be analytic and nonconstant in the domain D. Then $|f(z)|$ does not attain a maximum value at any point z_0 in D.*

Proof by Contradiction Assume the contrary, and suppose that there exists a point z_0 in D such that

(3) $$|f(z)| \leqq |f(z_0)| \quad \text{holds for all } z \text{ in } D.$$

If $C_0\colon |z - z_0| = R$ is any circle contained in D, then we can use identity (1) and property (22) of Section 5.2 to obtain

$$(4) \qquad |f(z_0)| = \left| \frac{1}{2\pi} \int_0^{2\pi} f(z_0 + re^{i\theta})\, d\theta \right| \leqq \frac{1}{2\pi} \int_0^{2\pi} |f(z_0 + re^{i\theta})|\, d\theta$$

for $0 \leqq r \leqq R$.

But in view of inequality (3) we can treat $|f(z)| = |f(z_0 + re^{i\theta})|$ as a real-valued function of the real variable θ and obtain

$$(5) \qquad \frac{1}{2\pi} \int_0^{2\pi} |f(z_0 + re^{i\theta})|\, d\theta \leqq \frac{1}{2\pi} \int_0^{2\pi} |f(z_0)|\, d\theta = |f(z_0)| \qquad \text{for } 0 \leqq r \leqq R.$$

If we combine inequalities (4) and (5), the result is the equation

$$|f(z_0)| = \frac{1}{2\pi} \int_0^{2\pi} |f(z_0 + re^{i\theta})|\, d\theta,$$

which can be written as

$$(6) \qquad \int_0^{2\pi} (|f(z_0)| - |f(z_0 + re^{i\theta})|)\, d\theta = 0 \qquad \text{for } 0 \leqq r \leqq R.$$

A theorem from the calculus of real-valued functions states that if the integral of a nonnegative continuous function taken over an interval is zero, then that function must be identically zero. Since the integrand in (6) is a nonnegative real-valued function, we conclude that it is identically zero; that is,

$$(7) \qquad |f(z_0)| = |f(z_0 + re^{i\theta})| \qquad \text{for } 0 \leqq r \leqq R \text{ and } 0 \leqq \theta \leqq 2\pi.$$

If the modulus of an analytic function is constant, then the results of Example 3.13 show that the function is constant. Therefore identity (7) implies that

$$(8) \qquad f(z) = f(z_0) \qquad \text{for all } z \text{ in the disk } D_0\colon |z - z_0| \leqq R.$$

Now let Z denote an arbitrary point in D, and let C be a contour in D that joins z_0 to Z. Let $2d$ denote the minimum distance from C to the boundary of D. Then we can find consecutive points $z_0, z_1, z_2, \ldots, z_n = Z$ along C with $|z_{k+1} - z_k| \leqq d$, such that the disks $D_k\colon |z - z_k| \leqq d$ for $k = 0, 1, \ldots, n$ are contained in D and cover C, as shown in Figure 5.29.

Since each disk D_k contains the center z_{k+1} of the next disk D_{k+1}, it follows that z_1 lies in D_0, and from equation (8) we see that $f(z_1) = f(z_0)$. Hence $|f(z)|$ also assumes its maximum value at z_1. An argument identical to the one given above will show that

$$(9) \qquad f(z) = f(z_1) = f(z_0) \qquad \text{for all } z \text{ in the disk } D_1.$$

We can proceed inductively and show that

$$(10) \qquad f(z) = f(z_{k+1}) = f(z_k) \qquad \text{for all } z \text{ in the disk } D_{k+1}.$$

By using (8), (9), and (10) it follows that $f(Z) = f(z_0)$. Therefore f is constant in D. With this contradiction the proof of the theorem is complete.

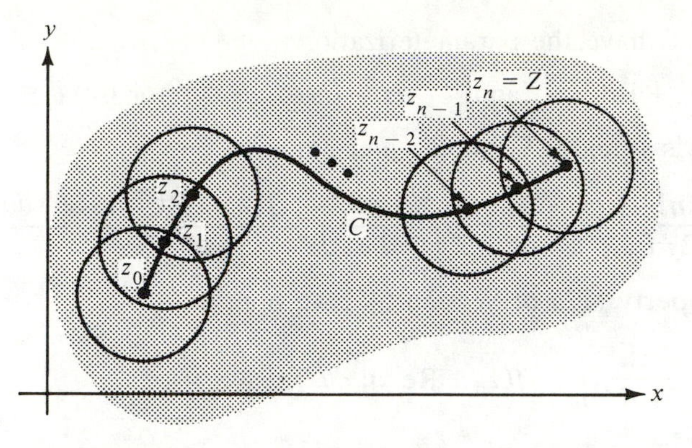

FIGURE 5.29 The "chain of disks" $D_0, D_1, \ldots, D_n$ that cover C.

The Maximum Modulus Principle is sometimes stated in the following form.

Theorem 5.12* (Maximum Modulus Principle) *Let f be analytic and nonconstant in the bounded domain D. If f is continuous on the closed region R that consists of D and all of its bondary points B, then $|f(z)|$ assumes its maximum value at some point z_0 on the boundary B.*

EXAMPLE 5.26 Let $f(z) = az + b$ where the domain is the disk $D = \{z: |z| < 1\}$. Then f is continuous on the closed region $R = \{z: |z| \leqq 1\}$. Prove that

$$\max_{|z| \leqq 1} |f(z)| = |a| + |b|$$

and that this value is assumed by f at a point $z_0 = e^{i\theta_0}$ on the boundary of D.

Solution From the triangle inequality and the fact that $|z| \leqq 1$ it follows that

$$|f(z)| = |az + b| \leqq |az| + |b| \leqq |a| + |b|.$$

If we choose $z_0 = e^{i\theta_0}$ where $\theta_0 = \arg b - \arg a$, then

$$\arg az_0 = \arg a + (\arg b - \arg a) = \arg b,$$

so the vectors az_0 and b lie on the same ray through the origin. Hence $|az_0 + b| = |az_0| + |b| = |a| + |b|$, and the result is established.

Theorem 5.13 (Cauchy's Inequalities) *Let f be analytic in the simply connected domain D that contains the circle $C: |z - z_0| = R$. If $|f(z)| \leqq M$ holds for all points z on C, then*

$$(11) \qquad |f^{(n)}(z_0)| \leqq \frac{n!M}{R^n} \qquad \text{for } n = 1, 2, \ldots.$$

Proof Let C have the parameterization

$$C: z(\theta) = z_0 + \mathrm{R}e^{i\theta} \qquad \text{and} \qquad dz = i\,\mathrm{R}e^{i\theta}\,d\theta \qquad \text{for } 0 \leqq \theta \leqq 2\pi.$$

We can use Cauchy's Integral Formulae and write

$$(12) \qquad f^{(n)}(z_0) = \frac{n!}{2\pi i}\int_C \frac{f(z)\,dz}{(z-z_0)^{n+1}} = \frac{n!}{2\pi i}\int_0^{2\pi} \frac{f(z_0 + \mathrm{R}e^{i\theta})\,i\,\mathrm{R}e^{i\theta}\,d\theta}{R^{n+1}e^{i(n+1)\theta}}.$$

Using (12) and property (22) of Section 5.2, we obtain

$$|f^{(n)}(z_0)| \leqq \frac{n!}{2\pi R^n}\int_0^{2\pi} |f(z_0 + \mathrm{R}e^{i\theta})|\,d\theta$$

$$\leqq \frac{n!}{2\pi R^n}\int_0^{2\pi} M\,d\theta = \frac{n!}{2\pi R^n} M 2\pi = \frac{n!M}{R^n},$$

and Theorem 5.13 is established.

The next result shows that a nonconstant entire function cannot be a bounded function.

Theorem 5.14 (Liouville's Theorem) *If f is an entire function and is bounded for all values of z in the complex plane, then f is constant.*

Proof Suppose that $|f(z)| \leqq M$ holds for all values of z. Let z_0 denote an arbitrary point. Then we can use the circle $C: |z - z_0| = R$, and Cauchy's Inequality with $n = 1$ implies that

$$(13) \qquad |f'(z_0)| \leqq \frac{M}{R}.$$

If we let $R \to \infty$ in (13), then we see that $f'(z_0) = 0$. Hence $f'(z) = 0$ for all z. If the derivative of an analytic function is zero for all z, then it must be a constant function. Therefore f is constant, and the theorem is proven.

EXAMPLE 5.27 The function $\sin z$ is *not* a bounded function.

Solution One way to see this is to observe that $\sin z$ is a nonconstant entire function, and therefore Liouville's Theorem implies that $\sin z$ cannot be bounded. Another way is to investigate the behavior of real and imaginary parts of $\sin z$. If we fix $x = \pi/2$ and let $y \to \infty$, then we see that

$$\lim_{y\to+\infty} \sin\left(\frac{\pi}{2} + iy\right) = \lim_{y\to+\infty} \sin\frac{\pi}{2}\cosh y + i\cos\frac{\pi}{2}\sinh y$$

$$= \lim_{y\to+\infty} \cosh y = +\infty.$$

Liouville's Theorem can be used to establish an important theorem of elementary algebra.

Theorem 5.15 (The Fundamental Theorem of Algebra) *If $P(z)$ is a polynomial of degree n, then P has at least one zero.*

Proof by Contradiction Assume the contrary and suppose that $P(z) \neq 0$ for all z. Then the function $f(z) = 1/P(z)$ is an entire function. We show that f is bounded as follows. First we write $P(z) = a_n z^n + a_{n-1} z^{n-1} + \cdots + a_1 z + a_0$ and consider the equation

$$(14) \qquad |f(z)| = \frac{1}{|P(z)|} = \frac{1}{|z|^n} \frac{1}{\left| a_n + \frac{a_{n-1}}{z} + \frac{a_{n-2}}{z^2} + \cdots + \frac{a_1}{z^{n-1}} + \frac{a_0}{z^n} \right|}.$$

Since $|a_k|/|z^{n-k}| = |a_k|/r^{n-k} \to 0$ as $|z| = r \to \infty$, it follows that

$$(15) \qquad a_n + \frac{a_{n-1}}{z} + \frac{a_{n-2}}{z^2} + \cdots + \frac{a_0}{z^n} \to a_n \qquad \text{as } |z| \to \infty.$$

If we use (15) in (14), then the result is

$$|f(z)| \to 0 \qquad \text{as } |z| \to \infty.$$

In particular, we can find a value of R such that

$$(16) \qquad |f(z)| \leqq 1 \qquad \text{for all } |z| \geqq R.$$

Consider

$$|f(z)| = [(u(x, y))^2 + (v(x, y))^2]^{1/2},$$

which is a continuous function of the two real variables x and y. A result from calculus regarding real functions says that a continuous function on a closed set is bounded. Hence $|f(z)|$ is bounded on the closed disk

$$x^2 + y^2 \leqq R^2;$$

that is, there exists a positive real number K such that

$$(17) \qquad |f(z)| \leqq K \qquad \text{for all } |z| \leqq R.$$

Combining (16) and (17), it follows that $|f(z)| \leqq M = \max\{K, 1\}$ holds for all z. Liouville's Theorem can now be used to conclude that f is constant. With this contradiction the proof of the theorem is complete.

Corollary 5.3 *Let P be a polynomial of degree n. Then P can be expressed as the product of linear factors. That is,*

$$P(z) = A(z - z_1)(z - z_2) \cdots (z - z_n)$$

where $z_1, z_2, \ldots, z_n$ are the zeros of P counted according to multiplicity and A is a constant.

EXERCISES FOR SECTION 5.6

For Exercises 1–4, express the given polynomial as a product of linear factors.

1. Factor $P(z) = z^4 + 4$.
2. Factor $P(z) = z^2 + (1+i)z + 5i$.
3. Factor $P(z) = z^4 - 4z^3 + 6z^2 - 4z + 5$.
4. Factor $P(z) = z^3 - (3+3i)z^2 + (-1+6i)z + 3 - i$. *Hint*: Show that $P(i) = 0$.
5. Let $f(z) = az^n + b$ where the region is the disk $R = \{z: |z| \leqq 1\}$. Show that
$$\max_{|z|\leqq 1} |f(z)| = |a| + |b|.$$
6. Show that $\cos z$ is *not* a bounded function.
7. Let $f(z) = z^2$ where the region is the rectangle $R = \{z = x + iy: 2 \leqq x \leqq 3$ and $1 \leqq y \leqq 3\}$.
Find the following:
(a) $\max_R |f(z)|$ **(b)** $\min_R |f(z)|$ **(c)** $\max_R \operatorname{Re} f(z)$ **(d)** $\min_R \operatorname{Im} f(z)$
Hint for (a) and (b): $|z|$ is the distance from 0 to z.
8. Let $F(z) = \sin z$ where the region is the rectangle
$$R = \left\{z = x + iy: 0 \leqq x \leqq \frac{\pi}{2} \text{ and } 0 \leqq y \leqq 2\right\}.$$
Find $\max_R |f(z)|$. *Hint*: $|\sin z|^2 = \sin^2 x + \sinh^2 y$.
9. Let f be analytic in the disk $|z| < 5$, and suppose that $|f(\xi)| \leqq 10$ for values of ξ on the circle $|\xi - 1| = 3$. **Find a bound for $|f^{(3)}(1)|$.**
10. Let f be analytic in the disk $|z| < 5$, and suppose that $|f(\xi)| \leqq 10$ for values of ξ on the circle $|\xi - 1| = 3$. Find a bound for $|f^{(3)}(0)|$.
11. Let f be an entire function such that $|f(z)| \leqq M|z|$ holds for all z.
(a) Show that $f''(z) = 0$ for all z, and **(b)** conclude that $f(z) = az + b$.
12. Establish the following *Minimum Modulus Principle*. Let f be analytic and nonconstant in the domain D. If $|f(z)| \geqq m$ where $m > 0$ holds for all z in D, then $|f(z)|$ does *not* attain a minimum value at any point z_0 in D.
13. Let $u(x, y)$ be harmonic for all (x, y). Show that
$$u(x_0, y_0) = \frac{1}{2\pi}\int_0^{2\pi} u(x_0 + R\cos\theta, y_0 + R\sin\theta)\, d\theta \qquad \text{where } R > 0.$$
Hint: Consider $f(z) = u(x, y) + iv(x, y)$.
14. Establish the following *Maximum Principle for harmonic functions*. Let $u(x, y)$ be harmonic and nonconstant in the simply connected domain D. Then u does not take on a maximum value at any point (x_0, y_0) in D. *Hint*: Let $f(z) = u(x, y) + iv(x, y)$ be analytic in D, and consider $F(z) = \exp(f(z))$ where $|F(z)| = e^{u(x,y)}$.
15. Let f be an entire function that has the property $|f(z)| \geqq 1$ for all z. Show that f is constant.
16. Let f be a nonconstant analytic function in the closed disk $R = \{z: |z| \leqq 1\}$. Suppose that $|f(z)| = K$ for all z on the circle $|z| = 1$. Show that f has a zero in D. *Hint*: Use both the Maximum and Minimum Modulus Principles.

5.7 Harmonic Functions and the Dirichlet Problem

The next result extends Theorem 3.7 to simply connected domains.

Theorem 5.16 *If $\phi(x, y)$ is a harmonic function in the simply connected domain D, then there exists an analytic function $F(z)$ defined on D such that*

(1) $$\phi(x, y) = \operatorname{Re} F(z) \qquad \text{for all } z \text{ in } D.$$

Proof For motivation, suppose that $\psi(x, y)$ is the harmonic conjugate of $\phi(x, y)$, and consider the analytic function

(2) $$F(z) = \phi(x, y) + i\psi(x, y).$$

Let the derivative be denoted as $f(z) = F'(z)$; then we define

(3) $$f(z) = u(x, y) + iv(x, y)$$

where

(4) $$u(x, y) = \phi_x(x, y) \qquad \text{and} \qquad v(x, y) = -\phi_y(x, y).$$

Now we see how to proceed. Since $\phi(x, y)$ is harmonic, Corollary 5.2 implies that both $\phi_x(x, y)$ and $\phi_y(x, y)$ exist; thus the definition of $f(z)$ given in (3) is justified. Now we show that the Cauchy-Riemann equations hold throughout D.

Since $\phi(x, y)$ is harmonic on D, $\phi_{xx}(x, y) + \phi_{yy}(x, y) = 0$ implies that

(5) $$\frac{\partial}{\partial x} \phi_x(x, y) = -\frac{\partial}{\partial y} \phi_y(x, y).$$

Partial differentiation of the functions in (4) can be used with (5) to obtain

(6) $$u_x(x, y) = v_y(x, y).$$

Since all the higher-order partial derivatives of $\phi(x, y)$ exist and are continuous, this implies that $\phi_{yx}(x, y) = \phi_{xy}(x, y)$. This in turn implies that

(7) $$\frac{\partial}{\partial y} \phi_x(x, y) = -\frac{\partial}{\partial x} [-\phi_y(x, y)].$$

The partial derivatives of the functions in (4) are used with (7) to obtain

(8) $$u_y(x, y) = -v_x(x, y).$$

Thus the Cauchy-Riemann equations hold, and we have proven that $f(z)$ is analytic throughout D. Theorem 5.5 assures us that $f(z)$ has an analytic antiderivative on D. The antiderivative is unique up to an additive constant. Hence the function $F(z)$ given in (2) is analytic on D, and the theorem is proven.

Dirichlet Problem for the Disk $|z| < 1$

Suppose that the value of an unknown harmonic function is known only at points on the boundary of the domain. Can this information be used to extend the function into the domain? Many applications are stated in this form of a boundary value problem. For example, if the steady state temperature is known at all points on the boundary of a region, then the temperature at all points of the interior are uniquely determined. If the boundary is the unit circle $|z| = 1$, the method outlined below can be used to define the function values inside the disk $|z| < 1$.

The Dirichlet problem for the closed unit disk can be stated as follows. Given the boundary values $U(\theta)$ for $-\pi < \theta \leqq \pi$, find a function $\phi(x, y)$ that is harmonic for $x^2 + y^2 < 1$ and such that

(9) $$\lim_{r \to 1} \phi(r \cos \theta, r \sin \theta) = U(\theta)$$

for each fixed θ where $U(\theta)$ is continuous (see Figure 5.30). The following result gives an integral solution to this problem.

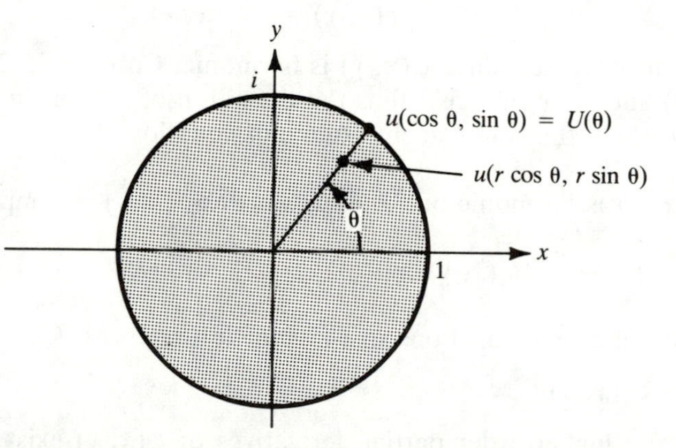

FIGURE 5.30 The Dirichlet problem for the unit disk $|z| < 1$.

Theorem 5.17 (Poisson Integral Formula) *Let $U(\theta)$ be a function that is piecewise continuous and bounded for $-\pi < \theta \leqq \pi$. Then there exists a unique function $\phi(x, y)$ that is harmonic in $|z| < 1$ that takes on the boundary values given in (9) at point of continuity of $U(\theta)$. Using polar coordinates, we can represent $\phi(x, y)$ by an integral:*

(10) $$\phi(r \cos \theta, r \sin \theta) = \frac{1}{2\pi} \int_{-\pi}^{\pi} \frac{(1 - r^2) U(t)\, dt}{1 + r^2 - 2r \cos(t - \theta)},$$

which is valid for $|z| < 1$. This formula is called the Poisson Integral Formula for the unit disk.

The proof of Theorem 5.17 is beyond the scope of this text. However, the following theorem conveys the meaning of the theorem, and its proof is built upon our development of complex integration.

Theorem 5.18 *Suppose that the function $u(x, y)$ is a harmonic in a simply connected domain that contains the closed unit disk $|z| \leqq 1$. If $u(x, y)$ takes on the boundary values stated in* (9), *then it has the integral representation* (10).

Proof Since $u(x, y)$ is harmonic in the simply connected domain, there exists a conjugate harmonic function $v(x, y)$ such that $f(z) = u(x, y) + iv(x, y)$ is an analytic function. Let C denote the contour consisting of the unit circle; then Cauchy's Integral Formula

$$(11) \qquad f(z) = \frac{1}{2\pi i} \int_C \frac{f(\xi)}{\xi - z}\, d\xi$$

expresses the value of $f(z)$ at any point z inside C in terms of the values of $f(\xi)$ at points ξ that lie on the circle C.

If we set $z^* = 1/\bar{z}$, then z^* lies outside the unit circle C, and the Cauchy-Goursat Theorem establishes the equation

$$(12) \qquad 0 = \frac{1}{2\pi i} \int_C \frac{f(\xi)}{\xi - z^*}\, d\xi.$$

Subtracting the expression for 0 in (12) from (11) and using the parameterization $\xi = e^{it}$, $d\xi = ie^{it}\, dt$ and the substitutions $z = re^{i\theta}$, $z^* = (1/r)e^{i\theta}$, we obtain

$$(13) \qquad f(z) = \frac{1}{2\pi} \int_{-\pi}^{\pi} \left[\frac{e^{it}}{e^{it} - re^{i\theta}} - \frac{e^{it}}{e^{it} - e^{i\theta}/r} \right] f(e^{it})\, dt.$$

The expression inside the brackets on the right side of (13) can be written

$$(14) \qquad \frac{e^{it}}{e^{it} - re^{i\theta}} - \frac{e^{it}}{e^{it} - e^{i\theta}/r} = \frac{1}{1 - re^{i(\theta - t)}} + \frac{re^{i(t-\theta)}}{1 - re^{i(t-\theta)}}$$
$$= \frac{1 - r^2}{1 + r^2 - 2r\cos(t - \theta)},$$

and it follows that

$$(15) \qquad f(z) = \frac{1}{2\pi} \int_{-\pi}^{\pi} \frac{(1 - r^2) f(e^{it})\, dt}{1 + r^2 - 2r\cos(t - \theta)}.$$

Since $u(x, y)$ is the real part of $f(z)$ and $U(t)$ is the real part of $f(e^{it})$, we can equate the real parts in this equation, and the proof is complete.

The real-valued function

$$(16)\qquad P(r, t-\theta) = \frac{1-r^2}{1+r^2-2r\cos(t-\theta)}$$

is known as the *Poisson kernal*. Using (14) and Lemma 6.1 in Chapter 6, we can express $P(r, t-\theta)$ in terms of trigonometric series:

$$(17)\qquad \begin{aligned} P(r, t-\theta) &= \frac{1}{1-re^{i(\theta-t)}} + \frac{re^{i(t-\theta)}}{1-re^{i(t-\theta)}} \\ &= \sum_{n=0}^{\infty} r^n e^{in(\theta-t)} + \sum_{n=1}^{\infty} r^n e^{in(t-\theta)} \\ &= 1 + \sum_{n=1}^{\infty} r^n [e^{in(\theta-t)} + e^{in(t-\theta)}] \\ &= 1 + 2\sum_{n=1}^{\infty} r^n \cos(n(\theta-t)) \\ &= 1 + 2\sum_{n=1}^{\infty} r^n [\cos n\theta \cos nt + \sin n\theta \sin nt]. \end{aligned}$$

Definition 5.2 (Fourier Series) *If $U(\theta)$ is periodic with period 2π and if $U(\theta)$ and $U'(\theta)$ are piecewise continuous, then the Fourier series for $U(\theta)$ is*

$$(18)\qquad \frac{a_0}{2} + \sum_{n=1}^{\infty} a_n \cos n\theta + \sum_{n=1}^{\infty} b_n \sin n\theta$$

where the coefficients $\{a_n\}$ and $\{b_n\}$ are given by the Euler formulas

$$(19)\qquad a_n = \frac{1}{\pi}\int_{-\pi}^{\pi} U(\theta)\cos n\theta \, d\theta \qquad \text{for } n = 0, 1, \ldots$$

and

$$(20)\qquad b_n = \frac{1}{\pi}\int_{-\pi}^{\pi} U(\theta)\sin n\theta \, d\theta \qquad \text{for } n = 1, 2, \ldots.$$

Theorem 5.19 *If $U(\theta)$ has the Fourier series representation*

$$(21)\qquad U(\theta) = \frac{a_0}{2} + \sum_{n=1}^{\infty} a_n \cos n\theta + \sum_{n=1}^{\infty} b_n \sin n\theta,$$

then the solution $u(x, y)$ to the Dirichlet problem in the unit disk is

$$(22)\qquad u(x, y) = \frac{a_0}{2} + \sum_{n=1}^{\infty} a_n r^n \cos n\theta + \sum_{n=1}^{\infty} b_n r^n \sin n\theta.$$

Proof It is easy to see that the series in (22) takes on the boundary values (9) when we substitute $r = 1$. Let us remark that Exercise 22 of Section

3.3 shows that the terms $r^n \cos n\theta$ and $r^n \sin n\theta$ are harmonic. Hence the right side of (22) defines a harmonic function. Now use (16) and (17) in equation (10) to get

$$\begin{aligned} u(x, y) &= \frac{1}{2\pi} \int_{-\pi}^{\pi} P(r, t-\theta) U(t)\, dt \\ &= \frac{1}{2\pi} \int_{-\pi}^{\pi} \left[1 + 2 \sum_{n=1}^{\infty} r^n \cos n\theta \cos nt \right. \\ &\quad \left. + 2 \sum_{n=1}^{\infty} r^n \sin n\theta \sin nt \right] U(t)\, dt \\ &= \frac{1}{2\pi} \int_{-\pi}^{\pi} U(t)\, dt + \sum_{n=1}^{\infty} r_n \cos n\theta \frac{1}{\pi} \int_{-\pi}^{\pi} U(t) \cos nt\, dt \\ &\quad + \sum_{n=1}^{\infty} r_n \sin n\theta \frac{1}{\pi} \int_{-\pi}^{\pi} U(t) \sin nt\, dt \\ &= \frac{a_0}{2} + \sum_{n=1}^{\infty} a_n r^n \cos n\theta + \sum_{n=1}^{\infty} b_n r^n \sin n\theta, \end{aligned}$$

and the proof of the theorem is complete.

EXAMPLE 5.28 Find the function $u(x, y)$ that is harmonic in the unit disk $|z| < 1$ and takes on the boundary values

(23) $$u(\cos\theta, \sin\theta) = U(\theta) = \theta/2 \qquad \text{for } -\pi < \theta < \pi.$$

Solution Using Euler's formulas (19) and (20) and integration by parts, we compute the coefficients of the Fourier series for $U(\theta)$:

(24) $$a_n = \frac{1}{\pi} \int_{-\pi}^{\pi} \frac{t}{2} \cos nt\, dt = \frac{1}{\pi} \left[\frac{t \sin nt}{2n} + \frac{\cos nt}{2n^2} \right] \Bigg|_{-\pi}^{\pi} = 0$$

for $n = 1, 2, \ldots$ and

(25) $$\begin{aligned} b_n &= \frac{1}{\pi} \int_{-\pi}^{\pi} \frac{t}{2} \sin nt\, dt = \frac{1}{\pi} \left[\frac{-t \cos nt}{2n} + \frac{\sin nt}{2n^2} \right] \Bigg|_{-\pi}^{\pi} \\ &= \frac{-\cos n\pi}{n} = (-1)^{n+1}/n \qquad \text{for } n = 1, 2, \ldots. \end{aligned}$$

The coefficient a_0 is obtained by a separate calculation:

(26) $$a_0 = \frac{1}{\pi} \int_{-\pi}^{\pi} \frac{t}{2}\, dt = \left[\frac{t^2}{4\pi} \right] \Bigg|_{-\pi}^{\pi} = 0.$$

Using these coefficients in formula (18), we obtain the Fourier series for $U(\theta)$:

$$U(\theta) = \theta/2 = \sum_{n=1}^{\infty} \frac{(-1)^{n+1}}{n} \sin n\theta \tag{27}$$

$$= \sin\theta - \frac{\sin 2\theta}{2} + \frac{\sin 3\theta}{3} - + \cdots.$$

According to Theorem 5.19, we can modify the series in (27) to obtain a representation for the solution $u(x, y)$:

$$u(x, y) = \sum_{n=1}^{\infty} \frac{(-1)^{n+1}}{n} r^n \sin n\theta$$

$$= r\sin\theta - \frac{r^2 \sin 2\theta}{2} + \frac{r^3 \sin 3\theta}{3} - + \cdots.$$

EXERCISES FOR SECTION 5.7

1. If $U(\theta) = K$ where K is a constant, then prove that $u(x, y) = K$.

2. Prove that

$$u(0, 0) = \frac{1}{2\pi} \int_{-\pi}^{\pi} U(\theta)\, d\theta,$$

that is, that the value at the center of the disk $|z| < 1$ is the average of the boundary values on the circle $|z| = 1$.

3. Extend the Poisson formula so that it can be used to define a harmonic function in a disk of arbitrary radius R.

4. Suppose that $U(\theta)$ is continuous. Prove that

$$m \leqq u(x, y) \leqq M \qquad \text{holds for all } |z| < 1$$

where m is the minimum value of $U(\theta)$ and M is the maximum value of $U(\theta)$.

For Exercises 5–9, find the solution $u(x, y)$ of the given Dirichlet problem in the unit disk $|z| < 1$. The boundary function $U(\theta)$ is defined on $-\pi < \theta \leqq \pi$. *Hint*: First find the Fourier series for $U(\theta)$ and then use Theorem 5.19.

5. $U(\theta) = |\theta|$

6. $U(\theta) = \theta|\theta|$

7. $U(\theta) = \theta^2$

8. $U(\theta) = \theta^3$

9. $U(\theta) = \begin{cases} \pi/2 - \theta & \text{for } 0 < \theta < \pi \\ \pi/2 + \theta & \text{for } -\pi < \theta < 0 \end{cases}$

Appendix A Goursat's Proof of the Cauchy-Goursat Theorem

Theorem 5.2 (Cauchy-Goursat Theorem) *Let f be analytic in a simply connected domain D. If C is a simple closed contour that lies in D, then*

$$\int_C f(z)\, dz = 0.$$

Proof We first establish the result for a triangular contour C with positive orientation. Construct four positively oriented contours C^1, C^2, C^3, C^4 that are the triangles obtained by joining the midpoints of the sides of C as shown in Figure 5.31.

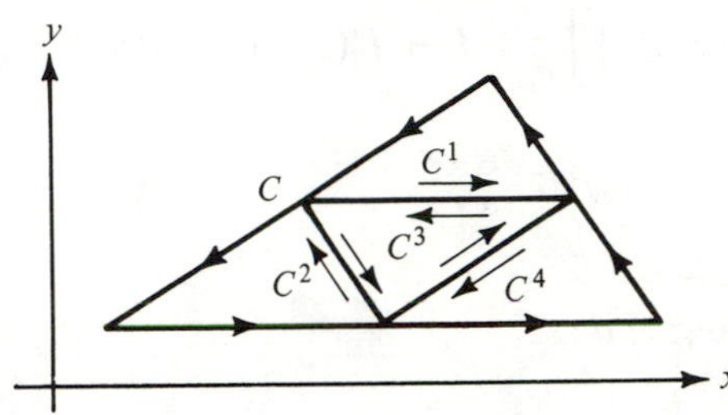

FIGURE 5.31 The triangular contours C and C^1, C^2, C^3, C^4.

Since each contour is positively oriented, if we sum the integrals along the four triangular contours, then the integrals along the segments interior to C cancel out in pairs. The result is

$$(1) \qquad \int_C f(z)\,dz = \sum_{k=1}^{4} \int_{C^k} f(z)\,dz.$$

Let C_1 be selected from C^1, C^2, C^3, C^4 so that the following relation holds true:

$$(2) \qquad \left|\int_C f(z)\,dz\right| \leqq \sum_{k=1}^{4} \left|\int_{C^k} f(z)\,dz\right| \leqq 4\left|\int_{C_1} f(z)\,dz\right|.$$

We can proceed inductively and carry out a similar subdivision process to obtain a sequence of triangular contours $\{C_n\}$ where the interior of C_{n+1} lies in the interior of C_n and the following inequality holds:

$$(3) \qquad \left|\int_{C_n} f(z)\,dz\right| \leqq 4\left|\int_{C_{n+1}} f(z)\,dz\right| \qquad \text{for } n = 1, 2, \ldots.$$

Let T_n denote the closed region that consists of C_n and its interior. Since the length of the sides of C_n go to zero as $n \to \infty$, there exists a unique point z_0 that belongs to all of the closed triangular regions T_n. Since f is analytic at the point z_0, there exists a function $\eta(z)$ with

$$(4) \qquad f(z) = f(z_0) + f'(z_0)(z - z_0) + \eta(z)(z - z_0).$$

Using (4) and integrating f along C_n, we find that

$$(5)\qquad \int_{C_n} f(z)\,dz = \int_{C_n} f(z_0)\,dz + \int_{C_n} f'(z_0)(z - z_0)\,dz$$

$$+ \int_{C_n} \eta(z)(z - z_0)\,dz$$

$$= [f(z_0) - f'(z_0)z_0] \int_{C_n} 1\,dz + f'(z_0) \int_{C_n} z\,dz$$

$$+ \int_{C_n} \eta(z)(z - z_0)\,dz$$

$$= \int_{C_n} \eta(z)(z - z_0)\,dz.$$

If $\varepsilon > 0$ is given, then a $\delta > 0$ can be found such that

$$(6)\qquad |z - z_0| < \delta \qquad \text{implies that } |\eta(z)| < \frac{\varepsilon}{L^2}$$

where L is the length of the original contour C. An integer n can now be chosen so that C_n lies in the neighborhood $|z - z_0| < \delta$, as shown in Figure 5.32.

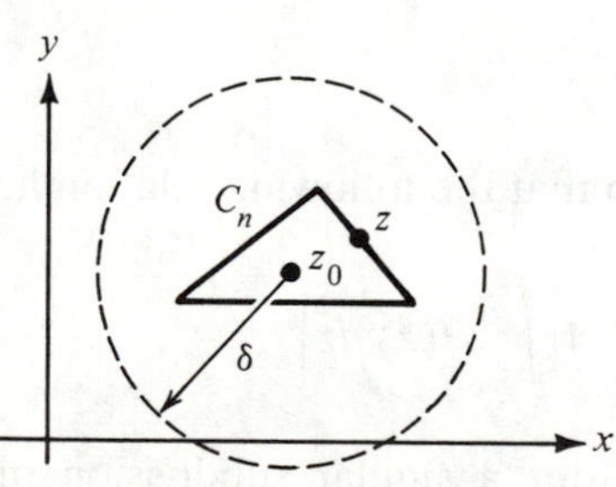

FIGURE 5.32 The contour C_n that lies in the neighborhood $|z - z_0| < \delta$.

Since the distance between a point z on a triangle and a point z_0 interior to the triangle is no greater than half the perimeter of the triangle, it follows that

$$(7)\qquad |z - z_0| < \tfrac{1}{2} L_n \qquad \text{for all } z \text{ on } C_n$$

where L_n is the length of the triangle C_n. From the above construction process, it follows that

$$(8)\qquad L_n = (\tfrac{1}{2})^n L \quad \text{and} \quad |z - z_0| < (\tfrac{1}{2})^{n+1} L \qquad \text{for } z \text{ on } C_n.$$

We can use (3), (6), (8) and equation (23) of Section 5.2 to obtain the following estimate:

$$\begin{aligned}\left|\int_C f(z)\,dz\right| &\leqq 4^n \int_{C_n} |\eta(z)(z - z_0)||dz| \\ &\leqq 4^n \int_{C_n} \frac{\varepsilon}{L^2}\left(\frac{1}{2}\right)^{n+1} L|dz| \\ &= \frac{2^{n-1}\varepsilon}{L}\int_{C_n} |dz| \\ &= \frac{2^{n-1}\varepsilon}{L}\left(\frac{1}{2}\right)^n L = \frac{\varepsilon}{2}.\end{aligned}$$

Since ε was arbitrary, it follows that (1) holds true for the triangular contour C. If C is a polygonal contour, then interior edges can be added until the interior is subdivided into a finite number of triangles. The integral around each triangle is zero, and the sum of all these integrals is equal to the integral around the polygonal contour C. Therefore equation (1) holds true for polygonal contours. The proof for an arbitrary simple closed contour is established by approximating the contour "sufficiently close" with a polygonal contour.

6

Series Representations

6.1 Convergence of Sequences and Series

This chapter will deal mainly with series representations of analytic functions. We start by developing the concepts of sequences and series of complex numbers.

Let $z_1, z_2, \ldots, z_n, \ldots$ be a sequence of complex numbers. We say that the sequence $\{z_n\}_1^\infty$ has the *limit* z_0 as $n \to \infty$, provided that the *terms* z_n are close to z_0 for all n sufficiently large, and we write

(1) $$\lim_{n\to\infty} z_n = z_0 \qquad \text{or} \qquad z_n \to z_0 \qquad \text{as } n \to \infty.$$

To be precise, we require that for any $\varepsilon > 0$ there corresponds a positive integer N such that

(2) $$|z_n - z_0| < \varepsilon \qquad \text{for all } n > N.$$

When the limit z_0 exists, the sequence is said to *converge* to z_0; otherwise, it has no limit and is said to *diverge*. Limits of complex sequences can be found by using properties of limits of real sequences.

Theorem 6.1 *Let $z_n = x_n + iy_n$ and $z_0 = x_0 + iy_0$. Then*

(3) $$\lim_{n\to\infty} z_n = z_0$$

if and only if

(4) $$\lim_{n\to\infty} x_n = x_0 \qquad \textit{and} \qquad \lim_{n\to\infty} y_n = y_0.$$

Proof Let us first assume that statement (3) is true and show that (4) is true. According to the definition of limit, for each $\varepsilon > 0$ there corresponds a positive integer N such that $|z_n - z_0| < \varepsilon$ whenever $n > N$. Since $z_n - z_0 = x_n - x_0 + i(y_n - y_0)$, we can use (2) of Section 1.3 to conclude that

$$|x_n - x_0| \leqq |z_n - z_0| < \varepsilon \qquad \text{and} \qquad |y_n - y_0| \leqq |z_n - z_0| < \varepsilon$$

whenever $n > N$, so (4) is true.

Conversely, let us now assume that (4) is true. Then for each $\varepsilon > 0$ there exist integers N_1 and N_2 such that

$$|x_n - x_0| < \frac{\varepsilon}{2} \quad \text{for } n > N_1 \qquad \text{and} \qquad |y_n - y_0| < \frac{\varepsilon}{2} \quad \text{for } n > N.$$

Let $N = \max\{N_1, N_2\}$. Then we can use the triangle inequality to conclude that

$$|z_n - z_0| \leqq |x_n - x_0| + |y_n - y_0| < \frac{\varepsilon}{2} + \frac{\varepsilon}{2} = \varepsilon \qquad \text{for all } n > N.$$

Hence the truth of (4) implies the truth of statement (3), and the proof of the theorem is complete.

It is left as an exercise to show that a given sequence $\{z_n\}$ can have at most one limit z_0; that is, the limit is unique.

EXAMPLE 6.1 Consider $z_n = (\sqrt{n} + i(n+1))/n$. Then we write

$$z_n = x_n + iy_n = \frac{1}{\sqrt{n}} + i\,\frac{n+1}{n}.$$

Using results about sequences of real numbers, which are studied in calculus, we find that

$$\lim_{n\to\infty} x_n = \lim_{n\to\infty} \frac{1}{\sqrt{n}} = 0 \qquad \text{and} \qquad \lim_{n\to\infty} y_n = \lim_{n\to\infty} \frac{n+1}{n} = 1.$$

Therefore

$$\lim_{n\to\infty} z_n = \lim_{n\to\infty} \frac{\sqrt{n} + i(n+1)}{n} = i.$$

EXAMPLE 6.2 Let us show that $\{(1+i)^n\}$ diverges. In this case we have

$$z_n = (1+i)^n = (\sqrt{2})^n \cos\frac{n\pi}{4} + i(\sqrt{2})^n \sin\frac{n\pi}{4}.$$

Since the real sequences $\{(\sqrt{2})^n \cos(n\pi/4)\}$ and $\{(\sqrt{2})^n \sin(n\pi/4)\}$ both diverge, we conclude that $\{(1+i)^n\}$ diverges.

Let $\{z_n\}$ be a sequence of complex numbers. We can form a new sequence by successively adding the terms of the sequence. Consider the sequence of *partial sums* $S_1, S_2, \ldots, S_n, \ldots$ defined as follows:

$$\begin{aligned}
(5) \qquad & S_1 = z_1, \\
& S_2 = z_1 + z_2, \\
& \;\vdots \\
& S_n = z_1 + z_2 + \cdots + z_n = \sum_{k=1}^{n} z_k, \\
& \;\vdots
\end{aligned}$$

The sum of all the terms z_n is called an *infinite series*, and we write

$$(6) \qquad z_1 + z_2 + \cdots + z_n + \cdots \qquad \text{or} \qquad \sum_{n=1}^{\infty} z_n.$$

The infinite series in (6) *converges*, if and only if there exists a complex number S such that

$$(7) \qquad S = \lim_{n\to\infty} S_n = \lim_{n\to\infty} \sum_{k=1}^{n} z_k.$$

The number S in (7) is called the *sum* of the series, and we write

$$(8) \qquad S = \sum_{n=1}^{\infty} z_n.$$

If a series does not converge, we say that it *diverges*. The sum of a series of complex numbers can be found by finding the sum of the real and imaginary parts. The series $\sum_{n=1}^{\infty} z_n$ is said to *converge absolutely* provided that the real series of magnitudes $\sum_{n=1}^{\infty} |z_n|$ converges.

Theorem 6.2 *Let $z_n = x_n + iy_n$ and $S = U + iV$. Then*

$$(9) \qquad S = \sum_{n=1}^{\infty} z_n = \sum_{n=1}^{\infty} (x_n + iy_n)$$

if and only if

$$(10) \qquad U = \sum_{n=1}^{\infty} x_n \qquad \text{and} \qquad V = \sum_{n=1}^{\infty} y_n.$$

Proof Let $U_n = \sum_{k=1}^{\infty} x_k$ and $V_n = \sum_{k=1}^{\infty} y_k$ and $S_n = U_n + iV_n$. We can use Theorem 6.1 to conclude that

$$(11) \qquad \lim_{n\to\infty} S_n = \lim_{n\to\infty} (U_n + iV_n) = U + iV = S$$

if and only if both $\lim_{n\to\infty} U_n = U$ and $\lim_{n\to\infty} V_n = V$, and the completion of the proof follows easily from definitions (7) and (8) and the results in (11).

Theorem 6.3 *If $\sum_{n=1}^{\infty} z_n$ is a convergent complex series, then*

$$(12) \qquad \lim_{n\to\infty} z_n = 0.$$

The proof of Theorem 6.3 is left as an exercise.

EXAMPLE 6.3 The series

$$\sum_{n=1}^{\infty} \frac{1 + in(-1)^n}{n^2} = \sum_{n=1}^{\infty} \left(\frac{1}{n^2} + i\,\frac{(-1)^2}{n} \right)$$

is convergent.

Solution From the calculus it is known that the series

$$\sum_{n=1}^{\infty} \frac{1}{n^2} \quad \text{and} \quad \sum_{n=1}^{\infty} \frac{(-1)^n}{n}$$

are convergent. Hence Theorem 6.2 implies that the given complex series is convergent.

EXAMPLE 6.4 The series

$$\sum_{n=1}^{\infty} \frac{(-1)^n + i}{n} = \sum_{n=1}^{\infty} \left(\frac{(-1)^n}{n} + i\,\frac{1}{n} \right)$$

is divergent.

Solution From the study of calculus it is known that the series $\sum_{n=1}^{\infty} (1/n)$ is divergent. Hence Theorem 6.2 implies that the given complex series is divergent.

EXAMPLE 6.5 The series $\sum_{n=1}^{\infty} (1+i)^n$ is divergent.

Solution Here we set $z_n = (1+i)^n$, and we observe that $\lim_{n\to\infty} |z_n| = \lim_{n\to\infty} (\sqrt{2})^n = \infty$. Hence $\lim_{n\to\infty} z_n \neq 0$, and Theorem 6.3 implies that the given series is not convergent; hence it is divergent.

We now state some results concerning complex series and leave the proofs for the reader.

Theorem 6.4 *Let $\sum_{n=1}^{\infty} z_n$ and $\sum_{n=1}^{\infty} w_n$ be convergent complex series, and let c be a complex constant. Then*

$$\sum_{n=1}^{\infty} cz_n = c \sum_{n=1}^{\infty} z_n \quad \textit{and} \tag{13}$$

$$\sum_{n=1}^{\infty} (z_n + w_n) = \sum_{n=1}^{\infty} z_n + \sum_{n=1}^{\infty} w_n. \tag{14}$$

Theorem 6.5 *Let $\sum_{n=0}^{\infty} a_n$ and $\sum_{n=0}^{\infty} b_n$ be convergent series, where a_n and b_n are complex numbers. The Cauchy product of the two series is defined to be*

$$\sum_{n=0}^{\infty} c_n \quad \textit{where} \quad c_n = \sum_{k=0}^{n} a_k b_{n-k}. \tag{15}$$

If the Cauchy product converges, then

$$\left(\sum_{n=0}^{\infty} a_n\right)\left(\sum_{n=0}^{\infty} b_n\right) = \sum_{n=0}^{\infty} c_n. \tag{16}$$

Proof The proof can be found in *Infinite Sequences and Series* by Konrad Knopp (translated by Frederick Bagemihl; New York: Dover, 1956).

To use series in the development of complex functions, we need to investigate the topic of convergence. We will need to understand pointwise convergence and uniform convergence.

Definition 6.1 (Pointwise Convergence) *Let $\{F_n(z)\}$ be a sequence of functions each of which is defined for all points in the set S. Suppose that $F(z)$ is also defined at all points in the set S. We say that $\{F_n(z)\}$ converges to $F(z)$ at the point $z = z_0$ provided that*

$$\lim_{n\to\infty} F_n(z_0) = F(z_0). \tag{17}$$

Also, the sequence $\{F_n(z)\}$ is said to converge pointwise to $F(z)$ on S provided that

$$\lim_{n\to\infty} F_n(z) = F(z) \qquad \textit{for each } z \textit{ in } S. \tag{18}$$

The connection between (17) and (18) must be understood. Statement (17) is concerned with the single point z_0 and says that for each $\varepsilon > 0$ there exists an N such that

$$|F_n(z_0) - F(z_0)| < \varepsilon \qquad \text{whenever } n \geqq N.$$

Statement (18) means that for each fixed z chosen in S we then consider any $\varepsilon > 0$. For this ε there now corresponds an $N = N_{z,\varepsilon}$, which depends on both z and ε, so that

$$|F_n(z) - F(z)| < \varepsilon \qquad \text{whenever } n \geqq N_{z,\varepsilon}.$$

Definition 6.2 (Uniform Convergence) *Let $\{F_n(z)\}$ be a sequence of functions each of which is defined for all points in the set S. Suppose that $F(z)$ is also defined at all points in the set S. We say that $\{F_n(z)\}$ converges uniformly to $F(z)$ on S provided that $\{F_n(z)\}$ converges pointwise to $F(z)$ on S and for any $\varepsilon > 0$ there exists an $N = N_\varepsilon$, which depends only on ε, such that*

$$n \geqq N_\varepsilon \qquad \text{implies that} \qquad |F_n(z) - F(z)| < \varepsilon \qquad \text{for all } z \text{ in } S. \tag{19}$$

Geometric series play an important role in the development of certain series representation of analytic functions, called Laurent Series. We state the following useful result.

Lemma 6.1 (Geometric Series) *The function $f(z) = 1/(1-z)$ has the following two representations:*

$$\frac{1}{1-z} = 1 + z + z^2 + \cdots + z^n + \cdots = \sum_{n=0}^{\infty} z^n \qquad \textit{for } |z| < 1. \tag{20}$$

(21) *Convergence in (20) is pointwise on the open disk $|z| < 1$ and*

(22) *Convergence in (20) is uniform on the closed disk $|z| \leqq R$ where $R < 1$,*

$$\frac{1}{1-z} = -z^{-1} - z^{-2} - z^{-3} - \cdots - z^{-n} - \cdots = -\sum_{n=1}^{\infty} z^{-n} \tag{23}$$

for $|z| > 1$.

(24) *Convergence in (23) is pointwise on the open region $|z| > 1$ and*

(25) *Convergence in (23) is uniform on the closed region $|z| \geqq R$ where $R > 1$*

The domains in which the series (20) and (23) converge are shown in Figure 6.1. Let us look at these series in detail and find out that convergence is faster at some points and slower at others.

To get a feel for the meaning of "uniform convergence," we could calculate some partial sums

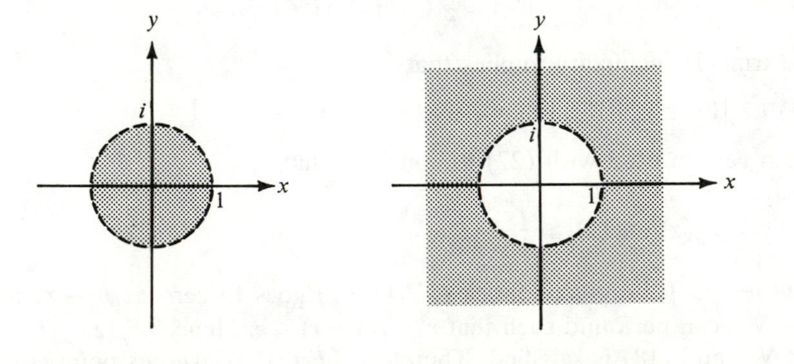

(a) The domain of convergence for the series in (20). **(b)** The domain of convergence for the series in (23).

FIGURE 6.1 Domains of convergence.

$$F_n(z) = \sum_{k=0}^{n} z^k = 1 + z + z^2 + z^3 + \cdots + z^n \tag{26}$$

of the geometric series. Suppose that we choose $\varepsilon = 1/100$ and consider values of $F_n(z)$ at $z = 1/2$ and $z = 9/10$. Calculation reveals that

$$F_7(1/2) = \sum_{k=0}^{7} (1/2)^7 = 1.9921875 \qquad \text{and} \qquad F(1/2) = 2.$$

Hence $|F_7(1/2) - F(1/2)| = 0.0078125 < \varepsilon$. However,

$$F_7(9/10) = \sum_{k=0}^{7} (9/10)^7 = 5.6953289 \qquad \text{and} \qquad F(9/10) = 10.$$

Thus $|F_7(9/10) - F(9/10)| = 4.3046711 > \varepsilon$. We see that summing eight terms does not produce "uniformly close" results. But if we are permitted to go out further in the summation, then we find that

$$F_{65}(9/10) = \sum_{k=0}^{65} (9/10)^7 = 9.99044995 \qquad \text{and} \qquad F(9/10) = 10.$$

Hence $|F_{65}(9/10) - F(9/10)| = 0.00955005 < \varepsilon$. Notice that convergence is slower at $z = 9/10$ than at $z = 1/2$.

Proof of (21) and (22) Let $F(z) = 1/(1 - z)$. First we consider pointwise convergence of the functions $\{F_n(z)\}$ given in (26). Fix the value $z = re^{i\theta}$, where $r < 1$. Then

$$|F_n(z) - F(z)| = \left|1 + z + z^2 + \cdots + z^n - \frac{1}{1 - z}\right| \tag{27}$$
$$= \left|\frac{1 - z^{n+1}}{1 - z} - \frac{1}{1 - z}\right| = \left|\frac{z^{n+1}}{1 - z}\right|.$$

The triangle inequality implies that

$$|1 - z| \geqq ||1| - |z|| = 1 - r, \tag{28}$$

which can be used with (27) to conclude that

$$|F_n(z) - F(z)| \leqq \frac{r^{n+1}}{1 - r}. \tag{29}$$

Now let $\varepsilon > 0$ be given. Since $r^{n+1}/(1 - r)$ goes to zero as $n \to \infty$, an integer $N = N_{z,\varepsilon}$ can be found such that $r^{N+1}/(1 - r) < \varepsilon$. Hence $|F_n(z) - F(z)| < \varepsilon$ for $n \geqq N_{z,\varepsilon}$ and (18) is satisfied. Therefore $\{F_n(z)\}$ converges pointwise to $F(z)$, and (21) is established.

Now we consider uniform convergence. Fix the number R, which satisfies $R < 1$. Let $z = re^{i\theta}$ be any complex number in the closed disk $|z| \leqq R$.

Then $r \leqq R$, and an argument similar to the one above can be used to derive the following relationship:

$$(30) \qquad |F_n(z) - F(z)| = \left|\frac{z^{n+1}}{1-z}\right| \leqq \frac{r^{n+1}}{1-r} \leqq \frac{R^{n+1}}{1-R}.$$

Let $\varepsilon > 0$ be given. Since $R^{n+1}/(1-R)$ goes to zero as $n \to \infty$, an integer $N = N_\varepsilon$ can be found such that $R^{N+1}/(1-R) < \varepsilon$. Hence $|F_n(z) - F(z)| < \varepsilon$ for $n \geqq N_\varepsilon$, and (19) is satisfied for all points z in the closed disk $|z| \leqq R$. Thus we have shown that $\{F_n(z)\}$ converges uniformly to $F(z)$ on $|z| \leqq R$, and the proof of (22) is complete.

One might expect that convergence would be uniform over the entire disk $|z| < 1$. However, this is not the case. Indeed, for each n the absolute error $|F_n(z) - F(z)|$ is unbounded, since

$$\lim_{z \to 1} |F_n(z) - F(z)| = \lim_{z \to 1} \left|\frac{z^{n+1}}{1-z}\right| = \infty.$$

This means that for each $\varepsilon > 0$ chosen in advance we can look at any function $F_n(z)$ and find a point z_n such that

$$(31) \qquad |F_n(z_n) - F(z_n)| > \varepsilon.$$

For example, the point $z_n = \varepsilon^{1/(n+1)}$ will cause (31) to occur. Therefore convergence is not uniform over the entire unit disk.

Proof of (24) Let $F(z) = 1/(1-z)$ and consider the following sequence of functions:

$$(32) \qquad F_n(z) = -\sum_{k=1}^{n} z^{-k} = -z^{-1} - z^{-2} - z^{-3} - \cdots - z^{-n}.$$

If we start with $|z| > 1$ and use the change of variable $Z = 1/z$, then (20) can be used with z replaced with Z to obtain

$$(33) \qquad \frac{1}{1-1/z} = \sum_{n=0}^{\infty} (1/z)^n = \sum_{n=0}^{\infty} z^{-n} = \lim_{n \to \infty} \sum_{k=0}^{n-1} z^{-k} \qquad \text{for } |z| > 1.$$

Now both sides of (33) are multiplied by $-1/z$, and we obtain

$$(34) \qquad \frac{1}{1-z} = \frac{-1}{z} \sum_{n=0}^{\infty} z^{-n} = -\sum_{n=1}^{\infty} z^{-n} = \lim_{n \to \infty} \sum_{k=1}^{n} z^{-k}.$$

Thus we have established (24). The proof of (25) is similar and is left as an exercise.

EXAMPLE 6.6 Show that $\sum_{n=0}^{\infty} ((1-i)^n/2^n) = 1 - i$.

Solution If we set $z = (1-i)/2$, then we see that $|z| = \sqrt{2}/2 < 1$, so we can use representation (17) for a geometric series. The sum is given by

$$\frac{1}{1 - \dfrac{1-i}{2}} = \frac{2}{2-1+i} = \frac{2}{1+i} = 1 - i.$$

We now state an important result that shows how to establish the convergence of complex series by comparison with a convergent series of real nonnegative terms. This result generalizes the comparison test, which is studied in calculus.

Theorem 6.6 (Comparison Test) *Let $\sum_{n=1}^{\infty} M_n$ be a convergent series of real nonnegative terms. If $\{z_n\}$ is a sequence of complex numbers and $|z_n| \leqq M_n$ holds for all n, then*

$$\sum_{n=1}^{\infty} z_n = \sum_{n=1}^{\infty} (x_n + iy_n)$$

converges.

Proof Using equations (2) of Section 1.3, we see that $|x_n| \leqq |z_n| \leqq M_n$ and $|y_n| \leqq |z_n| \leqq M_n$ holds for all n. The comparison test for real sequences can be used to conclude that

$$(35) \qquad \sum_{n=1}^{\infty} |x_n| \quad \text{and} \quad \sum_{n=1}^{\infty} |y_n|$$

are convergent. A result from calculus states that an absolutely convergent series is convergent. Hence (35) implies that

$$(36) \qquad \sum_{n=1}^{\infty} x_n \quad \text{and} \quad \sum_{n=1}^{\infty} y_n$$

are convergent. We can use the results in (36) together with Theorem 6.2 to conclude that $\sum_{n=1}^{\infty} z_n = \sum_{n=1}^{\infty} x_n + i \sum_{n=1}^{\infty} y_n$ is convergent.

Corollary 6.1 *If $\sum_{n=1}^{\infty} |z_n|$ converges, then $\sum_{n=1}^{\infty} z_n$ converges.*

EXAMPLE 6.7 Show that $\sum_{n=1}^{\infty} ((3+4i)^n/5^n n^2)$ converges.

Solution Calculating the modulus of the terms, we find that $|z_n| = |(3+4i)^n/5^n n^2| = 1/n^2 = M_n$. We can use the comparison test and the fact that $\sum_{n=1}^{\infty} (1/n^2)$ converges to conclude that $\sum_{n=1}^{\infty} ((3+4i)^n/5^n n^2)$ converges.

Other convergence tests for real series can be carried over to the complex case. We now state the ratio test. The proof is similar to those for real series and is left for the reader to establish.

Theorem 6.7 (d'Alembert's Ratio Test) *If* $\sum_{n=0}^{\infty} z_n$ *is a complex series with the property that*

$$\lim_{n\to\infty} \frac{|z_{n+1}|}{|z_n|} = L, \tag{37}$$

then the series converges if $L<1$ *and diverges if* $L>1$.

EXAMPLE 6.8 Show that $\sum_{n=0}^{\infty} ((1-i)^n/n!)$ converges.

Solution Using the ratio test, we find that

$$\lim_{n\to\infty} \frac{\left|\frac{(1-i)^{n+1}}{(n+1)!}\right|}{\left|\frac{(1-i)^n}{n!}\right|} = \lim_{n\to\infty} \frac{n!|1-i|}{(n+1)!} = \lim_{n\to\infty} \frac{|1-i|}{n+1} = \lim_{n\to\infty} \frac{\sqrt{2}}{n+1} = 0 = L.$$

Since $L<1$, the series converges.

EXAMPLE 6.9 Show that the series $\sum_{n=0}^{\infty} ((z-i)^n/2^n)$ converges for all values of z in the disk $|z-i|<2$ and diverges if $|z-i|>2$.

Solution Using the ratio test, we find that

$$\lim_{n\to\infty} \frac{\left|\frac{(z-i)^{n+1}}{2^{n+1}}\right|}{\left|\frac{(z-i)^n}{2^n}\right|} = \lim_{n\to\infty} \frac{|z-i|}{2} = \frac{|z-i|}{2} = L.$$

If $|z-i|<2$, then $L<1$ and the series converges. If $|z-i|>2$, then $L>1$, and the series diverges.

EXERCISES FOR SECTION 6.1

1. Find the following limits.

   (a) $\lim_{n\to\infty} \left(\frac{1}{2}+\frac{i}{4}\right)^n$ **(b)** $\lim_{n\to\infty} \frac{n+(i)^n}{n}$

   (c) $\lim_{n\to\infty} \frac{n^2+i2^n}{2^n}$ **(d)** $\lim_{n\to\infty} \frac{(n+i)(1+ni)}{n^2}$

2. Show that $\lim_{n\to\infty} (i)^{1/n} = 1$ where $(i)^{1/n}$ is the principal value of the nth root of i.
3. Let $\lim_{n\to\infty} z_n = z_0$. Show that $\lim_{n\to\infty} \overline{z}_n = \overline{z}_0$.

4. Let $\sum_{n=1}^{\infty} z_n = S$. Show that $\sum_{n=1}^{\infty} \overline{z}_n = \overline{S}$.

5. Show that $\displaystyle\sum_{n=0}^{\infty} \left(\frac{1}{2+i}\right)^n = \frac{3-i}{2}$.

6. Show that $\displaystyle\sum_{n=0}^{\infty} \left[\frac{1}{n+1+i} - \frac{1}{n+i}\right] = i$.

7. Show that $\displaystyle\sum_{n=1}^{\infty} \left(\frac{1}{n} + \frac{i}{2^n}\right)$ diverges.

8. Does $\displaystyle\lim_{n\to\infty} \left(\frac{1+i}{\sqrt{2}}\right)^n$ exist? Why?

9. Let $\{r_n\}$ and $\{\theta_n\}$ be two convergent sequences of real numbers such that

$$\lim_{n\to\infty} r_n = r_0 \qquad \text{and} \qquad \lim_{n\to\infty} \theta_n = \theta_0$$

Show that $\lim_{n\to\infty} r_n e^{i\theta_n} = r_0 e^{i\theta_0}$.

10. Show that $\displaystyle\sum_{n=0}^{\infty} \frac{(1+i)^n}{2^n} = 1 + i$.

11. Show that $\sum_{n=0}^{\infty} ((z+i)^n/2^n)$ converges for all values of z in the disk $|z+i| < 2$ and diverges if $|z+i| > 2$.

12. Is the series $\displaystyle\sum_{n=1}^{\infty} \frac{(4i)^n}{n!}$ convergent? Why?

13. Use the ratio test and show that the following series converge.

(a) $\displaystyle\sum_{n=0}^{\infty} \left(\frac{1+i}{2}\right)^n$ **(b)** $\displaystyle\sum_{n=1}^{\infty} \frac{(1+i)^n}{n2^n}$ **(c)** $\displaystyle\sum_{n=1}^{\infty} \frac{(1+i)^n}{n!}$ **(d)** $\displaystyle\sum_{n=0}^{\infty} \frac{(1+i)^{2n}}{(2n+1)!}$

14. Use the ratio test to find a disk in which the following series converge.

(a) $\displaystyle\sum_{n=0}^{\infty} (1+i)^n z^n$ **(b)** $\displaystyle\sum_{n=0}^{\infty} \frac{z^n}{(3+4i)^n}$ **(c)** $\displaystyle\sum_{n=0}^{\infty} \frac{(z-i)^n}{(3+4i)^n}$ **(d)** $\displaystyle\sum_{n=0}^{\infty} \frac{(z-3-4i)^n}{2^n}$

15. Show that if $\sum_{n=1}^{\infty} z_n$ converges, then $\lim_{n\to\infty} z_n = 0$. *Hint*: $z_n = S_n - S_{n-1}$.

16. Is the series $\displaystyle\sum_{n=1}^{\infty} \frac{(i)^n}{n}$ convergent? Why?

17. Let $\sum_{n=1}^{\infty} (x_n + iy_n) = U + iV$. If $c = a + ib$ is a complex constant, show that

$$\sum_{n=1}^{\infty} (a+ib)(x_n + iy_n) = (a+ib)(U+iV).$$

18. Let $f(z) = z + z^2 + z^4 + \cdots + z^{2^n} + \cdots$. Show that $f(z) = z + f(z^2)$.

19. Show that the limit of a sequence is unique.

20. Prove Theorem 6.5.

21. **(a)** Use the formula for geometric series with $z = re^{i\theta}$ where $r < 1$ and show that

$$\sum_{n=0}^{\infty} r^n e^{in\theta} = \frac{1 - r\cos\theta + ir\sin\theta}{1 + r^2 - 2r\cos\theta}.$$

(b) Use part (a) and obtain

$$\sum_{n=0}^{\infty} r^n \cos n\theta = \frac{1 - r\cos\theta}{1 + r^2 - 2r\cos\theta} \qquad \text{and}$$

$$\sum_{n=0}^{\infty} r^n \sin n\theta = \frac{r\sin\theta}{1 + r^2 - 2r\cos\theta}.$$

22. Show that $\sum_{n=0}^{\infty} e^{inz}$ converges for $\operatorname{Im} z > 0$.

23. Prove statement (25).

6.2 Taylor Series Representations

In Section 5.5 we saw that if f is analytic at z_0, then the derivatives $f^{(n)}(z_0)$ exist for all n. In this section we will see that if f is analytic at z_0, then f has a complex Taylor series representation that is valid in a disk with center z_0. The complex Maclaurin series expansions for the elementary functions $\exp z$, $\cos z$, $\sin z$, and $\text{Log}(1-z)$ can be obtained by using the coefficients of the real Maclaurin series for e^x, $\cos x$, $\sin x$, and $\ln(1-x)$ and replacing the real variable x with the complex variable z. In other words, when all the terms in the series are real, we obtain the familiar Taylor series that are studied in calculus.

Definition 6.3 *If $f(z)$ is analytic at z_0, then the series*

$$(1) \qquad f(z_0) + f'(z_0)(z-z_0) + \frac{f^{(2)}(z_0)}{2!}(z-z_0)^2 + \frac{f^{(3)}(z_0)}{3!}(z-z_0)^3 + \cdots$$

$$= \sum_{n=0}^{\infty} \frac{f^{(n)}(z_0)}{n!}(z-z_0)^n$$

is called the Taylor series for f centered around z_0. When the center is $z_0 = 0$, the series is called a Maclaurin series.

Theorem 6.8 (Taylor's Theorem) *If f is analytic in the simply connected domain D that contains the circle $|z-z_0| = R$, then the Taylor series (1) converges to $f(z)$ in the disk $|z-z_0| < R$; that is,*

$$(2) \qquad f(z) = \sum_{n=0}^{\infty} \frac{f^{(n)}(z_0)}{n!}(z-z_0)^n \qquad \textit{valid for } |z-z_0| < R.$$

Furthermore, this representation is valid in the largest disk with center z_0 that is contained in D, and convergence is uniform on any closed subdisk $|z-z_0| \leqq r$ where $r < R$.

Proof It will suffice to prove that convergence is uniform on every closed subdisk $|z-z_0| \leqq r$, since we can let r approach R and get pointwise convergence on the larger disk $|z-z_0| < R$. Let C be the circle $|\xi - z_0| = (r+R)/2$ as shown in Figure 6.2. Then we can use the Cauchy Integral Formula to write

$$(3) \qquad f(z) = \frac{1}{2\pi i}\int_C \frac{f(\xi)}{\xi - z}\, d\xi$$

where the variable ξ denotes a point on the circle C.

The term $1/(\xi - z)$ in the integrand in equation (3) can be replaced by using the formula for a finite geometric:

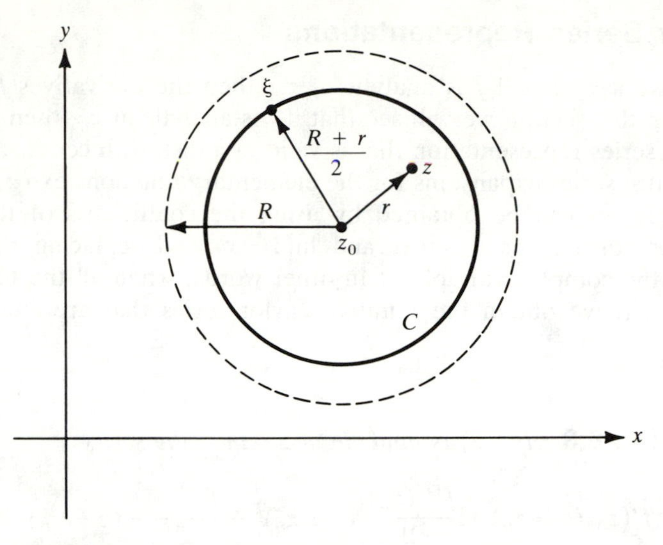

FIGURE 6.2 The circle of radius R and the contour C of radius $(R + r)/2$ in the proof of Taylor's Theorem.

$$(4)\qquad \frac{1}{\xi - z} = \frac{1}{(\xi - z_0) - (z - z_0)} = \frac{1}{\xi - z_0}\,\frac{1}{1 - (z - z_0)/(\xi - z_0)}$$

$$= \frac{1}{\xi - z_0} + \frac{z - z_0}{(\xi - z_0)^2} + \frac{(z - z_0)^2}{(\xi - z_0)^3} + \cdots + \frac{(z - z_0)^n}{(\xi - z_0)^{n+1}}$$

$$+ \frac{1}{\xi - z}\,\frac{(z - z_0)^{n+1}}{(\xi - z_0)^{n+1}}.$$

Substitution of the series in (4) into (3) results in

$$(5)\qquad f(z) = \frac{1}{2\pi i}\int_C \frac{f(\xi)\,d\xi}{\xi - z_0} + \frac{z - z_0}{2\pi i}\int_C \frac{f(\xi)\,d\xi}{(\xi - z_0)^2} + \frac{(z - z_0)^2}{2\pi i}\int_C \frac{f(\xi)\,d\xi}{(\xi - z_0)^3}$$

$$+ \cdots + \frac{(z - z_0)^n}{2\pi i}\int_C \frac{f(\xi)\,d\xi}{(\xi - z_0)^{n+1}} + E_n(z)$$

where the remainder term $E_n(z)$ can be expressed in the form

$$(6)\qquad E_n(z) = \frac{1}{2\pi i}\int_C \frac{f(\xi)(z - z_0)^{n+1}\,d\xi}{(\xi - z)(\xi - z_0)^{n+1}}.$$

Using Cauchy's Integral Formula, we see that

$$(7)\qquad f^{(k)}(z_0) = \frac{k!}{2\pi i}\int_C \frac{f(\xi)\,d\xi}{(\xi - z_0)^{k+1}} \qquad \text{for } k = 0, 1, \ldots, n.$$

Therefore we can use the results of equation (7) in (5) to conclude that

$$(8) \qquad f(z) = \sum_{k=0}^{n} \frac{f^{(k)}(z_0)}{k!} (z - z_0)^k + E_n(z).$$

The summation on the right side of (8) is the first $n + 1$ terms of the Taylor series. Hence all that is left to do is show that the remainder $E_n(z)$ can be made "uniformly small" for all z in the closed disk $|z - z_0| \leqq r$ by taking n sufficiently large.

We use the *ML* Inequality in Section 5.2 to bound the size of $|E_n(z)|$. According to our constructions shown in Figure 6.2, we have

$$(9) \qquad |z - z_0| \leqq r \quad \text{and} \quad |\xi - z_0| = (r + R)/2.$$

The triangle inequality and (9) can be used to see that

$$(10) \qquad |\xi - z| = |(\xi - z_0) - (z - z_0)| \geqq |(r + R)/2 - r| = (R - r)/2.$$

Let $L = 2\pi(r + R)/2$ be the length of the curve, and let $M = \max |f(\xi)|$ for ξ on C. When these values and (9) and (10) are used in the *ML* Inequality, we obtain

$$(11) \qquad |E_n(z)| \leqq \frac{1}{2\pi} M \frac{2}{R - r} \left(\frac{2r}{r + R}\right)^{n+1} 2\pi\left(\frac{r + R}{2}\right).$$

Since $2r/(r + R) < 1$, the right side of (11) goes to zero as n goes to infinity. Hence for any $\varepsilon > 0$ we can find an N_ε such that it is less than ε for all $n \geqq N$. Therefore the Taylor series (2) converges uniformly to $f(z)$ on the closed disk, and the proof is complete.

Corollary 6.2 *Suppose that f is analytic in the domain D that contains z_0. Let z_1 be a singular point of minimum distance to z_0. Then the Taylor series (2) converges for $|z - z_0| < |z_1 - z_0|$.*

If f is an entire function, then f has no singular points, and the radius of convergence of the Taylor series is $R = \infty$; for example, the Taylor series for $\exp z$, $\sin z$, and $\cos z$ have $R = \infty$. If $g(z) = 1/(1 + z^2)$ and $z_0 = 0$, then $z_1 = i$ is a singular of minimum distance to z_0. Hence the radius of convergence of the Maclaurin series for $g(z)$ is $R = 1$.

EXAMPLE 6.10

(a) Show that $\exp z = \sum_{n=0}^{\infty} (1/n!)z^n$ is valid for all z.

(b) Use the first five terms in the series and obtain an approximation to $\exp((1 + i)/8)$.

Solution Here we have $f(z) = \exp z$ and $f^{(n)}(z) = \exp z$ for all n. Hence the Maclaurin series for f is given by

$$\exp z = f(z) = \sum_{n=0}^{\infty} \frac{f^{(n)}(0)}{n!} z^n = \sum_{n=0}^{\infty} \frac{1}{n!} z^n$$

Since f is an entire function, the representation is valid for all z. If we use the first five terms to approximate $\exp((1+i)/8)$, then the result is as follows:

$$\begin{aligned}\exp\left(\frac{1+i}{8}\right) &\approx 1 + \frac{1+i}{8} + \frac{2i}{8^2 2!} + \frac{-2+2i}{8^3 3!} + \frac{-4}{8^4 4!} \\ &= 1 + \frac{1}{8} - \frac{1}{1536} - \frac{1}{24576} + i\left(\frac{1}{8} + \frac{1}{64} + \frac{1}{1536}\right) \\ &= 1.1243083 + i0.1412760.\end{aligned}$$

This is in agreement with the calculation

$$\exp\left(\frac{1+i}{8}\right) = e^{1/8} \cos \frac{1}{8} + ie^{1/8} \sin \frac{1}{8} \approx 1.1243073 + i0.1412750.$$

EXAMPLE 6.11 Show that $1/(1-z)^2 = \sum_{n=0}^{\infty} (n+1)z^n$ is valid in the disk $|z| < 1$.

Solution Here we have $f(z) = 1/(1-z)^2$ and $f^{(n)}(z) = (n+1)!/(1-z)^{n+2}$ for all n. Since $f^{(n)}(0) = (n+1)!$, the Maclaurin series for f is given by

$$\frac{1}{(1-z)^2} = f(z) = \sum_{n=0}^{\infty} \frac{f^{(n)}(0)}{n!} z^n = \sum_{n=0}^{\infty} \frac{(n+1)!}{n!} z^n = \sum_{n=0}^{\infty} (n+1)z^n.$$

Since f is analytic for all $|z| < 1$, Theorem 6.8 implies that the Maclaurin series is valid in any disk $|z| < R < 1$. Letting $R \to 1$, we conclude that the Maclaurin series is valid for $|z| < 1$.

Theorem 6.9 (Differentiation and Integration of Taylor Series) *Let f have a Taylor series representation given by equation* (1) *that is valid in the disk $|z - z_0| < R$. Then the Taylor series can be differentiated and integrated term by term to obtain*

(12) $$f'(z) = \sum_{n=1}^{\infty} \frac{f^{(n)}(z_0)}{(n-1)!} (z - z_0)^{n-1} \quad \textit{valid for } |z - z_0| < R \quad \textit{and}$$

(13) $$\int_{z_0}^{z} f(\xi)\, d\xi = \sum_{n=0}^{\infty} \frac{f^{(n)}(z_0)}{(n+1)!} (z - z_0)^{n+1} \quad \textit{valid for } |z - z_0| < R.$$

Proof of Equation (12) Let $g(z) = f'(z)$. Then g is analytic in the disk $|z - z_0| < R$ and has the Taylor series representation

$$(14) \qquad g(z) = \sum_{n=0}^{\infty} \frac{g^{(n)}(z_0)}{n!} (z - z_0)^n \qquad \text{for } |z - z_0| < R.$$

Since $g^{(n)}(z_0) = f^{(n+1)}(z_0)$, equation (14) can be written

$$f'(z) = g(z) = \sum_{n=0}^{\infty} \frac{f^{(n+1)}(z_0)}{n!} (z - z_0)^n = \sum_{n=1}^{\infty} \frac{f^{(n)}(z_0)}{(n-1)!} (z - z_0)^{n-1},$$

and equation (12) is established. The reader can verify equation (13).

EXAMPLE 6.12 Use the results of Theorem 6.9 to show that

$$(15) \qquad \frac{1}{(1-z)^2} = \sum_{n=1}^{\infty} nz^{n-1} \qquad \text{for } |z| < 1.$$

Solution Starting with $f(z) = 1/(1-z)$, $f^{(n)}(z) = n!/(1-z)^{n+1}$, and $f^{(n)}(0) = n!$, we find that the Maclaurin series for f is

$$(16) \qquad f(z) = \frac{1}{1-z} = \sum_{n=0}^{\infty} z^n \qquad \text{for } |z| < 1.$$

It is worthwhile to note that this is the same representation that was given in Lemma 6.1, where we discussed geometric series. Since $f'(z) = 1/(1-z)^2$, termwise differentiation of the series in equation (16) yields the desired result:

$$\frac{1}{(1-z)^2} = f'(z) = \sum_{n=1}^{\infty} nz^{n-1} \qquad \text{for } |z| < 1,$$

which is the same as the solution obtained in Example 6.11.

EXAMPLE 6.13 Use Theorem 6.9 to show that

$$(17) \qquad \text{Log}(1-z) = -\sum_{n=1}^{\infty} \frac{z^n}{n} \qquad \text{for } |z| < 1.$$

Solution If we start with $-\text{Log}(1-z) = \int_0^z d\xi/(1-\xi)$, which is valid in the left half-plane $\text{Re } z < 1$, then termwise integration of the series in equation (12) results in

$$\text{Log}(1-z) = -\int_0^z \sum_{n=0}^{\infty} \xi^n \, d\xi = -\sum_{n=1}^{\infty} \frac{z^n}{n} \qquad \text{valid for } |z| < 1.$$

EXAMPLE 6.14 Show that

$$\sin^3 z = \frac{3}{4} \sum_{n=0}^{\infty} \frac{(-1)^n (1-9^n) z^{2n+1}}{(2n+1)!}$$

is valid for all z.

Solution Since $\sin z$ is an entire function, it has a Maclaurin series representation

$$\sin z = \sum_{n=0}^{\infty} \frac{(-1)^n z^{2n+1}}{(2n+1)!} \qquad \text{valid for all } z.$$

We can use the trigonometric identity $\sin^3 z = \frac{3}{4} \sin z - \frac{1}{4} \sin 3z$ to obtain the following representation for $\sin^3 z$:

$$\begin{aligned}
\sin^3 z &= \frac{3}{4} \sum_{n=0}^{\infty} \frac{(-1)^n z^{2n+1}}{(2n+1)!} - \frac{1}{4} \sum_{n=0}^{\infty} \frac{(-1)^n (3z)^{2n+1}}{(2n+1)!} \\
&= \frac{3}{4} \sum_{n=0}^{\infty} \frac{(-1)^n z^{2n+1}}{(2n+1)!} - \frac{3}{4} \sum_{n=0}^{\infty} \frac{(-1)^n 9^n z^{2n+1}}{(2n+1)!} \\
&= \frac{3}{4} \sum_{n=0}^{\infty} \frac{(-1)^n (1-9^n) z^{2n+1}}{(2n+1)!}.
\end{aligned}$$

EXERCISES FOR SECTION 6.2

For Exercises 1–5, find the Maclaurin series expansions.

1. Show that $\sin z = \sum_{n=0}^{\infty} \frac{(-1)^n z^{2n+1}}{(2n+1)!}$ is valid for all z.

2. Show that $\cos z = \sum_{n=0}^{\infty} \frac{(-1)^n z^{2n}}{(2n)!}$ is valid for all z.

3. Show that $\sinh z = \sum_{n=0}^{\infty} \frac{z^{2n+1}}{(2n+1)!}$ is valid for all z.

4. Show that $\cosh z = \sum_{n=0}^{\infty} \frac{z^{2n}}{(2n)!}$ is valid for all z.

5. Show that $\mathrm{Log}(1+z) = \sum_{n=1}^{\infty} \frac{(-1)^{n-1}}{n} z^n$ is valid for $|z| < 1$.

6. Use the first three nonzero terms of the Maclaurin series and obtain an approximation to the following:

(a) $\cos\left(\frac{1+i}{8}\right)$ **(b)** $\cosh\left(\frac{1+i}{8}\right)$

7. Use the first two nonzero terms of the Maclaurin series and obtain an approximation to the following:

(a) $\sin\left(\frac{1+i}{10}\right)$ **(b)** $\sinh\left(\frac{1+i}{10}\right)$

8. Use Maclaurin series to establish the identity $\exp(iz) = \cos z + i \sin z$.

9. Differentiate the Maclaurin series for $1/(1-z)$ to show that

$$\frac{1}{(1-z)^3} = \sum_{n=2}^{\infty} \frac{n(n-1)}{2} z^{n-2} \qquad \text{valid for } |z| < 1.$$

10. **(a)** Use the substitution $Z = -z^2$ in the Maclaurin series for $1/(1-Z)$ to obtain the Maclaurin series for $1/(1+z^2)$.
(b) Integrate the Maclaurin series for $1/(1+z^2)$ term by term to obtain the Maclaurin series for $\arctan z$.

11. **(a)** Find the Maclaurin series for $\cos 3z$.
(b) Use the trigonometric identity $\cos 3z = 4\cos^3 z - 3\cos z$, and find the Maclaurin series for $\cos^3 z$.

12. Show that $f(z) = 1/(1-z)$ has a Taylor series representation about the point $z_0 = i$ given by

$$f(z) = \sum_{n=0}^{\infty} \frac{(z-i)^n}{(1-i)^{n+1}} \qquad \text{valid for } |z-i| < \sqrt{2}.$$

13. Find the Maclaurin series for $f(z) = (\sin z)/z$ if we define $f(0) = 1$.

14. Show that

$$\frac{1-z}{z-2} = \sum_{n=0}^{\infty} (z-1)^{n+1}$$

is valid for $|z-1| < 1$. *Hint*: Write

$$\frac{1-z}{z-2} = \frac{z-1}{1-(z-1)}.$$

15. Show that

$$\frac{1-z}{z-3} = \sum_{n=1}^{\infty} \frac{(z-1)^n}{2^n}$$

is valid for $|z-1| < 2$. *Hint*: Write

$$\frac{1-z}{z-3} = \frac{1}{2}\,\frac{z-1}{1-((z-1)/2)}.$$

16. Find the Maclaurin series for $F(z) = \int_0^z ((\sin \xi)/\xi)\, d\xi$.

17. Use the identity $\cos z = \frac{1}{2}(e^{iz} + e^{-iz})$, and find the Maclaurin series expansion for $f(z) = e^z \cos z = \frac{1}{2}e^{(1+i)z} + \frac{1}{2}e^{(1-i)z}$.

18. Find the Maclaurin series expansion for $f(z) = (z^2+1)\sin z$.

19. Let α denote a fixed complex number, and let $f(z) = (1+z)^\alpha = \exp(\alpha \operatorname{Log}(1+z))$ be the principal branch of $(1+z)^\alpha$. Establish the binomial expansion

$$(1+z)^\alpha = 1 + \alpha z + \frac{\alpha(\alpha-1)}{2!} z^2 + \frac{\alpha(\alpha-1)(\alpha-2)}{3!} z^3 + \cdots$$

that is valid for $|z| < 1$.

20. Suppose that $f(z)$ and $g(z)$ are analytic in a neighborhood of z_0 and $f(z_0) = 0$, $g/(z_0) = 0$ and $g'(z_0) \neq 0$. Use Taylor series to prove L'Hôpital's Rule:

$$\lim_{z \to z_0} \frac{f(z)}{g(z)} = \frac{f'(z_0)}{g'(z_0)}.$$

21. The Fresnel integrals $C(z)$ and $S(z)$ are defined by

$$C(z) = \int_0^z \cos(\xi^2)\, d\xi \qquad \text{and} \qquad S(z) = \int_0^z \sin(\xi^2)\, d\xi,$$

and $F(z)$ is defined by the equation $F(z) = C(z) + iS(z)$.

(a) Establish the identity

$$F(z) = \int_0^z \exp(i\xi^2)\, d\xi.$$

(b) Integrate the power series for $\exp(i\xi^2)$ and obtain the power series for $F(z)$.

(c) Use the partial sum involving terms up to z^9 to find approximations to $C(1.0)$ and $S(1.0)$.

22. Compute the Taylor series for the principal logarithm $f(z) = \text{Log}\, z$ expanded about the center $z_0 = -1 + i$. *Hint*: Use $f'(z) = [z - (-1+i) + (-1+i)]^{-1}$ and expand $f'(z)$ in powers of $[z - (-1+i)]$, then apply Theorem 6.9.

23. Let f be defined in a domain that contains the origin. The function f is said to be even if $f(-z) = f(z)$, and it is called odd if $f(-z) = -f(z)$. ***Hint.*** Use limits.

(a) Show that the derivative of an odd function is an even function.

(b) Show that the derivative of an even function is an odd function.

24. **(a)** If $f(z)$ is even, show that all the coefficients of the odd powers of z in the Maclaurin series are zero.

(b) If $f(z)$ is odd, show that all the coefficients of the even powers of z in the Maclaurin series are zero.

6.3 Laurent Series Representations

Suppose that f is analytic in a deleted neighborhood $0 < |z| < R$ but is not analytic at $z = 0$. Then f will *not* have a Maclaurin series representation. For example, $f(z) = z^{-2} \sin z$ is analytic for all $z \neq 0$. However, if we use the Maclaurin series for $\sin z$ and formally divide each term by z^2, then we obtain the representation

$$f(z) = z^{-2} \sin z = \frac{1}{z} - \frac{z}{6} + \frac{z^3}{120} - \frac{z^5}{5040} + \cdots$$

that is valid for $|z| > 0$. This is an example of a Laurent series representation with center $z_0 = 0$. This example generalizes the Taylor series method by permitting negative powers of z to be used in the representation. We now state the main result.

Theorem 6.10 (Laurent's Theorem) *Let $r \geqq 0$ and $R > r$, and suppose that f is analytic in the annulus $A = \{z\colon r < |z - z_0| < R\}$, which is shown in Figure 6.3. Then f has the Laurent series representation*

$$f(z) = \sum_{n=-\infty}^{\infty} a_n (z - z_0)^n \qquad \textit{valid for } r < |z - z_0| < R \tag{1}$$

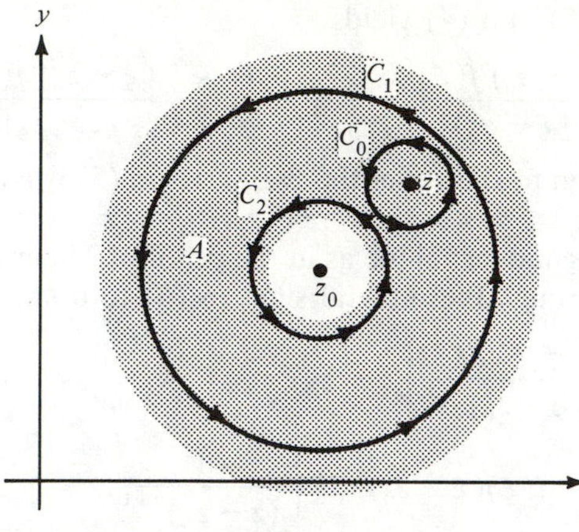

FIGURE 6.3 The circles C_0, C_1, C_2 that lie in the annulus A.

where the Laurent series coefficients a_n are given by

$$(2) \qquad a_n = \frac{1}{2\pi i}\int_C \frac{f(\xi)\,d\xi}{(\xi - z_0)^{n+1}} \qquad \textit{for } n = 0, \pm 1, \pm 2, \ldots$$

where C is any positively oriented circle with center z_0 and radius ρ where $r < \rho < R$.

Proof Let z be an arbitrary but fixed point in A. Let C_0: $|\xi - z| = r_0$ be a circle with center z such that the interior of C_0 is contained in A. Using the Cauchy Integral Formula, we write

$$(3) \qquad f(z) = \frac{1}{2\pi i}\int_{C_0} \frac{f(\xi)\,d\xi}{\xi - z}.$$

Let C_1: $|\xi - z_0| = r_1$ and C_2: $|\xi - z_0| = r_2$, where $r < r_2 < r_1 < R$, be two circles that lie in A such that C_0 lies in the region between C_1 and C_2 as shown in Figure 6.3. Using Theorem 5.4, the extended Cauchy-Goursat Theorem, we see that the integral of $f(\xi)/(\xi - z)$ taken around C_1 in the sum of the integrals of $f(\xi)/(\xi - z)$ taken around C_0 and C_2. This permits us to use equation (3) and express $f(z)$ in the following form:

$$(4) \qquad f(z) = \frac{1}{2\pi i}\int_{C_1} \frac{f(\xi)\,d\xi}{\xi - z} - \frac{1}{2\pi i}\int_{C_2} \frac{f(\xi)\,d\xi}{\xi - z}.$$

With the aid of Lemma 6.1 of Section 6.1 we obtain the following two expansions for $1/(\xi - z)$:

$$(5) \qquad \frac{1}{\xi - z} = \sum_{n=0}^{\infty} \frac{(z - z_0)^n}{(\xi - z_0)^{n+1}} \qquad \text{valid for } |z - z_0| < r_1 \text{ and } \xi \text{ on } C_1;$$

$$\frac{1}{\xi - z} = -\sum_{n=0}^{\infty} \frac{(\xi - z_0)^n}{(z - z_0)^{n+1}} \qquad \text{valid for } |z - z_0| > r_2 \text{ and } \xi \text{ on } C_2.$$

Substitution of the series in (5) into (4) yields

$$(6) \qquad f(z) = \frac{1}{2\pi i}\int_{C_1}\sum_{n=0}^{\infty}\frac{(z-z_0)^n f(\xi)\,d\xi}{(\xi-z_0)^{n+1}} + \frac{1}{2\pi i}\int_{C_2}\sum_{n=0}^{\infty}\frac{(\xi-z_0)^n f(\xi)\,d\xi}{(z-z_0)^{n+1}}.$$

Lemma 6.1 can be used to show that the two series in (5) will converge uniformly on the sets $|z-z_0| \leqq p_1 < r_1$ and $|z-z_0| \geqq p_2 > r_2$, respectively. This will enable us to use the same technique as in the proof of Theorem 6.8 to move the summation signs through the integral signs in (6) to obtain

$$(7) \qquad f(z) = \sum_{n=0}^{\infty}\left[\frac{1}{2\pi i}\int_{C_1}\frac{f(\xi)\,d\xi}{(\xi-z_0)^{n+1}}\right](z-z_0)^n + \sum_{n=0}^{\infty}\left[\frac{1}{2\pi i}\int_{C_2} f(\xi)(\xi-z_0)^n\,d\xi\right]\frac{1}{(z-z_0)^{n+1}}.$$

A final application of the extended Cauchy-Goursat Theorem will permit us to evaluate the integrals in (7) along the contour C. Thus we have

$$(8) \qquad f(z) = \sum_{n=0}^{\infty}\left[\frac{1}{2\pi i}\int_{C}\frac{f(\xi)\,d\xi}{(\xi-z_0)^{n+1}}\right](z-z_0)^n + \sum_{n=-\infty}^{-1}\left[\frac{1}{2\pi i}\int_{C}\frac{f(\xi)\,d\xi}{(\xi-z_0)^{n+1}}\right](z-z_0)^n.$$

The series involving the positive powers of $(z-z_0)$ in equation (8) is valid in the disk $|z-z_0|<r_1$, and the series involving the negative powers of $(z-z_0)$ is valid for $|z-z_0|>r_2$. Letting $r_1 \to R$ and $r_2 \to r$ and using equation (2), we find that $f(z)=\sum_{n=-\infty}^{\infty} a_n(z-z_0)^n$ is valid for $r<|z-z_0|<R$, which completes the proof of the theorem.

If f is analytic in the disk $D\colon |z-z_0|<R$, then the Laurent series in (1) reduces to the Taylor series of f with center z_0. This is easy to see if we look at equation (7). The coefficient for the positive power $(z-z_0)^n$ is seen to be $f^{(n)}(z_0)/n!$ by using Cauchy's integral formula for derivatives. Hence the series involving the positive powers of $(z-z_0)$ is actually the Taylor series for f. The coefficients for the negative powers of $(z-z_0)$ are shown to be zero by applying the Cauchy-Goursat Theorem. Therefore in this case there are no negative powers involved, and the Laurent series reduces to the Taylor series.

The Laurent series for an analytic function f in a specified annulus $r<|z-z_0|<R$ is unique. This is an important fact because the coefficients are seldom obtained by using equation (2). Some methods for finding the Laurent series coefficients are illustrated in the following examples.

EXAMPLE 6.15 Find three different Laurent series representations for $f(z) = 3/(2+z-z^2)$ involving powers of z.

Solution The function f is analytic in the disk $D: |z| < 1$, in the annulus $A: 1 < |z| < 2$, and in the region $R: |z| > 2$. We will find a different Laurent series for f in each of the three domains D, A, and R. We start by writing f in its partial fraction form:

$$(9) \qquad f(z) = \frac{3}{(1+z)(2-z)} = \frac{1}{1+z} + \frac{1}{2}\frac{1}{1-(z/2)}.$$

We can use Lemma 6.1 and obtain the following representations for the terms on the right side of equation (9):

$$(10) \qquad \frac{1}{1+z} = \sum_{n=0}^{\infty} (-1)^n z^n \qquad \text{valid for } |z| < 1,$$

$$(11) \qquad \frac{1}{1+z} = \sum_{n=1}^{\infty} \frac{(-1)^{n+1}}{z^n} \qquad \text{valid for } |z| > 1,$$

$$(12) \qquad \frac{1/2}{1-(z/2)} = \sum_{n=0}^{\infty} \frac{z^n}{2^{n+1}} \qquad \text{valid for } |z| < 2, \qquad \text{and}$$

$$(13) \qquad \frac{1/2}{1-(z/2)} = \sum_{n=1}^{\infty} \frac{-2^{n-1}}{z^n} \qquad \text{valid for } |z| > 2.$$

Representations (10) and (12) are both valid in the disk D, and we have

$$(14) \qquad f(z) = \sum_{n=0}^{\infty} \left[(-1)^n + \frac{1}{2^{n+1}}\right] z^n \qquad \text{valid for } |z| < 1,$$

which is a Laurent series that reduces to a Maclaurin series. In the annulus A, representations (11) and (12) are valid; hence

$$(15) \qquad f(z) = \sum_{n=1}^{\infty} \frac{(-1)^{n+1}}{z^n} + \sum_{n=0}^{\infty} \frac{z^n}{2^{n+1}} \qquad \text{valid for } 1 < |z| < 2.$$

Finally, in the region R we can use representations (11) and (13) to obtain

$$(16) \qquad f(z) = \sum_{n=1}^{\infty} \frac{(-1)^{n+1} - 2^{n-1}}{z^n} \qquad \text{valid for } |z| > 2.$$

EXAMPLE 6.16 Find the Laurent series representation for $f(z) = (\cos z - 1)/z^4$ that involves powers of z.

Solution We can use the Maclaurin series for $\cos z - 1$ to write

$$f(z) = \frac{\dfrac{-1}{2!}z^2 + \dfrac{1}{4!}z^4 - \dfrac{1}{6!}z^6 + \cdots}{z^4}.$$

We formally divide each term by z^4 to obtain the Laurent series

$$f(z) = \frac{-1}{2z^2} + \frac{1}{24} - \frac{z^2}{720} + \cdots \qquad \text{valid for } z \neq 0.$$

EXAMPLE 6.17 Find the Laurent series for $\exp(-1/z^2)$ centered at $z_0 = 0$.

Solution The Maclaurin series for $\exp Z$ is given by

$$(17) \qquad \exp Z = \sum_{n=0}^{\infty} \frac{Z^n}{n!} \qquad \text{valid for all } Z.$$

Using the substitution $Z = -z^{-2}$ in equation (17), we obtain

$$\exp\left(\frac{-1}{z^2}\right) = \sum_{n=0}^{\infty} \frac{(-1)^n}{n! z^{2n}} \qquad \text{valid for } |z| > 0.$$

Remark Consider the *real* function f defined by $f(x) = e^{-1/x^2}$ when $x \neq 0$ and $f(0) = 0$. Then $f'(0) = 0$ is shown by the following calculation:

$$\lim_{x \to 0} \frac{e^{-1/x^2}}{x} = \lim_{t \to \infty} \frac{t}{e^{t^2}} = 0 \qquad \text{where } t = \frac{1}{x}.$$

Further calculations show that $f^{(n)}(0) = 0$ for $n = 2, 3, \ldots$. Since f is not identically zero, it cannot have a Maclaurin series representation in powers of the real variable x. This shows that a real function can have derivatives of all orders at a point but fail to have a Taylor series representation. Let us compare this with the complex function in Example 6.17.

Consider the *complex* function F defined by $F(z) = \exp(-1/z^2)$ when $z \neq 0$ and $F(0) = 0$. Then F reduces to f when we set $z = x + 0i$; that is,

$$F(x + 0i) = \exp\left(\frac{-1}{(x + 0i)^2}\right) = e^{-1/x^2} = f(x).$$

The complex function F fails to have a Maclaurin series for a very simple reason: F is not continuous at $z = 0$. This is shown by finding the limits of F as $z \to 0$ along the coordinate axes:

$$\lim_{x \to 0} \exp\left(\frac{-1}{(x + 0i)^2}\right) = \lim_{x \to 0} e^{-1/x^2} = 0 \qquad \text{and}$$

$$\lim_{y \to 0} \exp\left(\frac{-1}{(0 + iy)^2}\right) = \lim_{y \to 0} e^{1/y^2} = \infty.$$

Since the limits are distinct, $\exp(-1/z^2)$ is not continuous at $z = 0$. Therefore it is not analytic at $z = 0$ and cannot have a Maclaurin series representation.

EXERCISES FOR SECTION 6.3

1. Find two Laurent series expansions for $f(z) = 1/(z^3 - z^4)$ that involve powers of z.
2. Find the Laurent series for $f(z) = (\sin 2z)/z^4$ that involves powers of z.

3. Show that

$$f(z) = \frac{1}{1-z} = \frac{1}{1-i}\,\frac{1}{1-\dfrac{z-i}{1-i}}$$

has a Laurent series representation about the point $z_0 = i$ given by

$$f(z) = \frac{1}{1-z} = -\sum_{n=1}^{\infty} \frac{(1-i)^{n-1}}{(z-i)^n} \qquad \text{valid for } |z-i| > \sqrt{2}.$$

4. Show that

$$\frac{1-z}{z-2} = -\sum_{n=0}^{\infty} \frac{1}{(z-1)^n}$$

is valid for $|z-1| > 1$. *Hint*: Use the hint for Exercise 14 in Section 6.2.

5. Show that

$$\frac{1-z}{z-3} = -\sum_{n=0}^{\infty} \frac{2^n}{(z-1)^n}$$

is valid for $|z-1| > 2$. *Hint*: Use the hint for Exercise 15 in Section 6.2.

6. Find the Laurent series for $\sin(1/z)$ centered at $z_0 = 0$.

7. Find the Laurent series for $f(z) = (\cosh z - \cos z)/z^5$ that involves powers of z.

8. Find the Laurent series for $f(z) = 1/(z^4(1-z)^2)$ that involves powers of z and is valid for $|z| > 1$. *Hint*: $1/(1-(1/z))^2 = z^2/(1-z)^2$.

9. Find two Laurent series for $z^{-1}(4-z)^{-2}$ involving powers of z. *Hint*: Use the result of Example 6.12.

10. Find three Laurent series for $(z^2 - 5z + 6)^{-1}$ centered at $z_0 = 0$.

11. Let a and b be positive real numbers with $b > a > 1$. Show that

$$\operatorname{Log} \frac{z-a}{z-b} = \sum_{n=1}^{\infty} \frac{b^n - a^n}{nz^n}$$

holds for $|z| > b$. *Hint*: $\operatorname{Log}(z-a)/(z-b) = \operatorname{Log}(1-(a/z)) - \operatorname{Log}(1-(b/z))$.

12. Use the Maclaurin series for $\sin z$ and then long division to show that the Laurent series for $\csc z$ with $z_0 = 0$ is

$$\csc z = \frac{1}{z} + \frac{z}{6} + \frac{7z^3}{360} + \cdots.$$

13. Can $\operatorname{Log} z$ be represented by a Maclaurin series or a Laurent series about the point $z_0 = 0$? Give a reason for your answer.

14. Show that $\cosh(z + (1/z)) = \sum_{n=-\infty}^{\infty} a_n z^n$ where

$$a_n = \frac{1}{2\pi}\int_0^{2\pi} \cos n\theta \cosh(2\cos\theta)\, d\theta.$$

Hint: Let the path of integration be the circle $C: |z| = 1$.

15. The *Bessel function* $J_n(z)$ is sometimes defined by the generating function

$$\exp\left(\frac{z}{2}\left(t - \frac{1}{t}\right)\right) = \sum_{n=-\infty}^{\infty} J_n(z) t^n.$$

Use the circle $C: |z| = 1$ as the contour of integration, and show that

$$J_n(z) = \frac{1}{\pi} \int_0^\pi \cos(n\theta - z \sin\theta)\, d\theta.$$

16. Consider the real-valued function $u(\theta) = 1/(5 - 4\cos\theta)$.
(a) Use the substitution $\cos\theta = (1/2)(z + 1/z)$ and obtain

$$u(\theta) = f(z) = \frac{-z}{(z-2)(2z-1)} = \frac{1}{3}\frac{1}{1 - z/2} - \frac{1}{3}\frac{1}{1-2z}.$$

(b) Expand the function $f(z)$ in part (a) in a Laurent series that is valid in the annulus $1/2 < |z| < 2$ and get

$$f(z) = \frac{1}{3} + \frac{1}{3}\sum_{n=1}^{\infty} 2^{-n}(z^n + z^{-n}).$$

(c) Use the substitutions $\cos(n\theta) = (1/2)(z^n + z^{-n})$ in part (b) and obtain the Fourier series for $u(\theta)$:

$$u(\theta) = \frac{1}{3} + \frac{1}{3}\sum_{n=1}^{\infty} 2^{-n+1}\cos(n\theta).$$

17. Suppose that the Laurent expansion $f(z) = \sum_{n=-\infty}^{\infty} a_n z^n$ converges in the annulus $r_1 < |z| < r_2$ where $r_1 < 1$ and $1 < r_2$. Consider the real-valued function $u(\theta) = f(e^{i\theta})$. Show that $u(\theta)$ has the Fourier series expansion

$$u(\theta) = f(e^{i\theta}) = \sum_{n=-\infty}^{\infty} a_n e^{i\theta} \qquad \text{where } a_n = \frac{1}{2\pi}\int_0^{2\pi} e^{-in\phi} f(e^{i\phi})\, d\phi.$$

18. *The Z-Transform*. Let $\{a_n\}$ be a sequence of complex numbers satisfying the growth condition $|a_n| \leqq MR^n$ for $n = 0, 1, \ldots$ for some fixed positive values M and R. Then the Z-transform of the sequence $\{a_n\}$ is the function $F(z)$ defined by

$$Z(\{a_n\}) = F(z) = \sum_{n=0}^{\infty} a_n z^{-n}.$$

Prove that $F(z)$ converges for $|z| > R$.

19. Find $Z(\{a_n\})$ for the following sequences:
(a) $a_n = 2$, **(b)** $a_n = 1/n!$, **(c)** $a_n = 1/(n+1)$,
(d) $a_n = 1$ when n is even, $a_n = 0$ when n is odd.

20. Prove the shifting property for the Z-transform:

$$Z(\{a_{n+1}\}) = z[Z(\{a_n\}) - a_0].$$

6.4 Power Series

We have seen that if f is analytic at the point z_0, then it has a Taylor series involving powers of $(z - z_0)$. We now state some general results concerning series. A *power series* in powers of $(z - z_0)$ is an infinite series of the form

$$\sum_{n=0}^{\infty} a_n (z - z_0)^n \tag{1}$$

where z is a complex variable, $a_0, a_1, \ldots$ are constants called coefficients, and the constant z_0 is called the center of the series. The power series in (1) converges for $z = z_0$, and the sum takes on the value a_0. Our first result shows that if the power series in (1) converges at some other value $z_1 \neq z_0$, then the series converges in an open disk with center z_0 and can be used to define a complex function.

Theorem 6.11

(i) *If the power series in (1) converges at the point $z_1 \neq z_0$, then it converges for all z in the disk $|z - z_0| < R$ where $R = |z_1 - z_0|$.*

(ii) *If the power series in (1) diverges at the point z_2, then it diverges for all z satisfying $|z - z_0| > \rho$ where $\rho = |z_2 - z_0|$.*

(iii) *Furthermore, a power series that converges at $z_1 \neq z_0$ defines a complex function $S(z)$ that represents the sum of the series. We write*

$$S(z) = \sum_{n=0}^{\infty} a_n (z - z_0)^n \qquad \text{for } |z - z_0| < R. \tag{2}$$

The situation is illustrated in Figure 6.4.

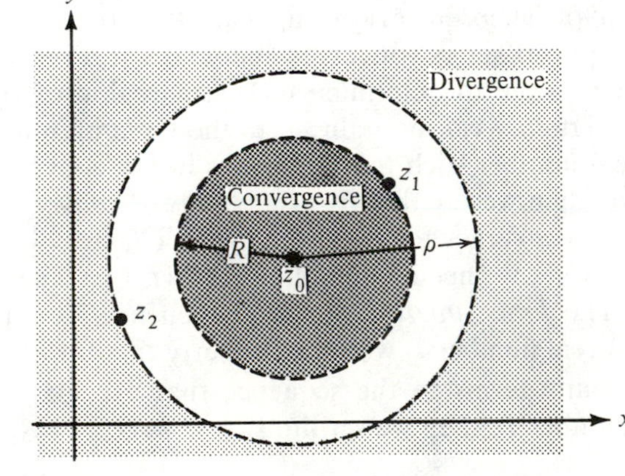

FIGURE 6.4 The regions where a power series can converge and diverge according to the statement of Theorem 6.11.

Proof of Part (i) Let r be any real number satisfying $0 < r < R$. Now we will show that the series in (1) converges at each point in the closed disk $D_r: |z - z_0| \leqq r$. Since the series converges at z_1, according to Exercise 15 of Section 6.1, we have $\lim_{n\to\infty} a_n(z_1 - z_0)^n = 0$. Therefore there exists an integer N such that $|a_n(z_1 - z_0)^n| \leqq 1$ for $n \geqq N$. If we set $M = \max\{1, |a_0|, \ldots, |a_{N-1}(z_1 - z_0)^{N-1}|\}$, then we see that

$$|a_n||z_1 - z_0|^n \leqq M \qquad \text{holds for all } n = 0, 1, \ldots. \tag{3}$$

We now consider the series of absolute values:

$$(4)\qquad \sum_{n=0}^{\infty} |a_n||z - z_0|^n = \sum_{n=0}^{\infty} |a_n||z_1 - z_0|^n \frac{|z - z_0|^n}{|z_1 - z_0|^n}.$$

If we use the fact that $|z - z_0|^n/|z_1 - z_0|^n \leqq r^n/R^n = K^n$ where $0 < K < 1$, then using (3) and (4), we obtain

$$(5)\qquad \sum_{n=0}^{\infty} |a_n||z - z_0|^n \leqq \sum_{n=0}^{\infty} MK^n = \frac{M}{1 - K}.$$

The result in (5) can be used with the comparison test in Section 6.1 to conclude that the power series in (1) converges for $|z - z_0| \leqq r$. Letting $r \to R$ establishes part (i) of Theorem 6.11.

Proof by Contradiction of Part (ii) Assume the contrary, and suppose that series (1) diverges at z_2. Also suppose that there exists a value z_3 with $|z_3 - z_0| > |z_2 - z_0|$ at which the series converges. Then part (i) implies that the series converges for all z in the disk $|z - z_0| < |z_3 - z_0|$. In particular, the series would converge at the point z_2. With this contradiction, part (ii) is established.

From the statement of Theorem 6.11 it appears that there is a gap between the region where a power series converges and where it diverges. For an arbitrary series it is not possible to determine the exact region of convergence, but the region of uncertainty can be reduced to the circumference of a circle. If R is the largest real number such that the series in (1) converges for all z in the disk $|z - z_0| < R$, then we say that R is the *radius of convergence* of the power series. If the series converges for all z, then we set $R = \infty$. Theorem 6.12 states how to find R, but first we need a definition. Let $\{t_n\}$ be a sequence of positive real numbers. The *limit superior* of the sequence is denoted by $\limsup t_n$ and is the smallest real number L with the property that for any $\varepsilon > 0$ there are at most finitely many terms in the sequence that are larger than $L + \epsilon$. If there is no such number L, then we set $\limsup t_n = \infty$.

Theorem 6.12 (Radius of Convergence) *For the power series in* (1) *there exists a number R with $0 \leqq R \leqq +\infty$ called the radius of convergence of the series, such that*

$$S(z) = \sum_{n=0}^{\infty} a_n(z - z_0)^n \qquad \textit{converges for } |z - z_0| < R$$

and the series diverges when $|z - z_0| > R$. (See Figure 6.5.) The value R is given by the equation $R = 1/L$ ($\infty =$ "$1/0^+$") where L can be found by the following methods.

(6) *d'Alembert's Ratio Test:* $L = \lim_{n\to\infty} |a_{n+1}|/|a_n|$ *if the limit exists.*

(7) *Cauchy's Root Test:* $L = \lim_{n\to\infty} \sqrt[n]{|a_n|}$ *if the limit exists.*

(8) *Cauchy-Hadamard formula:* $L = \limsup \sqrt[n]{|a_n|}$, *which always exists.*

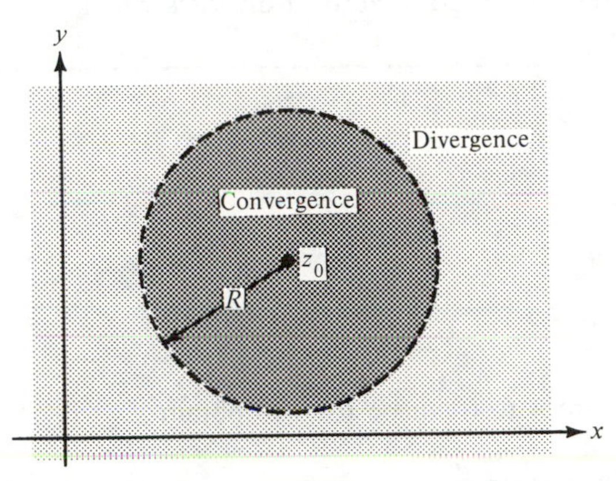

FIGURE 6.5 The radius of convergence of a power series.

Proof We will establish the method in statement (7) and leave the others for the reader. Suppose that the number L is given by the limit in (7) and that $|z - z_0| < R$. We can choose a value r such that $|z - z_0| \leqq r < R$. Since $2/(R + r) > L$, there exists an integer N such that

$$(9) \qquad \sqrt[n]{|a_n|} < \frac{2}{R+r} \qquad \text{or} \qquad |a_n| < \left(\frac{2}{R+r}\right)^n$$

holds for all $n \geqq N$. If we consider the series of terms taken in absolute value, then by using the inequalities in (9) we find that

$$(10) \qquad \sum_{n=N}^{\infty} |a_n||z - z_0|^n \leqq \sum_{n=N}^{\infty} \left(\frac{2}{R+r}\right)^n r^n = \sum_{n=N}^{\infty} \left(\frac{2r}{R+r}\right)^n.$$

Since $2r/(R + r) < 1$, the series on the right side of (10) is a convergent geometric series, and the comparison test of Section 6.1 implies that

$$(11) \qquad \sum_{n=N}^{\infty} a_n(z - z_0)^n \qquad \text{converges for } |z - z_0| \leqq r.$$

We can add the first N terms of the series to (11) and conclude that the power series in (1) converges for $|z - z_0| \leqq r$. Letting $r \to R$ establishes the first part of Theorem 6.12.

Now suppose that $|z - z_0| > R$. Let r be chosen so that $|z - z_0| > r > R$. Since $1/r < L$, there exists an integer N such that

$$(12) \qquad \sqrt[n]{|a_n|} > \frac{1}{r} \qquad \text{holds for all } n \geqq N.$$

Using inequality (12), we find that

$$\lim_{n\to\infty} |a_n||z - z_0|^n > \lim_{n\to\infty} \frac{r^n}{r^n} = 1.$$

Since the terms of the series do not go to zero, it diverges when $|z - z_0| > R$, and the proof is complete.

EXAMPLE 6.18 The power series

$$\sum_{n=0}^{\infty} \left(\frac{n+3}{2n+1}\right)^n z^n$$

has radius of convergence $R = 2$.

Solution Cauchy's root test can be used to obtain

$$L = \lim_{n\to\infty} \sqrt[n]{|a_n|} = \lim_{n\to\infty} \frac{n+3}{2n+1} = \frac{1}{2}.$$

Therefore the radius of convergence is $R = 1/L = 2$.

EXAMPLE 6.19 The limit superior of the sequence

$$\{t_n\} = \{1, 4, 5, 4, 5, 4, 5, \ldots\}$$

is $\limsup t_n = 5$.

Solution If we set $L = 5$, then for any $\varepsilon > 0$ there are at most finitely many (actually there are none) terms in the sequence that are larger than $L + \varepsilon = 5 + \varepsilon$. Therefore $\limsup t_n = 5$.

EXAMPLE 6.20 The limit superior of the sequence

$$\{t_n\} = \{5, 4.1, 5.1, 4.01, 5.01, 4.001, 5.001, \ldots\}$$

is $\limsup t_n = 5$.

Solution If we set $L = 5$, then for any $\varepsilon > 0$ there are finitely many terms in the sequence larger than $L + \varepsilon = 5 + \varepsilon$. Therefore $\limsup t_n = 5$.

EXAMPLE 6.21 The power series

$$\sum_{n=0}^{\infty} a_n z^n = 1 + 4z + 5^2z^2 + 4^3z^3 + 5^4z^4 + 4^5z^5 + 5^6z^6 + \cdots$$

has radius of convergence $R = 1/5$.

Solution According to Example 6.19, the sequence $\{t_n\} = \{\sqrt[n]{|a_n|}\}$ has the limit superior given by $\limsup t_n = 5$. Using the Cauchy-Hadamard Formula (8), we have $L = \limsup \sqrt[n]{|a_n|} = 5$. Therefore the radius of convergence is $R = 1/L = 1/5$.

Important results concerning power series are the facts that they define continuous functions that are differentiable and hence they can be used to define an analytic function.

Theorem 6.13 *If the power series* $S(z) = \sum_{n=0}^{\infty} a_n(z - z_0)^n$ *has radius of convergence R, then*

(13) *The series is uniformly convergent on any subdisk* $|z - z_0| \leqq r$ *where* $r < R$ *and*

(14) *The function* $S(z)$ *is continuous on* $|z| < R$.

Proof of (13) The validity of (13) follows by a close examination of the proof of Theorem 6.12. In the proof of statement (7) we assumed that $r < R$ and worked with values of z in the closed disk $|z - z_0| \leqq r$. For any $\varepsilon > 0$ an integer N can be found such that the geometric series on the right side of (10) can be made less than ε. This in turn can be used to show that error term $E_n(z)$ defined by

$$E_n(z) = S(z) - \sum_{k=1}^{n} a_k(z - z_0)^k$$

will satisfy

$$|E_n(z)| \leqq \varepsilon \qquad \text{for all } n \geqq N \text{ and for all } |z - z_0| \leqq r.$$

Therefore the series converges uniformly to $S(z)$ on $|z - z_0| \leqq r$.

Proof of (14) We will show that $S(z)$ is continuous at $z = z_1$. From uniform convergence, for any $\varepsilon > 0$ there exists an integer $N = N_\varepsilon$ such that

(15) $|E_n(z)| < \varepsilon/3$

for $n \geqq N$ and all z in the closed disk $|z - z_0| \leqq (R + |z_1 - z_0|)/2$. In particular, the Nth partial sum

$$(16)\qquad S_N(z) = \sum_{k=0}^{N} a_k(z - z_0)^k$$

is a polynomial of degree N and hence is continuous at z_1. Hence we can find a δ that is smaller than $(R - |z_1 - z_0|)/2$ so that

$$(17)\qquad |S_N(z) - S_N(z_1)| < \varepsilon/3 \qquad \text{whenever } |z - z_1| < \delta.$$

Notice that the relation

$$S(z) = S_N(z) + E_N(z)$$

implies that

$$|S(z) - S(z_1)| = |S_N(z) - S_N(z_1) + E_N(z) - E_N(z_1)|$$

and also

$$(18)\qquad |S(z) - S(z_1)| \leqq |S_N(z) - S_N(z_1)| + |E_N(z)| + |E_N(z_1)|.$$

Using (15) and (17) in (18) results in

$$|S(z) - S(z_1)| \leqq \varepsilon/3 + \varepsilon/3 + \varepsilon/3 = \varepsilon \qquad \text{for all } |z - z_1| < \delta.$$

Therefore $S(z)$ is continuous at z_1, and statement (14) is proven.

Theorem 6.14 *Let $S(z) = \sum_{n=0}^{\infty} a_n(z - z_0)^n$ be a power series with radius of convergence $R > 0$. Then $S(z)$ is analytic in the disk $|z - z_0| < R$, and its derivative is given by*

$$(19)\qquad S'(z) = \sum_{n=1}^{\infty} na_n(z - z_0)^{n-1} \qquad \textit{valid for } |z - z_0| < R.$$

Proof We first show that the series on the right side of equation (19) has radius of convergence R. Let $z \neq z_0$ denote a fixed value in the disk $|z - z_0| < R$; then we can find another value z_1 that lies in this disk and is farther away from z_0, that is, $|z - z_0| < |z_1 - z_0| < R$. The power series for $S(z_1)$ converges, and the details in the proof of Theorem 6.11 show that the series of terms taken in absolute value converge. That is,

$$(20)\qquad \sum_{n=0}^{\infty} |a_n||z_1 - z_0|^n < \infty.$$

Let us compare this series with the differentiated series. If we set $r = |z - z_0|/|z_1 - z_0|$, then we have

$$(21)\qquad |na_n(z - z_0)^{n-1}| = \frac{n}{|z - z_0|}\frac{|z - z_0|^n}{|z_1 - z_0|^n}|a_n||z_1 - z_0|^n$$

$$= \frac{nr^n}{|z - z_0|}|a_n||z_1 - z_0|^n.$$

Since $0 < r < 1$, it follows that $nr^n \to 0$ as $n \to \infty$. Hence it is possible to find a positive real number M such that

$$(22) \qquad \frac{nr^n}{|z - z_0|} < M \qquad \text{for } n = 0, 1, 2, \ldots .$$

Using (22) in (21), we obtain the inequality

$$(23) \qquad |na_n(z - z_0)^{n-1}| < M|a_n||z_1 - z_0|^n \qquad \text{for all } n.$$

By using (20) and (23) and Theorem 6.6 (the comparison test) it follows that the series in (19) converges for any fixed value of z satisfying $|z - z_0| < R$.

We now show that $S(z)$ is analytic and that equality in (19) holds. For simplicity we will prove the result for the special case $z_0 = 0$:

$$(24) \qquad S(x) = \sum_{n=0}^{\infty} a_n z^n \qquad \text{and} \qquad S'(z) = \sum_{n=1}^{\infty} na_n z^{n-1}.$$

To continue the proof, we need to observe that if we differentiate the series in (24) again, we will obtain the series

$$(25) \qquad h(z) = \sum_{n=2}^{\infty} n(n-1)a_n z^{n-2},$$

and this series converges absolutely for $|z| < R$ and defines a continuous function.

Suppose that z is fixed and $|z| = r$ where $r < R$. Let us also assume that Δz satisfies $|\Delta z| < (R - r)/2$ so that $|z + \Delta z| \leqq |z| + |\Delta z| \leqq (r + R)/2$. Hence the series $S(z + \Delta z)$, $S'(z + \Delta z)$, and $h(|z| + |\Delta z|)$ all converge. Define $g(\Delta z)$ by

$$(26) \qquad g(\Delta z) = \frac{S(z + \Delta z) - S(z)}{\Delta z} - S'(z) = \sum_{n=2}^{\infty} a_n\left(\frac{(z + \Delta z)^n - z^n}{\Delta z} - nz^{n-1}\right).$$

The binomial formula gives us

$$\begin{aligned}\frac{(z + \Delta z)^n - z^n}{\Delta z} - nz^{n-1} &= \binom{n}{2} z^{n-2}\Delta z + \binom{n}{3} z^{n-3}(\Delta z)^2 + \cdots \\ &\quad + \binom{n}{n-1} z(\Delta z)^{n-2} + \binom{n}{n}(\Delta z)^{n-1} \\ &= \frac{n(n-1)}{2}\,\Delta z\left[z^{n-2} + \frac{n-2}{3}\, z^{n-3}\Delta z + \cdots \right. \\ &\quad \left. + \frac{n-1}{2}\, z(\Delta z)^{n-3} + \frac{2}{n(n-1)}\,(\Delta z)^{n-2}\right].\end{aligned}$$

The modulus of the expression in brackets is less than or equal to the following expression:

$$\begin{aligned}&|z|^{n-2} + (n-2)|z|^{n-3}|\Delta z| + \cdots + (n-2)|z||\Delta z|^{n-3} + |\Delta z|^{n-2} \\ &\quad = (|z| + |\Delta z|)^{n-2}.\end{aligned}$$

Thus we obtain the following estimate:

$$\left|\frac{(z+\Delta z)^n - z^n}{\Delta z} - nz^{n-1}\right| \leqq \frac{1}{2}\,|\Delta z|n(n-1)(|z|+|\Delta z|)^{n-2},$$

which can be used with (26) to obtain

$$\text{(27)}\qquad |g(\Delta z)| \leqq \frac{1}{2}\,|\Delta z| \sum_{n=2}^{\infty} n(n-1)|a_n|(|z|+|\Delta z|)^{n-2}$$

$$\leqq \frac{1}{2}\,|\Delta z||h(|z|+|\Delta z|)|.$$

Since the series in (25) is continuous on the closed disk $|z| \leqq (r+R)/2$, it is bounded on this disk; thus there exists an M such that

$$\text{(28)}\qquad |h(|z|+|\Delta z|)| \leqq M \qquad \text{for all } |\Delta z| < (R-|z|)/2.$$

Combining (27) and (28), we see that $g(\Delta z)\to 0$ as $\Delta z \to 0$. This combined with (26) gives

$$\lim_{\Delta z\to 0} \frac{S(z+\Delta z)-S(z)}{\Delta z} - S'(z) = 0,$$

and the proof of Theorem 6.14 is complete.

EXAMPLE 6.22 The Bessel function of order zero is given by

$$J_0(z) = \sum_{n=0}^{\infty} \frac{(-1)^n}{(n!)^2}\left(\frac{z}{2}\right)^{2n} = 1 - \frac{z^2}{2^2} + \frac{z^4}{2^2 4^2} - \frac{z^6}{2^2 4^2 6^2} + \cdots,$$

and termwise differentiation shows that its derivative is given by

$$J_0'(z) = \sum_{n=0}^{\infty} \frac{(-1)^{n+1}}{n!(n+1)!}\left(\frac{z}{2}\right)^{2n+1} = \frac{-z}{2} + \frac{1}{1!2!}\left(\frac{z}{2}\right)^3 - \frac{1}{2!3!}\left(\frac{z}{2}\right)^5 + \cdots.$$

The representations are valid for all z.

Solution The ratio test can be used to see that

$$L = \lim_{n\to\infty} \frac{|a_{n+1}|}{|a_n|} = \lim_{n\to\infty} \frac{(n!)^2 2^{2n}}{((n+1)!)^2 2^{2n+2}} = \lim_{n\to\infty} \frac{1}{4(n+1)^2} = 0.$$

Therefore the radius of convergence is $R=\infty$, and Theorem 6.14 can be used to obtain $J_0'(z)$ by termwise differentiation. We remark that the Bessel function $J_1(z)$ of order one is known to satisfy the relation $J_1(z) = -J_0'(z)$.

We now state a result concerning the uniqueness of power series. This will permit us to obtain a power series for an analytic function by any means that is convenient. In fact, we show that the power series representation is actually the Taylor series representation.

Theorem 6.15 (Uniqueness of Series) *Let f be represented by a power series*

$$f(z) = \sum_{n=0}^{\infty} a_n(z - z_0)^n \qquad \text{for } |z - z_0| < R. \tag{29}$$

Then the coefficients a_n satisfy the identity

$$a_n = \frac{f^{(n)}(z_0)}{n!} \qquad \text{for all } n. \tag{30}$$

Hence the power series is actually the Taylor series for f with center z_0.

Proof If we set $z = z_0$ in equation (29), then we see that $a_0 = f(z_0)$. Using Theorem 6.13, we differentiate the series in (29) to obtain

$$f'(z) = a_1 + 2a_2(z - z_0) + 3a_3(z - z_0)^2 + \cdots. \tag{31}$$

If we set $z = z_0$ in equation (31), then we see that $a_1 = f'(z_0)$. Continuing in this fashion, we can differentiate the series in (29) n times, and the result will be

$$f^{(n)}(z) = n!a_n + \frac{(n+1)!}{1!} a_{n+1}(z - z_0) + \frac{(n+2)!}{2!} a_{n+2}(z - z_0)^2 + \cdots. \tag{32}$$

Setting $z = z_0$ in equation (32) and solving for a_n yields $a_n = f^{(n)}(z_0)/n!$, which proves Theorem 6.15.

EXAMPLE 6.23 Show that

$$\exp\left(\frac{-z^2}{2}\right) = \sum_{n=0}^{\infty} \frac{(-1)^n z^{2n}}{2^n n!}$$

is valid for all z.

Solution Using the substitution $Z = -z^2/2$ and the expansion for $\exp Z$ given in Example 6.10, we obtain

$$\exp\left(\frac{-z^2}{2}\right) = \sum_{n=0}^{\infty} \frac{(-1)^n z^{2n}}{2^n n!} \qquad \text{for all } z.$$

We conclude this section with a result concerning the sum and product of series.

Theorem 6.16 *Let f and g have the power series representations*

$$f(z) = \sum_{n=0}^{\infty} a_n(z - z_0)^n \qquad \text{and} \qquad g(z) = \sum_{n=0}^{\infty} b_n(z - z_0)^n$$

that have the radii of convergence R_1 and R_2, respectively. If c is a complex constant, then

$$(33)\qquad cf(z) = \sum_{n=0}^{\infty} ca_n(z - z_0)^n \qquad \text{for } |z - z_0| < R_1.$$

If $R = \min\{R_1, R_2\}$, then

$$(34)\qquad f(z) + g(z) = \sum_{n=0}^{\infty} (a_n + b_n)(z - z_0)^n \qquad \text{for } |z - z_0| < R.$$

The Cauchy product *of the two series is given by*

$$(35)\qquad f(z)g(z) = \sum_{n=0}^{\infty} \left(\sum_{k=0}^{n} a_k b_{n-k}\right)(z - z_0)^n \qquad \text{for } |z - z_0| < R.$$

Proof The proofs of equations (33) and (34) are left as exercises. To establish equation (35), we observe that the function $h(z) = f(z)g(z)$ is analytic for $|z - z_0| < R$. The derivatives of h are given by

$$h' = f'g + fg', \qquad h'' = f''g + 2f'g' + fg''.$$

The nth derivative of h is given by Leibniz's formula for the derivative of a product of functions

$$(36)\qquad h^{(n)}(z) = \sum_{k=0}^{n} \frac{n!}{k!(n-k)!} f^{(k)}(z) g^{(n-k)}(z).$$

We can use identities (30) and (36) to conclude that

$$(37)\qquad \frac{h^{(n)}(z_0)}{n!} = \sum_{k=0}^{n} \frac{f^{(k)}(z_0)}{k!} \frac{g^{(n-k)}(z_0)}{(n-k)!} = \sum_{k=0}^{n} a_k b_{n-k}.$$

Using equation (37) and Theorem 6.15, we obtain

$$h(z) = \sum_{n=0}^{\infty} \frac{h^{(n)}(z_0)}{n!}(z - z_0)^n = \sum_{n=0}^{\infty}\left(\sum_{k=0}^{n} a_k b_{n-k}\right)(z - z_0)^n,$$

and the proof is complete.

EXAMPLE 6.24 Use the Cauchy product of series to show that

$$\frac{1}{(1-z)^2} = \sum_{n=0}^{\infty} (n+1)z^n \qquad \text{for } |z| < 1.$$

Solution Let $f(z) = g(z) = 1/(1 - z) = \sum_{n=0}^{\infty} z^n$ for $|z| < 1$. Then $a_n = b_n = 1$ for all n. Using equation (35), we obtain

$$\frac{1}{(1-z)^2} = f(z)g(z) = \sum_{n=0}^{\infty}\left(\sum_{k=0}^{n} a_k b_{n-k}\right) z^n = \sum_{n=0}^{\infty} (n+1)z^n,$$

which is in agreement with Example 6.11.

EXERCISES FOR SECTION 6.4

1. Find the radius of convergence of $\sum_{n=0}^{\infty} \frac{z^n}{(1+i)^n}$.

2. Find the radius of convergence of $\sum_{n=2}^{\infty} \frac{n(n-1)z^n}{(3+4i)^n}$.

3. Show that $\sum_{n=0}^{\infty} (n^n/n!)z^n$ has radius of convergence $R = 1/e$. *Hint*: Use the fact that $\lim_{n\to\infty} (1+(1/n))^n = e$.

4. Show that the series $\sum_{n=0}^{\infty} ((n+7)/(4n+2))^n z^n$ has radius of convergence $R = 4$. *Hint*: Use the root test.

5. Find the radius of convergence of $\sum_{n=0}^{\infty} \left(\frac{3n+7}{4n+2}\right)^n z^n$.

6. Find the radius of convergence of $\sum_{n=0}^{\infty} \frac{2^n}{1+3^n} z^n$.

7. **(a)** Show that if $\{t_n\} = \{1, 3, 1, 3, 1, 3, \ldots\}$, then $\limsup t_n = 3$.

   (b) Show that $\sum_{n=0}^{\infty} (2-(-1)^n)^n z^n$ has radius of convergence $R = \frac{1}{3}$.

8. Use the Cauchy-Hadamard Formula (8) to show that the series $\sum_{n=0}^{\infty} z^{2^n}$ has radius of convergence $R = 1$.

9. Find the radius of convergence of the following power series representation:

$$\sinh z + \frac{4}{4-z^2} = 1 + z + \frac{z^2}{2^2} + \frac{z^3}{3!} + \frac{z^4}{2^4} + \frac{z^5}{5!} + \frac{z^6}{2^6} + \frac{z^7}{7!} + \cdots.$$

10. Let $f(z) = \sum_{n=0}^{\infty} a_n z^n$ be an entire function.
    (a) Find the series representation for $\overline{f(z)}$ using powers of $\overline{z}$.
    (b) Show that $\overline{f(\overline{z})}$ is an entire function.
    (c) Does $\overline{f(\overline{z})} = f(z)$? Why?

11. Suppose that $\sum_{n=0}^{\infty} a_n z^n$ has radius of convergence R. Show that $\sum_{n=0}^{\infty} a_n^2 z^n$ has radius of convergence R^2. *Hint:* Use (8).

12. Suppose that $\sum_{n=0}^{\infty} a_n z^n$ converges for $z = 1, 2, 3, \ldots$. Show that $\sum_{n=0}^{\infty} a_n z^n$ converges for all z.

13. Does there exist a power series $\sum_{n=0}^{\infty} a_n z^n$ that converges at $z_1 = 4 - i$ and diverges at $z_2 = 2 + 3i$? Why?

14. Find $f^{(3)}(0)$ for the following functions.

    (a) $f(z) = \sum_{n=0}^{\infty} (3+(-1)^n)^n z^n$ **(b)** $f(z) = \sum_{n=0}^{\infty} \frac{(1+i)^n}{n} z^n$

    (c) $f(z) = \sum_{n=0}^{\infty} \frac{z^n}{(\sqrt{3}+i)^n}$

15. Let $f(z)$ have the power series representation

$$f(z) = \sum_{n=0}^{\infty} a_n z^n = 1 + z + 2z^2 + 3z^3 + 5z^4 + \cdots$$

where the coefficients a_n are the Fibonacci numbers

$$a_0 = 1, \quad a_1 = 1, \quad \text{and} \quad a_n = a_{n-1} + a_{n-2} \quad \text{for } n \geqq 2.$$

(a) Show that f satisfies the equation

$$f(z) = 1 + zf(z) + z^2 f(z).$$

(b) Solve the equation in part (a) for f, and show that

$$f(z) = \frac{1}{1 - z - z^2}.$$

16. Establish identities (33) and (34).

17. Use Maclaurin series and the Cauchy product in (35) to verify the identity $\sin 2z = 2\cos z \sin z$ up to terms involving z^5.

18. *Division of Power Series*: Let f and g be analytic at z_0 and have the power series representations

$$f(z) = \sum_{n=0}^{\infty} a_n (z - z_0)^n \qquad \text{and} \qquad g(z) = \sum_{n=0}^{\infty} b_n (z - z_0)^n.$$

If $g(z_0) = b_0 \neq 0$, then show that the quotient f/g has the power series representation

$$\frac{f(z)}{g(z)} = \sum_{n=0}^{\infty} c_n (z - z_0)^n$$

where the coefficients c_n satisfy the equations

$$a_n = b_0 c_n + b_1 c_{n-1} + \cdots + b_{n-1} c_1 + b_n c_0 \qquad \text{for } n = 0, 1, 2, \ldots.$$

19. Use the result of Exercise 18 to show that the first few terms in the Maclaurin series for $\sec z$ are given by

$$\sec z = \frac{1}{\cos z} = \frac{1}{1 - \dfrac{z^2}{2!} + \dfrac{z^4}{4!} - \dfrac{z^6}{6!} + \cdots} = 1 + \frac{1}{2} z^2 + \frac{5}{24} z^4 + \cdots$$

for $|z| < \dfrac{\pi}{2}$.

Hint: Use $f(z) = 1$ and $g(z) = \cos z$.

20. Use the result of Exercise 18 to show that the first few terms in the Maclaurin series for $\tan z$ are given by

$$\tan z = \frac{\sin z}{\cos z} = \frac{z - \dfrac{z^3}{3!} + \dfrac{z^5}{5!} - \cdots}{1 - \dfrac{z^2}{2!} + \dfrac{z^4}{4!} - \cdots} = z + \frac{1}{3} z^3 + \frac{2}{15} z^5 + \cdots \qquad \text{for } |z| < \frac{\pi}{2}.$$

Hint: Use $f(z) = \sin z$ and $g(z) = \cos z$.

21. Suppose that

$$\sum_{n=0}^{\infty} a_n z^n = \sum_{n=0}^{\infty} b_n z^n$$

holds for $|z| < R$. Prove that $a_n = b_n$ for all n.

6.5 Singularities, Zeros, and Poles

The point z_0 is called a *singular point*, or singularity, of the complex function f if f is not analytic at z_0; but every neighborhood of z_0 contains at least one point at which f is analytic. For example, the function $f(z) = 1/(1 - z)$ is analytic for all $z \neq 1$, but f is not analytic at $z_0 = 1$. Hence the point $z_0 = 1$ is a singular point of f. Another example is $g(z) = \text{Log } z$, the principal branch of the logarithm. We saw in Section 4.4 that g is analytic for all z except at the origin and at the points on the negative x axis. Hence the origin and each point on the negative x-axis is a singularity of g.

The point z_0 is called an *isolated singularity* of the complex function f, if f is not analytic at z_0, but is analytic in the punctured disk $0 < |z - z_0| < R$. We can look at the special case of Laurent's Theorem with $r = 0$ and obtain the following classification of isolated singularities.

Definition 6.4 *Let f have an isolated singularity at z_0 and have the Laurent series representation*

(1) $$f(z) = \sum_{n=-\infty}^{\infty} a_n(z - z_0)^n \qquad \textit{valid for } 0 < |z - z_0| < R.$$

Then we distinguish the following types of singularities at z_0.

(i) *If $a_n = 0$ for $n = -1, -2, -3, \ldots$, then f has a* removable singularity *at z_0.*

(ii) *If $a_{-k} \neq 0$ and $a_n = 0$ for $n = -k - 1, -k - 2, -k - 3, \ldots$, then f has a* pole of order k *at z_0.*

(iii) *If $a_n \neq 0$ for an infinitely many negative integers n, then f has an* essential singularity *at z_0.*

Let us investigate the three cases that arise. If f has a removable singularity at z_0, then it has a Laurent series representation:

(2) $$f(z) = \sum_{n=0}^{\infty} a_n(z - z_0)^n \qquad \text{valid for } 0 < |z - z_0| < R.$$

Using the result of Theorem 6.13, we see that the power series in (2) defines an analytic function in the disk $|z - z_0| < R$. If we use this series to define $f(z_0) = a_0$, then the function f becomes analytic at z_0, and the singularity is "removed." For example, the function $f(z) = z^{-1} \sin z$ has an isolated singularity at 0. The Laurent series for f is given by

$$f(z) = \frac{\sin z}{z} = \frac{1}{z}\left(z - \frac{z^3}{3!} + \frac{z^5}{5!} - \frac{z^7}{7!} + \cdots\right)$$

$$= 1 - \frac{z^2}{3!} + \frac{z^4}{5!} - \frac{z^6}{7!} + \cdots \qquad \text{valid for } |z| > 0.$$

If we define $f(0) = 1$, then f is analytic at $z_0 = 0$, and the singularity is "removed." Another example is $g(z) = z^{-2}(\cos z - 1)$, which has a singularity at $z_0 = 0$. The Laurent series for g is given by

$$g(z) = \frac{1}{z^2}\left(-\frac{z^2}{2!} + \frac{z^4}{4!} - \frac{z^6}{6!} + \cdots\right)$$

$$= -\frac{1}{2} + \frac{z^2}{4!} - \frac{z^4}{6!} + \cdots \qquad \text{valid for } |z| > 0.$$

If we define $g(0) = -1/2$, then g is analytic for all z.

Suppose that f has a pole of order k at z_0. Then the Laurent series for f is given by

$$(3) \qquad f(z) = \sum_{n=-k}^{\infty} a_n(z - z_0)^n \qquad \text{valid for } 0 < |z - z_0| < R$$

where $a_{-k} \neq 0$. For example,

$$f(z) = z^{-3} \sin z = \frac{1}{z^2} - \frac{1}{3!} + \frac{z^2}{5!} - \frac{z^4}{7!} + \cdots$$

has a pole of order 2 at $z_0 = 0$. If f has a pole of order 1 at z_0, then we say that f has a *simple pole* at z_0. An example is

$$f(z) = z^{-1} \exp z = \frac{1}{z} + 1 + \frac{z}{2!} + \frac{z^2}{3!} + \cdots,$$

which has a simple pole at $z_0 = 0$.

If an infinite number of negative powers of $(z - z_0)$ occur in the Laurent series (1), then f cannot have a pole at z_0. In this case, z_0 is an essential singularity. For example

$$f(z) = z^2 \sin \frac{1}{z} = z - \frac{1}{3!z} - \frac{1}{5!z^3} + \frac{1}{7!z^5} + \cdots$$

has an essential singularity at the origin.

Definition 6.5 *Consider a function f that is analytic at z_0 and has the power series representation*

$$(4) \qquad f(z) = \sum_{n=0}^{\infty} a_n(z - z_0)^n \qquad \textit{valid for } |z - z_0| < R$$

where $a_0 = f(z_0)$ and $a_n = f^{(n)}(z_0)/n!$. If $a_0 = 0$, then $f(z_0) = 0$ so that z_0 is a zero of f. If, in addition,

$$a_0 = a_1 = \cdots = a_{k-1} = 0 \qquad \textit{and} \qquad a_k \neq 0,$$

then we say that f has a zero of order k *at z_0. A zero of order one is sometimes called a simple zero.*

For example,

$$f(z) = z \sin(z^2) = z^3 - \frac{z^7}{3!} + \frac{z^{11}}{5!} - \frac{z^{15}}{7!} + \cdots$$

has a zero of order 3 at $z_0 = 0$.

Consider the function $f(z) = -1 - z + \exp z$. The first two derivatives are $f'(z) = -1 + \exp z$ and $f''(z) = \exp z$. Calculation reveals that $f(0) = 0$, $f'(0) = 0$, and $f''(0) = 1$. Applying Definition 6.5, we see that $f(z)$ has a zero of order $k = 2$ at $z_0 = 0$.

Theorem 6.17 *Let f be analytic at the point z_0. Then the function f has a zero of order k at z_0 if and only if f can be expressed in the form*

$$f(z) = (z - z_0)^k g(z) \tag{5}$$

where g is analytic at z_0 and $g(z_0) \neq 0$.

Proof Suppose that f has a zero of order k at z_0. Then using equation (4), we write

$$f(z) = (z - z_0)^k \sum_{n=0}^{\infty} a_{k+n}(z - z_0)^n \qquad \text{valid for } |z - z_0| < R \tag{6}$$

where $a_k \neq 0$. The function g is defined to be the series on the right side of equation (6). That is,

$$g(z) = \sum_{n=0}^{\infty} a_{k+n}(z - z_0)^n \qquad \text{valid for } |z - z_0| < R.$$

Theorem 6.14 implies that g is analytic at z_0, and $g(z_0) = a_k \neq 0$.

Conversely, suppose that f has the form given by equation (5). Since g is analytic at z_0, it has the power series representation

$$g(z) = \sum_{n=0}^{\infty} b_n(z - z_0)^n \tag{7}$$

where $g(z_0) = b_0 \neq 0$. We can multiply both sides of equation (7) by $(z - z_0)^k$ and obtain the following power series representation for f:

$$f(z) = \sum_{n=0}^{\infty} b_n(z - z_0)^{n+k} = \sum_{n=k}^{\infty} b_{n-k}(z - z_0)^n.$$

Therefore f has a zero of order k at z_0, and the proof is complete.

An immediate consequence of Theorem 6.17 is the following result. The proof is left as an exercise.

Corollary 6.3 *If f and g are analytic at z_0 and have zeros of order m and n, respectively, at z_0, then their product $h(z) = f(z)g(z)$ has a zero of order $k = m + n$ at z_0.*

For example, let $f(z) = z^3 \sin z$. Then $f(z)$ can be factored as the product of z^3 and $\sin z$, which have zeros of order $m = 3$ and $n = 1$ at $z_0 = 0$. Hence $z_0 = 0$ is a zero of order $k = 4$ of $f(z)$.

Our next result gives us a useful way to characterize a pole.

Theorem 6.18 *Let f be analytic in the punctured disk $0 < |z - z_0| < R$. Then f has a pole of order k at z_0 if and only if f can be expressed in the form*

$$f(z) = \frac{h(z)}{(z - z_0)^k} \tag{8}$$

where h is analytic at z_0 and $h(z_0) \neq 0$.

Proof Suppose that f has a pole of order k at z_0. The the Laurent series for f given in equation (3) can be written as

$$f(z) = \frac{1}{(z - z_0)^k} \sum_{n=0}^{\infty} a_{n-k}(z - z_0)^n \qquad \text{for } 0 < |z - z_0| < R \tag{9}$$

where $a_{-k} \neq 0$. The function h is defined to be the series on the right side of equation (9); that is,

$$h(z) = \sum_{n=0}^{\infty} a_{n-k}(z - z_0)^n \qquad \text{for } 0 < |z - z_0| < R. \tag{10}$$

The power series for h given in (10) is valid in the disk $|z - z_0| < R$. Hence h is analytic at z_0, and $h(z_0) = a_{-k} \neq 0$.

Conversely, suppose that f has the form given by equation (8). Since h is analytic at z_0, it has a power series representation

$$h(z) = \sum_{n=0}^{\infty} b_n(z - z_0)^n \qquad \text{where } b_0 \neq 0. \tag{11}$$

We can divide both sides of equation (11) by $(z - z_0)^k$ and obtain the following Laurent series representation for f:

$$f(z) = \sum_{n=0}^{\infty} b_n(z - z_0)^{n-k} = \sum_{n=-k}^{\infty} b_{n+k}(z - z_0)^n.$$

Therefore f has a pole of order k at z_0, and the theorem is proven.

The following results will be useful in determining the order of a zero or a pole. The proofs follow easily from Theorems 6.17 and 6.18 and are left as exercises.

Theorem 6.19

(i) *If f is analytic and has a zero of order k at z_0, then $g(z) = 1/f(z)$ has a pole of order k at z_0.*

(ii) *Conversely, if f has a pole of order k at z_0, then $h(z) = 1/f(z)$ has a removable singularity at z_0. If we define $h(z_0) = 0$, then h has a zero of order k at z_0.*

Corollary 6.4 *If f and g have poles of order m and n, respectively, at z_0, then their product $h(z) = f(z)g(z)$ has a pole of order $k = m + n$ at z_0.*

Corollary 6.5 *Let f and g be analytic at z_0 and have zeros of order m and n, respectively, at z_0. Then their quotient $h(z) = f(z)/g(z)$ has the following behavior.*

(i) *If $m > n$, then h has a removable singularity at z_0. Defining $h(z_0) = 0$, we find that h has a zero of order $m - n$ at z_0.*

(ii) *If $m < n$, then h has a pole of order $n - m$ at z_0.*

(iii) *If $m = n$, then h has a removable singularity at z_0 and can be defined so that h is analytic at z_0 by $h(z_0) = \lim_{z \to z_0} h(z)$.*

EXAMPLE 6.25 Locate the zeros and poles of $h(z) = (\tan z)/z$ and determine their order.

Solution In Section 4.3 we saw that the zeros of $f(z) = \sin z$ occur at the points $z = n\pi$ where n is an integer. Since $f'(n\pi) = \cos(n\pi) \neq 0$, the zeros of f are simple. In a similar fashion it can be shown that the function $g(z) = z \cos z$ has simple zeros at the points $z = 0$ and $z = (n + \frac{1}{2})\pi$ where n is an integer. From the information given we find that $h(z) = f(z)/g(z)$ has the following behavior.

(i) h has simple zeros at $z = n\pi$ where $n = \pm 1, \pm 2, \ldots$.

(ii) h has simple poles at $z = (n + \frac{1}{2})\pi$ where n is an integer.

(iii) h is analytic at 0 and $h(0) \neq 0$.

EXAMPLE 6.26 Locate the poles of $g(z) = 1/(5z^4 + 26z^2 + 5)$, and specify their order.

Solution The roots of the quadratic equation $5Z^2 + 26Z + 5 = 0$ occur at the points $Z_1 = -5$ and $Z_2 = -1/5$. If we use the substitution $Z = z^2$, then we see that the function $f(z) = 5z^4 + 26z^2 + 5$ has simple zeros at the points $\pm i\sqrt{5}$ and $\pm i/\sqrt{5}$. Theorem 6.19 implies that g has simple poles at $\pm i\sqrt{5}$ and $\pm i/\sqrt{5}$.

EXAMPLE 6.27 Locate the poles of $g(z) = (\pi \cot(\pi z))/z^2$, and specify their order.

Solution The functions $z^2 \sin(\pi z)$ and $f(z) = (z^2 \sin(\pi z))/(\pi \cos(\pi z))$ have a zero of order 3 at $z = 0$ and simple zeros at the points $z = \pm 1, \pm 2, \ldots$. Using Theorem 6.19, we see that g has a pole of order 3 at $z = 0$ and simple poles at the points $z = \pm 1, \pm 2, \ldots$.

EXERCISES FOR SECTION 6.5

Locate the zeros of the functions in Exercises 1 and 2, and determine their order.

1. **(a)** $(1+z^2)^4$ **(b)** $\sin^2 z$ **(c)** z^2+2z+2
(d) $\sin(z^2)$ **(e)** z^4+10z^2+9 **(f)** $1+\exp z$

2. **(a)** z^6+1 **(b)** $z^3\exp(z-1)$ **(c)** z^6+2z^3+1
(d) $z^3\cos^2 z$ **(e)** z^8+z^4 **(f)** $z^2\cosh z$

Locate the poles of the functions in Exercises 3 and 4, and determine their order.

3. **(a)** $(z^2+1)^{-3}(z-1)^{-4}$ **(b)** $z^{-1}(z^2-2z+2)^{-2}$
(c) $(z^6+1)^{-1}$ **(d)** $(z^4+z^3-2z^2)^{-1}$
(e) $(3z^4+10z^2+3)^{-1}$ **(f)** $(i+2/z)^{-1}(3+4/z)^{-1}$

4. **(a)** $z\cot z$ **(b)** $z^{-5}\sin z$ **(c)** $(z^2\sin z)^{-1}$
(d) $z^{-1}\csc z$ **(e)** $(1-\exp z)^{-1}$ **(f)** $z^{-5}\sinh z$

Locate the singularities of the functions in Exercises 5 and 6, and determine their type.

5. **(a)** $z^{-2}(z-\sin z)$ **(b)** $\sin(1/z)$
(c) $z\exp(1/z)$ **(d)** $\tan z$

6. **(a)** $(z^2+z)^{-1}\sin z$ **(b)** $z/\sin z$
(c) $(\exp z-1)/z$ **(d)** $(\cos z-\cos 2z)/z^4$

For Exercises 7–10, use L'Hôpital's Rule to find the limit.

7. $\displaystyle\lim_{z\to 1+i}\frac{z-1-i}{z^4+4}$ **8.** $\displaystyle\lim_{z\to i}\frac{z^2-2iz-1}{z^4+2z^2+1}$

9. $\displaystyle\lim_{z\to i}\frac{1+z^6}{1+z^2}$ **10.** $\displaystyle\lim_{z\to 0}\frac{\sin z+\sinh z-2z}{z^3}$

11. Let f be analytic and have a zero of order k at z_0. Show that $f'(z)$ has a zero of order $k-1$ at z_0.

12. Let f and g be analytic at z_0 and have zeros of order m and n, respectively, at z_0. What can you say about the zero of $f+g$ at z_0?

13. Let f and g have poles of order m and n, respectively, at z_0. Show that $f+g$ has either a pole or removable singularity at z_0.

14. Let f be analytic and have a zero of order k at z_0. Show that $f'(z)/f(z)$ has a simple pole at z_0.

15. Let f have a pole of order k at z_0. Show that $f'(z)$ has a pole of order $k+1$ at z_0.

16. Establish Corollary 6.3. **17.** Establish Corollary 6.4.

18. Establish Theorem 6.19. **19.** Find the singularities of $\cot z-1/z$.

20. Find the singularities of $\mathrm{Log}(z^2)$. **21.** Find the singularities of $\dfrac{1}{\sin(1/z)}$.

22. If $f(z)$ has a removable singularity at z_0, then prove that $1/f(z)$ has either a removable singularity or a pole at z_0.

6.6 Some Theoretical Results

In this section we mention a few results concerning the nature of analytic functions. A point z_0 of a set S is called *isolated* if there exists a neighborhood of z_0 that does not contain any other points of S. The first result shows that the zeros of an analytic function are isolated.

Theorem 6.20 *Let f be analytic and have a zero at z_0. If f is not identically zero, then there exists a punctured disk $0 < |z - z_0| < R$ in which f has no other zeros.*

Proof If all the Taylor coefficients $a_n = f^{(n)}(z_0)/n!$ of f at z_0 are zero, then f is identically zero on some disk centered at z_0. A proof that is similar to the proof of the Maximum Modulus Principle given in Section 6.1 will show that f is identically zero.

If f is not identically zero, then let k be the smallest integer such that $f^{(k)}(z_0) \neq 0$. Then f has a zero of order k at z_0, and f can be written in the form

$$f(z) = (z - z_0)^k g(z)$$

where g is analytic and $g(z_0) \neq 0$. Since g is a continuous function, there exists a disk $|z - z_0| < R$ throughout which g is nonzero. Therefore $f(z) \neq 0$ in the punctured disk $0 < |z - z_0| < R$, and the theorem is proven.

An immediate consequence is the following result.

Theorem 6.21 (Identity Theorem) *Let f be analytic in the domain D, and let z_0 be a point in D. If there exists a sequence of points z_1, $z_2, \ldots$ in D such that $z_n \to z_0$ and $f(z_n) = 0$ for $n = 1, 2, \ldots$, then $f(z) = 0$ for all points in D.*

Corollary 6.6 *Let f and g be analytic in the domain D, and let z_0 be a point in D. If there exists a sequence $z_1, z_2, \ldots$ in D, such that $z_n \to z_0$ and $f(z_n) = g(z_n)$ for $n = 1, 2, \ldots$, then $f(z) = g(z)$ for all points in D.*

The behavior of a complex function near a removable singularity is shown in the following result.

Theorem 6.22 (Riemann) *Let f be analytic in the punctured disk $D: 0 < |z - z_0| < R$. If f is bounded in D, then f either is analytic at z_0 or has a removable singularity at z_0.*

Proof Consider the function g defined as follows:

$$(1) \qquad g(z) = \begin{cases} (z - z_0)^2 f(z) & \text{for } z \neq z_0, \\ 0 & \text{when } z = z_0. \end{cases}$$

Then g is analytic in D. Since f is bounded, we can compute $g'(z_0)$ to be 0 as follows:

$$(2) \qquad g'(z_0) = \lim_{z \to z_0} \frac{g(z) - g(z_0)}{z - z_0} = \lim_{z \to z_0} (z - z_0) f(z) = 0.$$

Hence g is analytic at z_0 and has the Taylor series representation

$$(3) \qquad g(z) = \sum_{n=2}^{\infty} \frac{g^{(n)}(z_0)}{n!} (z - z_0)^n.$$

We can divide both sides of equation (3) by $(z - z_0)^2$ and use (1) to obtain the following power series representation for f:

$$f(z) = \sum_{n=2}^{\infty} \frac{g^{(n)}(z_0)}{n!} (z - z_0)^{n-2}.$$

Therefore f is analytic at z_0, and the proof is complete.

Corollary 6.7 *Let f be analytic in the punctured disk $0 < |z - z_0| < R$. If $\lim_{z \to z_0} f(z)$ exists, then f can be defined at z_0 to be analytic at z_0.*

The behavior of a complex function near a pole is demonstrated by the following result.

Theorem 6.23 *If f has a pole of order k at z_0, then*

$$\lim_{z \to z_0} |f(z)| = \infty.$$

Proof Using Theorem 6.18, we write

$$f(z) = \frac{h(z)}{(z - z_0)^k}$$

where h is analytic and $h(z_0) \neq 0$. Since

$$\lim_{z \to z_0} |h(z)| = |h(z_0)| \qquad \text{and} \qquad \lim_{z \to z_0} |(z - z_0)^k| = 0,$$

we conclude that

$$\lim_{z \to z_0} f(z) = \lim_{z \to z_0} |h(z)| \lim_{z \to z_0} \frac{1}{|(z - z_0)^k|} = \infty,$$

and Theorem 6.23 is proven.

EXERCISES FOR SECTION 6.6

1. Consider the function $f(z) = z \sin(1/z)$.

(a) Show that there is a sequence $\{z_n\}$ of points converging to $z = 0$ such that $f(z_n) = 0$ for $n = 1, 2, 3, \ldots$.

(b) Does this contradict the Identity Theorem?

2. Determine whether there exists a function $f(z)$ that is analytic at $z = 0$ such that

$$f\left(\frac{1}{2n}\right) = 0 \quad \text{and} \quad f\left(\frac{1}{2n-1}\right) = 1 \quad \text{for } n = 1, 2, \ldots .$$

3. Determine whether there exists a function $f(z)$ that is analytic at $z = 0$ such that

$$f\left(\frac{1}{n}\right) = f\left(\frac{-1}{n}\right) = \frac{1}{n^2} \quad \text{for } n = 1, 2, \ldots .$$

4. Determine whether there exists a function $f(z)$ that is analytic at $z = 0$ such that

$$f\left(\frac{1}{n}\right) = f\left(\frac{-1}{n}\right) = \frac{1}{n^3} \quad \text{for } n = 1, 2, \ldots .$$

7

Residue Theory

7.1 The Residue Theorem

The Cauchy Integral Formulae in Section 5.5 are useful in evaluating contour integrals over a simple closed contour C where the integrand has the form $f(z)/(z - z_0)^k$ and f is an analytic function. In this case the singularity of the integrand is at worst a pole of order k at z_0. In this section we extend this result to integrals that have a finite number of isolated singularities and lie inside the contour C. The new method can be used in cases where the integrand has an essential singularity at z_0 and is an important extension of the previous method.

Let f have a nonremovable isolated singularity at the point z_0. Then f has the Laurent series representation

$$(1) \qquad f(z) = \sum_{n=-\infty}^{\infty} a_n (z - z_0)^n \qquad \text{valid for } 0 < |z - z_0| < R.$$

The coefficient a_{-1} of $1/(z - z_0)$ is called the *residue* of f at z_0, and we use the notation

$$(2) \qquad \text{Res}[f, z_0] = a_{-1}.$$

EXAMPLE 7.1 If $f(z) = \exp(2/z)$, then the Laurent series (1) has the form

$$f(z) = \exp\left(\frac{2}{z}\right) = 1 + \frac{2}{z} + \frac{2^2}{2!z^2} + \frac{2^3}{3!z^3} + \cdots,$$

and we see that $\text{Res}[f, 0] = 2$.

EXAMPLE 7.2 If $g(z) = \dfrac{3}{2z + z^2 - z^3}$, then $\text{Res}[g, 0] = \frac{3}{2}$.

Solution Using Example 6.15, we find that g has three Laurent series representations involving powers of z. The Laurent series of the form (1) is given by

$$g(z) = \sum_{n=0}^{\infty} \left[(-1)^n + \frac{1}{2^{n+1}} \right] z^{n-1} \qquad \text{valid for } 0 < |z| < 1.$$

Computing the first few coefficients, we obtain

$$g(z) = \frac{3}{2}\frac{1}{z} - \frac{3}{4} + \frac{9}{8}z - \frac{15}{16}z^2 + \cdots.$$

Therefore $\text{Res}[g, 0] = \frac{3}{2}$.

Let us recall that the Laurent series coefficients in (1) are given by

$$(3) \qquad a_n = \frac{1}{2\pi i} \int_C \frac{f(\xi)\, d\xi}{(\xi - z_0)^{n+1}} \qquad \text{for } n = 0, \pm 1, \pm 2, \ldots$$

where C is any positively oriented circle $|z - z_0| = r$ with $0 < r < R$. This gives us an important fact concerning $\text{Res}[f, z_0]$. If we set $n = -1$ in equation (3), then we obtain

$$(4) \qquad \int_C f(\xi)\, d\xi = 2\pi i a_{-1} = 2\pi i \,\text{Res}[f, z_0]$$

where z_0 is the only singularity of f that lies inside C. If we are able to find the Laurent series expansion for f given in equation (1), then equation (4) give us an important tool for evaluating contour integrals.

EXAMPLE 7.3 $\int_C \exp(2/z)\, dz = 4\pi i$ where C is the unit circle $|z| = 1$ taken with positive orientation.

Solution We have seen that the residue of $f(z) = \exp(2/z)$ at $z_0 = 0$ is $\text{Res}[f, 0] = 2$. Using equation (4), we find that

$$\int_C \exp\left(\frac{2}{z}\right) dz = 2\pi i \,\text{Res}[f, 0] = 4\pi i.$$

Theorem 7.1 (Cauchy's Residue Theorem) *Let D be a simply connected domain, and let C be a simple closed positively oriented contour that lies in D. If f is analytic inside C and on C, except at the points $z_1, z_2, \ldots, z_n$ that lie inside C, then*

$$(5) \qquad \int_C f(z)\, dz = 2\pi i \sum_{k=1}^{n} \text{Res}[f, z_k].$$

The situation is illustrated in Figure 7.1.

Proof Since there are a finite number of singular points inside C, there exists an $r > 0$ such that the circles $C_k: |z - z_k| = r$ (for $k = 1, 2, \ldots, n$)

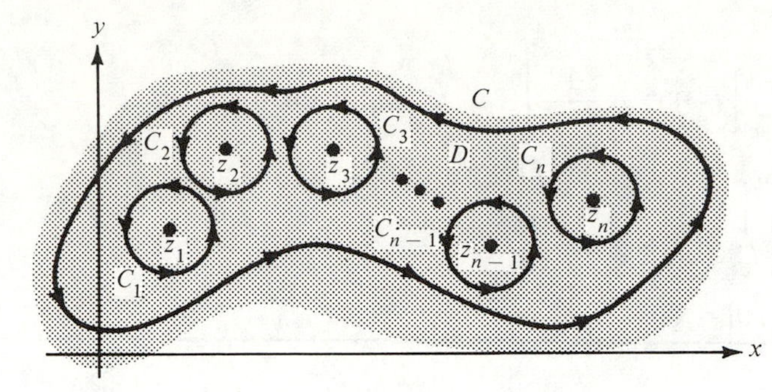

FIGURE 7.1 The domain D and contour C and the singular points z_1, $z_2, \ldots, z_n$ in the statement of Cauchy's Residue Theorem.

are mutually disjoint and all lie inside C. Using Theorem 5.4, the extended Cauchy-Goursat theorem, it follows that

$$(6) \qquad \int_C f(z)\,dz = \sum_{k=1}^{n} \int_{C_k} f(z)\,dz.$$

Since f is analytic in a punctured disk with center z_k that contains the circle C_k, equation (4) can be used to obtain

$$(7) \qquad \int_{C_k} f(z)\,dz = 2\pi i \operatorname{Res}[f, z_k] \qquad \text{for } k = 1, 2, \ldots, n.$$

Using (7) in (6) results in

$$\int_C f(z)\,dz = 2\pi i \sum_{k=1}^{n} \operatorname{Res}[f, z_k],$$

and the theorem is proven.

7.2 Calculation of Residues

The calculation of a Laurent series expansion is tedious in most circumstances. Since the residue at z_0 involves only the coefficient a_{-1} in the Laurent expansion, we seek a method to calculate the residue from special information about the nature of the singularity at z_0.

If f has a removable singularity at z_0, then our work in Section 6.5 showed that $a_{-n} = 0$ for $n = 1, 2, \ldots$. Therefore if z_0 is a removable singularity, then $\operatorname{Res}[f, z_0] = 0$.

Theorem 7.2 (Residues at Poles)

(i) *If f has a simple pole at z_0, then*

$$\text{Res}[f, z_0] = \lim_{z \to z_0} (z - z_0) f(z). \tag{1}$$

(ii) *If f has a pole of order 2 at z_0, then*

$$\text{Res}[f, z_0] = \lim_{z \to z_0} \frac{d}{dz} (z - z_0)^2 f(z). \tag{2}$$

(iii) *If f has a pole of order k at z_0, then*

$$\text{Res}[f, z_0] = \frac{1}{(k-1)!} \lim_{z \to z_0} \frac{d^{k-1}}{dz^{k-1}} (z - z_0)^k f(z). \tag{3}$$

Proof If f has a simple pole at z_0, then we write

$$f(z) = \frac{a_{-1}}{z - z_0} + a_0 + a_1(z - z_0) + a_2(z - z_0)^2 + \cdots. \tag{4}$$

If we multiply both sides of (4) by $(z - z_0)$ and take the limit as $z \to z_0$, then we obtain

$$\lim_{z \to z_0} (z - z_0) f(z) = \lim_{z \to z_0} [a_{-1} + a_0(z - z_0) + a_1(z - z_0)^2 + \cdots]$$
$$= a_{-1} = \text{Res}[f, z_0],$$

and (1) is established.

Since equation (2) is a special case of equation (3), let us suppose that f has a pole of order k at z_0. Then f can be written as

$$f(z) = \frac{a_{-k}}{(z - z_0)^k} + \frac{a_{-k+1}}{(z - z_0)^{k-1}} + \cdots + \frac{a_{-1}}{z - z_0} + a_0 + a_1(z - z_0) + \cdots. \tag{5}$$

If we multiply both sides of equation (5) by $(z - z_0)^k$, then the result is

$$(z - z_0)^k f(z) = a_{-k} + \cdots + a_{-1}(z - z_0)^{k-1} + a_0(z - z_0)^k + \cdots. \tag{6}$$

We can differentiate both sides of equation (6) $k - 1$ times to obtain

$$\frac{d^{k-1}}{dz^{k-1}} (z - z_0)^k f(z) = (k-1)! a_{-1} + k! a_0 (z - z_0) + \frac{(k+1)!}{2} a_1 (z - z_0)^2 + \cdots. \tag{7}$$

If we let $z \to z_0$ in (7), then

$$\lim_{z \to z_0} \frac{d^{k-1}}{dz^{k-1}} (z - z_0)^k f(z) = (k-1)! a_{-1} = (k-1)! \text{Res}[f, z_0],$$

and equation (3) is established.

EXAMPLE 7.4 Find the residue of $f(z) = \dfrac{\pi \cot(\pi z)}{z^2}$ at $z_0 = 0$.

Solution We can express $f(z) = (\pi \cos(\pi z))/(z^2 \sin(\pi z))$. Since $z^2 \sin(\pi z)$ has a zero of order 3 at $z_0 = 0$, we see that f has a pole of order 3 at $z_0 = 0$. Therefore using equation (3), we find that

$$\begin{aligned}
\operatorname{Res}[f, 0] &= \frac{1}{2!} \lim_{z \to 0} \frac{d^2}{dz^2} \pi z \cot(\pi z) \\
&= \frac{1}{2} \lim_{z \to 0} \frac{d}{dz} (\pi \cot(\pi z) - \pi^2 z \csc^2(\pi z)) \\
&= \pi^2 \lim_{z \to 0} (\pi z \cot(\pi z) - 1) \csc^2(\pi z) \\
&= \pi^2 \lim_{z \to 0} \frac{\pi z \cos(\pi z) - \sin(\pi z)}{\sin^3(\pi z)}.
\end{aligned}$$

This last limit involves an indeterminate form and can be evaluated by using L'Hôpital's Rule:

$$\begin{aligned}
\operatorname{Res}[f, 0] &= \pi^2 \lim_{z \to 0} \frac{-\pi^2 z \sin(\pi z)}{3\pi \sin^2(\pi z) \cos(\pi z)} \\
&= \frac{-\pi^2}{3} \lim_{z \to 0} \frac{\pi z}{\sin(\pi z)} \lim_{z \to 0} \frac{1}{\cos(\pi z)} = \frac{-\pi^2}{3}.
\end{aligned}$$

EXAMPLE 7.5 Find $\int_C (dz/(z^4 + z^3 - 2z^2))$ where C is the circle $|z| = 3$ taken with the positive orientation.

Solution The integrand can be written as $f(z) = 1/(z^2(z+2)(z-1))$. The singularities of f that lie inside C are simple poles at the points 1 and -2 and a pole of order 2 at the origin. We compute the residues as follows:

$$\operatorname{Res}[f, 0] = \lim_{z \to 0} \frac{d}{dz} z^2 f(z) = \lim_{z \to 0} \frac{-2z - 1}{(z^2 + z - 2)^2} = \frac{-1}{4},$$

$$\operatorname{Res}[f, 1] = \lim_{z \to 1} (z - 1) f(z) = \lim_{z \to 1} \frac{1}{z^2(z+2)} = \frac{1}{3},$$

$$\operatorname{Res}[f, -2] = \lim_{z \to -2} (z + 2) f(z) = \lim_{z \to -2} \frac{1}{z^2(z-1)} = \frac{-1}{12}.$$

The value of the integral is now found by using the Residue Theorem.

$$\int_C \frac{dz}{z^4 + z^3 - 2z^2} = 2\pi i \left[\frac{-1}{4} + \frac{1}{3} - \frac{1}{12} \right] = 0.$$

The value 0 for the integral is not an obvious answer, and all of the above calculations are required to find it.

EXAMPLE 7.6 Find $\int_C (z^4+4)^{-1}\,dz$ where C is the circle $|z-1|=2$ taken with the positive orientation.

Solution The singularities of the integrand $f(z)=1/(z^4+4)$ that lie inside C are simple poles that occur at the points $1 \pm i$. (The points $-1 \pm i$ lie outside C.) It is tedious to factor the denominator, so we use a different approach. If z_0 is any one of the singularities of f, then L'Hôpital's Rule can be used to compute $\operatorname{Res}[f, z_0]$ as follows:

$$\operatorname{Res}[f, z_0] = \lim_{z \to z_0} \frac{z - z_0}{z^4 + 4} = \lim_{z \to z_0} \frac{1}{4z^3} = \frac{1}{4z_0^3}.$$

Since $z_0^4 = -4$, this can be further simplified to yield $\operatorname{Res}[f, z_0] = (-1/16)z_0$. Hence $\operatorname{Res}[f, 1+i] = (-1-i)/16$, and $\operatorname{Res}[f, 1-i] = (-1+i)/16$. The Residue Theorem can now be used to obtain

$$\int_C \frac{dz}{z^4+4} = 2\pi i\left(\frac{-1-i}{16} + \frac{-1+i}{16}\right) = \frac{-\pi i}{4}.$$

The theory of residues can be used to expand the quotient of two polynomials into its *partial fraction* representation.

Lemma 7.1 *Let $P(z)$ be a polynomial of degree at most 2. If a, b, and c are distinct complex numbers, then*

$$(8) \qquad f(z) = \frac{P(z)}{(z-a)(z-b)(z-c)} = \frac{A}{z-a} + \frac{B}{z-b} + \frac{C}{z-c}$$

where

$$A = \operatorname{Res}[f, a] = \frac{P(a)}{(a-b)(a-c)},$$

$$B = \operatorname{Res}[f, b] = \frac{P(b)}{(b-a)(b-c)},$$

$$C = \operatorname{Res}[f, c] = \frac{P(c)}{(c-a)(c-b)}.$$

Proof It will suffice to prove that $A = \operatorname{Res}[f, a]$. We can expand f in its Laurent series about the point $z=a$ by expanding the three terms on the right side of (8) in their Laurent series about $z=a$ and adding them. The term $A/(z-a)$ is itself a one-term Laurent series. The term $B/(z-b)$ is analytic at $z=a$, and its Laurent series is actually a Taylor series,

$$(9) \qquad \frac{B}{z-b} = \frac{-B}{b-a}\,\frac{1}{1-\dfrac{z-a}{b-a}} = -\sum_{n=0}^{\infty} \frac{B}{(b-a)^{n+1}}(z-a)^n.$$

The expansion for the term $C/(z-c)$ is given by

$$(10)\qquad \frac{C}{z-c} = -\sum_{n=0}^{\infty} \frac{C}{(c-a)^{n+1}}(z-a)^n.$$

We can substitute equations (9) and (10) into (8) to obtain

$$f(z) = \frac{A}{z-a} - \sum_{n=0}^{\infty}\left[\frac{B}{(b-a)^{n+1}} + \frac{C}{(c-a)^{n+1}}\right](z-a)^n.$$

Therefore $A = \text{Res}[f, a]$, and calculation reveals that

$$\text{Res}[f, a] = \lim_{z\to a} \frac{P(z)}{(z-b)(z-c)} = \frac{P(a)}{(a-b)(a-c)}.$$

EXAMPLE 7.7 Express $f(z) = \dfrac{3z+2}{z(z-1)(z-2)}$ in partial fractions.

Solution Computing the residues, we obtain

$$\text{Res}[f, 0] = 1, \qquad \text{Res}[f, 1] = -5, \qquad \text{Res}[f, 2] = 4.$$

Therefore

$$\frac{3z+2}{z(z-1)(z-2)} = \frac{1}{z} - \frac{5}{z-1} + \frac{4}{z-2}.$$

If a repeated root occurs, then the process is similar.

Lemma 7.2 *If* $P(z)$ *has degree at most* 2, *then*

$$(11)\qquad f(z) = \frac{P(z)}{(z-a)^2(z-b)} = \frac{A}{(z-a)^2} + \frac{B}{z-a} + \frac{C}{z-b}$$

where $A = \text{Res}[(z-a)f(z), a]$, $B = \text{Res}[f, a]$, *and* $C = \text{Res}[f, b]$.

EXAMPLE 7.8 Express $f(z) = \dfrac{z^2+3z+2}{z^2(z-1)}$ in partial fractions.

Solution Calculating the residues we find that

$$\text{Res}[zf(z), 0] = \lim_{z\to 0} \frac{z^2+3z+2}{z-1} = -2,$$

$$\text{Res}[f, 0] = \lim_{z\to 0} \frac{d}{dz}\frac{z^2+3z+2}{z-1}$$

$$= \lim_{z\to 0} \frac{(2z+3)(z-1)-(z^2+3z+2)}{(z-1)^2} = -5,$$

$$\text{Res}[f, 1] = \lim_{z\to 1} \frac{z^2+3z+2}{z^2} = 6.$$

Hence $A = -2$, $B = -5$, and $C = 6$, and we can use (11) to obtain

$$\frac{z^2 + 3z + 2}{z^2(z-1)} = \frac{-2}{z^2} - \frac{5}{z} + \frac{6}{z-1}.$$

EXERCISES FOR SECTION 7.2

Find $\text{Res}[f, 0]$ for the functions in Exercises 1–4.

1. **(a)** $z^{-1} \exp z$ **(b)** $z^{-3} \cosh 4z$ **(c)** $\csc z$ **(d)** $(z^2 + 4z + 5)/(z^2 + z)$
2. **(a)** $\cot z$ **(b)** $z^{-3} \cos z$ **(c)** $z^{-1} \sin z$ **(d)** $(z^2 + 4z + 5)/z^3$
3. **(a)** $\exp(1 + 1/z)$ **(b)** $z^4 \sin(1/z)$ **(c)** $z^{-1} \csc z$
4. **(a)** $z^{-2} \csc z$ **(b)** $(\exp 4z - 1)/\sin^2 z$ **(c)** $z^{-1} \csc^2 z$

For Exercises 5–15, assume that the contour C has positive orientation.

5. Find $\int_C \frac{dz}{z^4 + 4}$ where C is the circle $|z + 1 - i| = 1$.
6. Find $\int_C \frac{dz}{z(z^2 - 2z + 2)}$ where C is the circle $|z - i| = 2$.
7. Find $\int_C \frac{\exp z \, dz}{z^3 + z}$ where C is the circle $|z| = 2$.
8. Find $\int_C \frac{\sin z \, dz}{4z^2 - \pi^2}$ where C is the circle $|z| = 2$.
9. Find $\int_C \frac{\sin z \, dz}{z^2 + 1}$ where C is the circle $|z| = 2$.
10. Find $\int_C (z-1)^{-2}(z^2 + 4)^{-1} \, dz$ along the following contours:
    (a) the circle $|z| = 4$ **(b)** the circle $|z - 1| = 1$
11. Find $\int_C (z^6 + 1)^{-1} \, dz$ along the following contours:
    (a) the circle $|z - i| = \frac{1}{2}$ **(b)** the circle $|z - (1 + i)/2| = 1$
    Hint: If z_0 is a singularity of $f(z) = 1/(z^6 + 1)$, then show that $\text{Res}[f, z_0] = (-1/6)z_0$.
12. Find $\int_C (3z^4 + 10z^2 + 3)^{-1} \, dz$ along the following contours:
    (a) the circle $|z - i\sqrt{3}| = 1$ **(b)** the circle $|z - i/\sqrt{3}| = 1$
13. Find $\int_C (z^4 - z^3 - 2z^2)^{-1} \, dz$ along the following contours:
    (a) the circle $|z| = \frac{1}{2}$ **(b)** the circle $|z| = \frac{3}{2}$
14. Find $\int_C \frac{dz}{z^2 \sin z}$ where C is the circle $|z| = 1$.
15. Find $\int_C \frac{dz}{z \sin^2 z}$ where C is the circle $|z| = 1$.
16. Let f and g have an isolated singularity at z_0. Show that $\text{Res}[f + g, z_0] = \text{Res}[f, z_0] + \text{Res}[g, z_0]$.

17. Let f and g be analytic at z_0. If $f(z_0) \neq 0$ and g has a simple zero at z_0, then show that

$$\text{Res}\left[\frac{f}{g}, z_0\right] = \frac{f(z_0)}{g'(z_0)}.$$

18. Use residues to find the partial fraction representations of the following functions.

(a) $\dfrac{1}{z^2 + 3z + 2}$ **(b)** $\dfrac{3z - 3}{z^2 - z - 2}$ **(c)** $\dfrac{z^2 - 7z + 4}{z^2(z + 4)}$

(d) $\dfrac{10z}{(z^2 + 4)(z^2 + 9)}$ **(e)** $\dfrac{2z^2 - 3z - 1}{(z - 1)^3}$ **(f)** $\dfrac{z^3 + 3z^2 - z + 1}{z(z + 1)^2(z^2 + 1)}$

19. Let f be analytic in a simply connected domain D, and let C be a simply closed positively oriented contour in D. If z_0 is the only zero of f in D and z_0 lies interior to C, then show that

$$\frac{1}{2\pi i} \int_C \frac{f'(z)}{f(z)}\, dz = k$$

where k is the order of the zero at z_0.

20. Let f be analytic at the points $z = 0, \pm 1, \pm 2, \ldots$. If $g(z) = \pi f(z) \cot(\pi z)$, then show that

$$\text{Res}[g, n] = f(n) \qquad \text{for } n = 0, \pm 1, \pm 2, \ldots.$$

7.3 Trigonometric Integrals

The evaluation of certain definite integrals can be accomplished with the aid of the Residue Theorem. If the definite integral can be interpreted as the parametric form of a contour integral of an analytic function along a simple closed contour, then the Residue Theorem can be used to evaluate the equivalent complex integral.

The method in this section can be used to evaluate integrals of the form

$$(1) \qquad \int_0^{2\pi} F(\cos\theta, \sin\theta)\, d\theta$$

where $F(u, v)$ is a function of the two real variables u and v. Let us consider the contour C that consists of the unit circle $|z| = 1$, taken with the parameterization

$$(2) \qquad C\colon z = \cos\theta + i\sin\theta, \quad dz = (-\sin\theta + i\cos\theta)\, d\theta \qquad \text{for } 0 \leqq \theta \leqq 2\pi.$$

Using $1/z = \cos\theta - i\sin\theta$ and (2), we can obtain

$$(3) \qquad \cos\theta = \frac{1}{2}\left(z + \frac{1}{z}\right), \qquad \sin\theta = \frac{1}{2i}\left(z - \frac{1}{z}\right), \qquad \text{and} \qquad d\theta = \frac{dz}{iz}.$$

If we use the substitutions (3) in expression (1), then the definite integral is transformed into a contour integral

$$(4) \qquad \int_0^{2\pi} F(\cos\theta, \sin\theta)\, d\theta = \int_C f(z)\, dz$$

where the new integrand is

$$(5) \qquad f(z) = \frac{F\left(\frac{1}{2}\left(z + \frac{1}{z}\right), \frac{1}{2i}\left(z - \frac{1}{z}\right)\right)}{iz}.$$

Suppose that f is analytic for $|z| \leqq 1$, except at the points z_1, $z_2, \ldots, z_n$ that lie interior to C. Then the Residue Theorem can be used to conclude that

$$(6) \qquad \int_0^{2\pi} F(\cos\theta, \sin\theta)\, d\theta = 2\pi i \sum_{k=1}^{n} \operatorname{Res}[f, z_k].$$

The situation is illustrated in Figure 7.2.

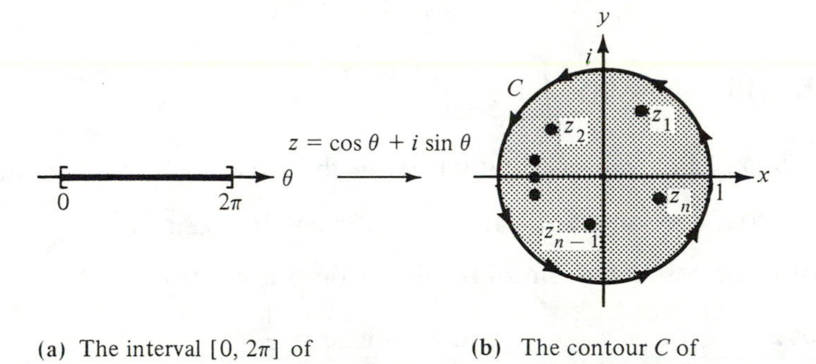

(a) The interval $[0, 2\pi]$ of integration for $F(\cos\theta, \sin\theta)$.

(b) The contour C of integration for $f(z)$.

FIGURE 7.2 The change of variables from a definite integral on $[0, 2\pi]$ to a contour integral around C.

EXAMPLE 7.9 Show that $\displaystyle\int_0^{2\pi} \frac{d\theta}{1 + 3\cos^2\theta} = \pi.$

Solution The complex integrand f of equation (5) is given by

$$f(z) = \frac{1}{iz[1 + \frac{3}{4}(z + z^{-1})^2]} = \frac{-i4z}{3z^4 + 10z^2 + 3}.$$

The singularities of f are poles that are located at the points where $3(z^2)^2 + 10(z^2) + 3 = 0$. The quadratic formula can be used to see that the singular points of f satisfy the relation $z^2 = (-10 \pm \sqrt{100 - 36})/6 = (-5 \pm 4)/3$. Hence the only singularities of f that lie inside the circle C: $|z| = 1$ are simple poles located at the two points $z_1 = i/\sqrt{3}$ and $z_2 = -i/\sqrt{3}$. Theorem 7.2 with the aid of L'Hôpital's Rule can be used to calculate the residues of f at z_k (for $k = 1, 2$) as follows:

$$\begin{aligned}\text{Res}[f, z_k] &= \lim_{z \to z_k} \frac{-i4z(z - z_k)}{3z^4 + 10z^2 + 3} = \frac{0}{0} \\ &= \lim_{z \to z_k} \frac{-i4(2z - z_k)}{12z^3 + 20z} \\ &= \frac{-i4z_k}{12z_k^3 + 20z_k} = \frac{-i}{3z_k^2 + 5}.\end{aligned}$$

Since $z_k = \pm i/\sqrt{3}$ and $z_k^2 = -1/3$, we see that the residues are given by $\text{Res}[f, z_k] = -i/(3(-1/3) + 5) = -i/4$. Equation (6) can now be used to compute the value of the integral

$$\int_0^{2\pi} \frac{d\theta}{1 + 3\cos^2\theta} = 2\pi i\left[\frac{-i}{4} + \frac{-i}{4}\right] = \pi.$$

EXAMPLE 7.10 Show that $\displaystyle\int_0^{2\pi} \frac{\cos 2\theta \, d\theta}{5 - 4\cos\theta} = \frac{\pi}{6}$.

Solution For values of z that lie on the circle $C: |z| = 1$, we have

$$z^2 = \cos 2\theta + i \sin 2\theta \qquad \text{and} \qquad z^{-2} = \cos 2\theta - i \sin 2\theta.$$

We can solve for $\cos 2\theta$ and $\sin 2\theta$ to obtain the substitutions

(7) $$\cos 2\theta = \frac{1}{2}(z^2 + z^{-2}) \qquad \text{and} \qquad \sin 2\theta = \frac{1}{2i}(z^2 - z^{-2}).$$

Using the substitutions in (3) and (7), we find that the complex integrand f in (5) can be written as

$$f(z) = \frac{\frac{1}{2}(z^2 + z^{-2})}{iz[5 - 2(z + z^{-1})]} = \frac{i(z^4 + 1)}{2z^2(z - 2)(2z - 1)}.$$

The singularities of f that lie inside C are poles that are located at the points $z_1 = 0$ and $z_2 = \frac{1}{2}$. Using Theorem 7.2 to calculate the residues results in

$$\begin{aligned}\text{Res}[f, 0] &= \lim_{z \to 0} \frac{d}{dz} z^2 f(z) = \lim_{z \to 0} \frac{d}{dz} \frac{i(z^4 + 1)}{2(2z^2 - 5z + 2)} \\ &= \lim_{z \to 0} i \frac{4z^3(2z^2 - 5z + 2) - (4z - 5)(z^4 + 1)}{2(2z^2 - 5z + 2)^2} = \frac{5i}{8}\end{aligned}$$

and

$$\text{Res}[f, \tfrac{1}{2}] = \lim_{z \to 1/2} (z - \tfrac{1}{2}) f(z) = \lim_{z \to 1/2} \frac{i(z^4 + 1)}{4z^2(z - 2)} = \frac{-17i}{24}.$$

Therefore using equation (6), we conclude that

$$\int_0^{2\pi} \frac{\cos 2\theta \, d\theta}{5 - 4\cos\theta} = 2\pi i \left[\frac{5i}{8} - \frac{17i}{24} \right] = \frac{\pi}{6}.$$

EXERCISES FOR SECTION 7.3

Use residues to find the following:

1. $\displaystyle\int_0^{2\pi} \frac{d\theta}{3\cos\theta + 5} = \frac{\pi}{2}$

2. $\displaystyle\int_0^{2\pi} \frac{d\theta}{4\sin\theta + 5}$

3. $\displaystyle\int_0^{2\pi} \frac{d\theta}{15\sin^2\theta + 1} = \frac{\pi}{2}$

4. $\displaystyle\int_0^{2\pi} \frac{d\theta}{5\cos^2\theta + 4}$

5. $\displaystyle\int_0^{2\pi} \frac{\sin^2\theta \, d\theta}{5 + 4\cos\theta} = \frac{\pi}{4}$

6. $\displaystyle\int_0^{2\pi} \frac{\sin^2\theta \, d\theta}{5 - 3\cos\theta}$

7. $\displaystyle\int_0^{2\pi} \frac{d\theta}{(5 + 3\cos\theta)^2} = \frac{5\pi}{32}$

8. $\displaystyle\int_0^{2\pi} \frac{d\theta}{(5 + 4\cos\theta)^2}$

9. $\displaystyle\int_0^{2\pi} \frac{\cos(2\theta)\, d\theta}{5 + 3\cos\theta} = \frac{\pi}{18}$

10 $\displaystyle\int_0^{2\pi} \frac{\cos(2\theta)\, d\theta}{13 - 12\cos\theta}$

11. $\displaystyle\int_0^{2\pi} \frac{d\theta}{(1 + 3\cos^2\theta)^2} = \frac{5\pi}{8}$

12. $\displaystyle\int_0^{2\pi} \frac{d\theta}{(1 + 8\cos^2\theta)^2}$

13. $\displaystyle\int_0^{2\pi} \frac{\cos^2(3\theta)\, d\theta}{5 - 4\cos(2\theta)} = \frac{3\pi}{8}$

14. $\displaystyle\int_0^{2\pi} \frac{\cos^2(3\theta)\, d\theta}{5 - 3\cos(2\theta)}$

15. $\displaystyle\int_0^{2\pi} \frac{d\theta}{a\cos\theta + b\sin\theta + d} = \frac{2\pi}{\sqrt{d^2 - a^2 - b^2}}$ where a, b, d are real and $a^2 + b^2 < d^2$

16. $\displaystyle\int_0^{2\pi} \frac{d\theta}{a\cos^2\theta + b\sin^2\theta + d} = \frac{2\pi}{\sqrt{(a + d)(b + d)}}$ where a, b, d are real and $a > d$ and $b > d$

7.4 Improper Integrals of Rational Functions

An important application of the theory of residues is the evaluation of certain types of improper integrals. Let $f(x)$ be a continuous function of the real variable x on the interval $0 \leqq x < \infty$. Recall from calculus that the improper integral of f over $[0, \infty)$ is defined by

$$(1) \qquad \int_0^{\infty} f(x)\, dx = \lim_{b \to \infty} \int_0^{b} f(x)\, dx$$

provided that the limit exists. If f is defined for all real x, then the integral of f

over $(-\infty, \infty)$ is defined by

$$(2) \qquad \int_{-\infty}^{\infty} f(x)\,dx = \lim_{a\to-\infty} \int_a^0 f(x)\,dx + \lim_{b\to\infty} \int_0^b f(x)\,dx$$

provided that both limits exist. If the integral in (2) exists, then its value can be obtained by taking a single limit as follows:

$$(3) \qquad \int_{-\infty}^{\infty} f(x)\,dx = \lim_{R\to\infty} \int_{-R}^{R} f(x)\,dx.$$

However, for some functions the limit on the right side of equation (3) exists when definition (2) does not exist.

EXAMPLE 7.11 $\lim_{R\to\infty} \int_{-R}^{R} x\,dx = \lim_{R\to\infty} [R^2/2 - (-R)^2/2] = 0$, but the improper integral of $f(x) = x$ over $(-\infty, \infty)$ does not exist. Therefore equation (3) can be used to extend the notion of the value of an improper integral and motivates us to make the following definition.

Let $f(x)$ be a continuous real valued function for all x. The *Cauchy principal value* of the integral (2) is defined by

$$(4) \qquad \text{P.V.} \int_{-\infty}^{\infty} f(x)\,dx = \lim_{R\to\infty} \int_{-R}^{R} f(x)\,dx$$

provided that the limit exists. Therefore Example 7.11 shows that

$$\text{P.V.} \int_{-\infty}^{\infty} x\,dx = 0.$$

EXAMPLE 7.12

$$\int_{-\infty}^{\infty} \frac{dx}{x^2+1} = \lim_{R\to\infty} \int_{-R}^{R} \frac{dx}{x^2+1}$$
$$= \lim_{R\to\infty} [\arctan R - \arctan(-R)]$$
$$= \frac{\pi}{2} - \frac{-\pi}{2} = \pi.$$

If $f(x) = P(x)/Q(x)$ where P and Q are polynomials, then f is called a *rational function*. Techniques in calculus were developed to integrate rational functions. We now show how the Residue Theorem can be used to obtain the Cauchy principal value of the integral of f over $(-\infty, \infty)$.

Theorem 7.3 *Let $f(z) = P(z)/Q(z)$ where P and Q are polynomials of degree m and n, respectively. If $Q(x) \neq 0$ for all real x and $n \geqq m+2$, then*

$$(5) \qquad \text{P.V.} \int_{-\infty}^{\infty} \frac{P(x)}{Q(x)}\,dx = 2\pi i \sum_{j=1}^{k} \text{Res}\left[\frac{P}{Q}, z_j\right]$$

where $z_1, z_2, \ldots, z_{k-1}$, and z_k are the poles of P/Q that lie in the upper half-plane. The situation is illustrated in Figure 7.3.

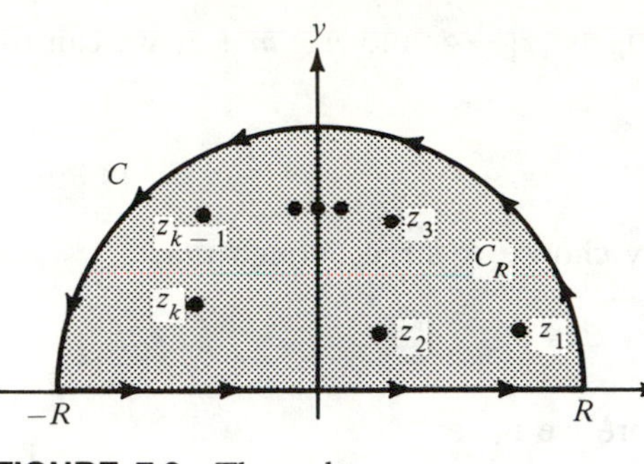

FIGURE 7.3 The poles $z_1, z_2, \ldots, z_{k-1}, z_k$ of P/Q that lie in the upper half-plane.

Proof Since there are a finite number of poles of P/Q that lie in the upper half-plane, a real number R can be found such that the poles all lie inside the contour C, which consists of the segment $-R \leqq x \leqq R$ of the x axis together with the upper semicircle C_R of radius R shown in Figure 7.3. Property (17) in Section 5.2 can be used to write

$$(6) \qquad \int_{-R}^{R} \frac{P(x)}{Q(x)}\, dx = \int_C \frac{P(z)}{Q(z)}\, dz - \int_{C_R} \frac{P(z)}{Q(z)}\, dz.$$

The Residue Theorem can be used to express (6) in the form

$$(7) \qquad \int_{-R}^{R} \frac{P(x)}{Q(x)}\, dx = 2\pi i \sum_{j=1}^{k} \operatorname{Res}\left[\frac{P}{Q}, z_j\right] - \int_{C_R} \frac{P(z)}{Q(z)}\, dz.$$

The result will be established if we can show that the integral of $P(z)/Q(z)$ along C_R on the right side of equation (7) goes to zero as $R \to \infty$. Since $n \geqq m + 2$, the degree of the polynomial $Q(z)$ is greater than the degree of $zP(z)$. Suppose that

$$P(z) = a_m z^m + a_{m-1} z^{m-1} + \cdots + a_1 z + a_0$$

and

$$Q(z) = b_n z^n + b_{n-1} z^{n-1} + \cdots + b_1 z + b_0.$$

Then

$$P(z) = z^m P_1(z) \qquad \text{where}$$

$$P_1(z) = a_m + a_{m-1} z^{-1} + \cdots + a_1 z^{-m+1} + a_0 z^{-m}$$

and

$$Q(z) = z^n Q_1(z) \qquad \text{where}$$

$$Q_1(z) = b_n + b_{n-1} z^{-1} + \cdots + b_1 z^{-n+1} + b_0 z^{-n}.$$

Therefore we have

$$(8) \qquad \frac{zP(z)}{Q(z)} = \frac{z^{m+1} P_1(z)}{z^n Q_1(z)}.$$

Since $P_1(z) \to a_m$ and $Q_1(z) \to b_n$ as $|z| \to \infty$ and $n \geqq m + 2$, we can use (8) to see that

$$\left|\frac{zP(z)}{Q(z)}\right| \to 0 \qquad \text{as } |z| \to \infty.$$

Therefore for any $\varepsilon > 0$ we may choose R large enough that

$$\left|\frac{zP(z)}{Q(z)}\right| < \frac{\varepsilon}{\pi}$$

whenever z lies on C_R. Therefore we have

$$(9) \qquad \left|\frac{P(z)}{Q(z)}\right| < \frac{\varepsilon}{\pi|z|} = \frac{\varepsilon}{\pi R} \qquad \text{whenever } z \text{ lies on } C_R.$$

Using inequality (22) of Section 5.2 and the result of (9), we obtain the estimate

$$(10) \qquad \left|\int_{C_R} \frac{P(z)}{Q(z)}\, dz\right| \leqq \int_{C_R} \frac{\varepsilon}{\pi R} |dz| = \frac{\varepsilon}{\pi R} \pi R = \varepsilon.$$

Since $\varepsilon > 0$ was arbitrary, (10) shows that

$$(11) \qquad \lim_{R \to \infty} \int_{C_R} \frac{P(z)}{Q(z)}\, dz = 0.$$

We can use equation (11) in equation (7) and use definition (4) to conclude that

$$\text{P.V.} \int_{-\infty}^{\infty} \frac{P(x)}{Q(x)}\, dx = \lim_{R \to \infty} \int_{-R}^{R} \frac{P(x)}{Q(x)}\, dx = 2\pi i \sum_{j=1}^{k} \text{Res}\left[\frac{P}{Q}, z_j\right],$$

and the theorem is proven.

EXAMPLE 7.13 Show that $\displaystyle\int_{-\infty}^{\infty} \frac{dx}{(x^2+1)(x^2+4)} = \frac{\pi}{6}$.

Solution The integrand can be written in the form

$$f(z) = \frac{1}{(z+i)(z-i)(z+2i)(z-2i)}.$$

We see that f has simple poles at the points $z_1 = i$ and $z_2 = 2i$ in the upper half-plane. Computing the residues, we obtain

$$\text{Res}[f, i] = \frac{-i}{6} \qquad \text{and} \qquad \text{Res}[f, 2i] = \frac{i}{12}.$$

Using Theorem 7.3, we conclude that

$$\int_{-\infty}^{\infty} \frac{dx}{(x^2+1)(x^2+4)} = 2\pi i\left[\frac{-i}{6} + \frac{i}{12}\right] = \frac{\pi}{6}.$$

EXAMPLE 7.14 Show that $\displaystyle\int_{-\infty}^{\infty} \frac{dx}{(x^2+4)^3} = \frac{3\pi}{256}$.

Solution The integrand $f(z) = 1/(z^2+4)^3$ has a pole of order 3 at the point $z_1 = 2i$. Computing the residue, we find that

$$\begin{aligned}\text{Res}[f, 2i] &= \frac{1}{2} \lim_{z\to 2i} \frac{d^2}{dz^2} \frac{1}{(z+2i)^3} \\ &= \frac{1}{2} \lim_{z\to 2i} \frac{d}{dz} \frac{-3}{(z+2i)^4} \\ &= \frac{1}{2} \lim_{z\to 2i} \frac{12}{(z+2i)^5} = \frac{-3i}{512}.\end{aligned}$$

Therefore $\displaystyle\int_{-\infty}^{\infty} \frac{dx}{(x^2+4)^3} = 2\pi i \frac{-3i}{512} = \frac{3\pi}{256}$.

EXERCISES FOR SECTION 7.4

Use residues to find the values of the integrals in Exercises 1–15.

1. $\displaystyle\int_{-\infty}^{\infty} \frac{x^2\,dx}{(x^2+16)^2} = \frac{\pi}{8}$

2. $\displaystyle\int_{-\infty}^{\infty} \frac{dx}{x^2+16}$

3. $\displaystyle\int_{-\infty}^{\infty} \frac{x\,dx}{(x^2+9)^2} = 0$

4. $\displaystyle\int_{-\infty}^{\infty} \frac{x+3}{(x^2+9)^2}\,dx$

5. $\displaystyle\int_{-\infty}^{\infty} \frac{2x^2+3}{(x^2+9)^2}\,dx = \frac{7\pi}{18}$

6. $\displaystyle\int_{-\infty}^{\infty} \frac{dx}{x^4+4}$

7. $\displaystyle\int_{-\infty}^{\infty} \frac{x^2\,dx}{x^4+4} = \frac{\pi}{2}$

8. $\displaystyle\int_{-\infty}^{\infty} \frac{x^2\,dx}{(x^2+4)^3}$

9. $\displaystyle\int_{-\infty}^{\infty} \frac{dx}{(x^2+1)^2(x^2+4)} = \frac{\pi}{9}$

10. $\displaystyle\int_{-\infty}^{\infty} \frac{x+2}{(x^2+4)(x^2+9)}\,dx$

11. $\displaystyle\int_{-\infty}^{\infty} \frac{3x^2+2}{(x^2+4)(x^2+9)}\,dx = \frac{2\pi}{3}$

12. $\displaystyle\int_{-\infty}^{\infty} \frac{dx}{x^6+1}$

13. $\displaystyle\int_{-\infty}^{\infty} \frac{x^4\,dx}{x^6+1} = \frac{2\pi}{3}$

14. $\displaystyle\int_{-\infty}^{\infty} \frac{dx}{(x^2+a^2)(x^2+b^2)} = \frac{\pi}{ab(a+b)}$ where $a>0$ and $b>0$

15. $\displaystyle\int_{-\infty}^{\infty} \frac{x^2\,dx}{(x^2+a^2)^3} = \frac{\pi}{8a^3}$ where $a>0$

7.5 Improper Integrals Involving Trigonometric Functions

Let P and Q be polynomials of degree m and n, respectively, where $n \geqq m + 1$. If $Q(x) \neq 0$ for all real x, then

$$\text{P.V.} \int_{-\infty}^{\infty} \frac{P(x)}{Q(x)} \cos x \, dx \qquad \text{and} \qquad \text{P.V.} \int_{-\infty}^{\infty} \frac{P(x)}{Q(x)} \sin x \, dx$$

are convergent improper integrals. Integrals of this type are sometimes encountered in the study of Fourier Transforms and Fourier Integrals. We now show how these improper integrals can be evaluated.

It is of particular importance to observe that we will be using identities

$$(1) \qquad \cos(\alpha x) = \text{Re} \exp(i\alpha x) \qquad \text{and} \qquad \sin(\alpha x) = \text{Im} \exp(i\alpha x)$$

where α is a positive real number. The crucial step in the proof of Theorem 7.4 will not hold if $\cos(\alpha z)$ and $\sin(\alpha z)$ are used instead of $\exp(i\alpha z)$. Lemma 7.3 will give the details.

Theorem 7.4 *Let P and Q be polynomials of degree m and n, respectively, where $n \geqq m + 1$, and $Q(x) \neq 0$ for all real x. If $\alpha > 0$ and*

$$(2) \qquad f(z) = \frac{\exp(i\alpha z) P(z)}{Q(z)},$$

then

$$(3) \qquad \text{P.V.} \int_{-\infty}^{\infty} \frac{P(x)}{Q(x)} \cos(\alpha x) \, dx = -2\pi \sum_{j=1}^{k} \text{Im Res}[f, z_j] \qquad \textit{and}$$

$$(4) \qquad \text{P.V.} \int_{-\infty}^{\infty} \frac{P(x)}{Q(x)} \sin(\alpha x) \, dx = 2\pi \sum_{j=1}^{k} \text{Re Res}[f, z_j]$$

where $z_1, z_2, \ldots, z_{k-1}, z_k$ are the poles of f that lie in the upper half-plane and where Re Res$[f, z_j]$ *and* Im Res$[f, z_j]$ *are the real and imaginary parts of* Res$[f, z_j]$, *respectively.*

The proof of the theorem is similar to the proof of Theorem 7.3. Before we turn to the proof, let us first give some examples.

EXAMPLE 7.15 Show that $\text{P.V.} \int_{-\infty}^{\infty} \frac{x \sin x \, dx}{x^2 + 4} = \frac{\pi}{e^2}$.

Solution The complex function f in equation (2) is $f(z) = z \exp(iz)/(z^2 + 4)$ and has a simple pole at the point $z_1 = 2i$ in the upper half-plane. Calculating the residue results in

$$\text{Res}[f, 2i] = \lim_{z \to 2i} \frac{z \exp(iz)}{z + 2i} = \frac{2ie^{-2}}{4i} = \frac{1}{2e^2}.$$

Using equation (4), we find that

$$\text{P.V.} \int_{-\infty}^{\infty} \frac{x \sin x \, dx}{x^2 + 4} = 2\pi \text{ Re Res}[f, 2i] = \frac{\pi}{e^2}.$$

EXAMPLE 7.16 Show that $\displaystyle\int_{-\infty}^{\infty} \frac{\cos x \, dx}{x^4 + 4} = \frac{\pi(\cos 1 + \sin 1)}{4e}$.

Solution The complex function f in equation (2) is $f(z) = \exp(iz)/(z^4 + 4)$ and has simple poles at the points $z_1 = 1 + i$ and $z_2 = -1 + i$ in the upper half-plane. The residues are found with the aid of L'Hôpital's Rule.

$$\text{Res}[f, 1+i] = \lim_{z \to 1+i} \frac{(z - 1 - i)\exp(iz)}{z^4 + 4} = \frac{0}{0}$$

$$= \lim_{z \to 1+i} \frac{[1 + i(z - 1 - i)]\exp(iz)}{4z^3}$$

$$= \frac{\exp(-1 + i)}{4(1 + i)^3} = \frac{\sin 1 - \cos 1 - i(\cos 1 + \sin 1)}{16e}.$$

Similarly,

$$\text{Res}[f, -1+i] = \frac{\cos 1 - \sin 1 - i(\cos 1 + \sin 1)}{16\text{e}}.$$

Using equation (3), we find that

$$\int_{-\infty}^{\infty} \frac{\cos x \, dx}{x^4 + 4} = -2\pi(\text{Im Res}[f, 1+i] + \text{Im Res}[f, -1+i])$$

$$= \frac{\pi(\cos 1 + \sin 1)}{4e}.$$

We now turn to the proof of Theorem 7.4, a theorem that depends on the following result.

Lemma 7.3 (Jordan's Lemma) *Let P and Q be polynomials of degree m and n, respectively, where $n \geqq m + 1$. If C_R is the upper semicircle $z = \text{R}e^{i\theta}$ for $0 \leqq \theta \leqq \pi$, then*

$$(5) \qquad \lim_{R \to \infty} \int_{C_R} \frac{\exp(iz)P(z)}{Q(z)} \, dz = 0.$$

Proof Since $n \geqq m+1$, it follows that $|P(z)/Q(z)| \to 0$ as $|z| \to \infty$. Therefore for $\varepsilon > 0$ given there exists an $R_\varepsilon > 0$ such that

$$(6) \qquad \left|\frac{P(z)}{Q(z)}\right| < \frac{\varepsilon}{\pi} \qquad \text{whenever } |z| \geqq R_\varepsilon.$$

Using inequality (22) of Section 5.2 together with (6), we obtain the estimate

$$(7) \qquad \left|\int_{C_R} \frac{\exp(iz)P(z)}{Q(z)}\,dz\right| \leqq \int_{C_R} \frac{\varepsilon}{\pi}|e^{iz}||dz| \qquad \text{where } R \geqq R_\varepsilon.$$

The parameterization of C_R leads to the equations

$$(8) \qquad |dz| = R\,d\theta \qquad \text{and} \qquad |e^{iz}| = e^{-y} = e^{-R\sin\theta}.$$

Using the trigonometric identity $\sin(\pi-\theta) = \sin\theta$ and (8), we can express the integral on the right side of (7) as

$$(9) \qquad \int_{C_R} \frac{\varepsilon}{\pi}|e^{iz}||dz| = \frac{\varepsilon}{\pi}\int_0^\pi e^{-R\sin\theta}R\,d\theta = \frac{2\varepsilon}{\pi}\int_0^{\pi/2} e^{-R\sin\theta}R\,d\theta.$$

On the interval $0 \leqq \theta \leqq \pi/2$ we can use the inequality

$$(10) \qquad 0 \leqq \frac{2\theta}{\pi} \leqq \sin\theta.$$

We can combine the results of (7), (9), and (10) to conclude that, for $R \geqq R_\varepsilon$,

$$\left|\int_{C_R} \frac{\exp(iz)P(z)\,dz}{Q(z)}\right| \leqq \frac{2\varepsilon}{\pi}\int_0^{\pi/2} e^{-2R\theta/\pi}R\,d\theta$$
$$= -\varepsilon e^{-2R\theta/\pi}\Big|_0^{\pi/2} < \varepsilon.$$

Since $\varepsilon > 0$ is arbitrary, Lemma 7.3 is proven.

Proof of Theorem 7.4 Let C be the contour that consists of the segment $-R \leqq x \leqq R$ of the real axis together with the semicircle C_R of Lemma 7.3. Property (17) of Section 5.2 can be used to write

$$(11) \qquad \int_{-R}^{R} \frac{\exp(i\alpha x)P(x)\,dx}{Q(x)} = \int_C \frac{\exp(i\alpha z)P(z)\,dz}{Q(z)} - \int_{C_R} \frac{\exp(i\alpha z)P(z)\,dz}{Q(z)}.$$

If R is sufficiently large, then all the poles $z_1, z_2, \ldots, z_k$ of f will lie inside C, and we can use the Residue Theorem to obtain

$$(12) \qquad \int_{-R}^{R} \frac{\exp(i\alpha x)P(x)\,dx}{Q(x)} = 2\pi i \sum_{j=1}^{k} \operatorname{Res}[f, z_j] - \int_{C_R} \frac{\exp(i\alpha z)P(z)\,dz}{Q(z)}.$$

Since α is a positive real number, the change of variables $Z = \alpha z$ shows that Jordan's Lemma holds true for the integrand $\exp(i\alpha z)P(z)/Q(z)$. Hence we can let $R \to \infty$ in equation (12) to obtain

$$(13) \qquad \text{P.V.} \int_{-\infty}^{\infty} \frac{[\cos(\alpha x) + i \sin(\alpha x)]P(x)\, dx}{Q(x)} = 2\pi i \sum_{j=1}^{k} \text{Res}[f, z_j]$$
$$= -2\pi \sum_{j=1}^{k} \text{Im Res}[f, z_j] + 2\pi i \sum_{j=1}^{k} \text{Re Res}[f, z_j].$$

Equating the real and imaginary parts of (13) results in equations (3) and (4), respectively, and Theorem 7.4 is proven.

EXERCISES FOR SECTION 7.5

Use residues to find the integrals in Exercises 1–12.

1. $\displaystyle\int_{-\infty}^{\infty} \frac{\cos x\, dx}{x^2+9} = \frac{\pi}{3e^3}$ and $\displaystyle\int_{-\infty}^{\infty} \frac{\sin x\, dx}{x^2+9} = 0$

2. P.V. $\displaystyle\int_{-\infty}^{\infty} \frac{x \cos x\, dx}{x^2+9}$ and P.V. $\displaystyle\int_{-\infty}^{\infty} \frac{x \sin x\, dx}{x^2+9}$

3. $\displaystyle\int_{-\infty}^{\infty} \frac{x \sin x\, dx}{(x^2+4)^2} = \frac{\pi}{4e^2}$

4. $\displaystyle\int_{-\infty}^{\infty} \frac{\cos x\, dx}{(x^2+4)^2}$

5. $\displaystyle\int_{-\infty}^{\infty} \frac{\cos x\, dx}{(x^2+4)(x^2+9)} = \frac{\pi}{5}\left(\frac{1}{2e^2} - \frac{1}{3e^3}\right)$

6. $\displaystyle\int_{-\infty}^{\infty} \frac{\cos x\, dx}{(x^2+1)(x^2+4)}$

7. $\displaystyle\int_{-\infty}^{\infty} \frac{\cos x\, dx}{x^2-2x+5} = \frac{\pi \cos 1}{2e^2}$

8. $\displaystyle\int_{-\infty}^{\infty} \frac{\cos x\, dx}{x^2-4x+5}$

9. $\displaystyle\int_{-\infty}^{\infty} \frac{x \sin x\, dx}{x^4+4} = \frac{\pi \sin 1}{2e}$

10. P.V. $\displaystyle\int_{-\infty}^{\infty} \frac{x^3 \sin x\, dx}{x^4+4}$

11. $\displaystyle\int_{-\infty}^{\infty} \frac{\cos 2x\, dx}{x^2+2x+2} = \frac{\pi \cos 2}{e^2}$

12. P.V. $\displaystyle\int_{-\infty}^{\infty} \frac{x^3 \sin 2x\, dx}{x^4+4}$

7.6 Indented Contour Integrals

If f is continuous on the interval $b < x \leqq c$, then the improper integral of f over $(b, c]$ is defined by

$$(1) \qquad \int_b^c f(x)\, dx = \lim_{r \to b^+} \int_r^c f(x)\, dx$$

provided that the limit exists. Similarly, if f is continuous on the interval $a \leqq x < b$, then the improper integral of f over $[a, b)$ is defined by

$$(2) \qquad \int_a^b f(x)\,dx = \lim_{R \to b^-} \int_a^R f(x)\,dx$$

provided that the limit exists. For example,

$$\int_0^9 \frac{dx}{2\sqrt{x}} = \lim_{r \to 0^+} \int_r^9 \frac{dx}{2\sqrt{x}} = \lim_{r \to 0^+} [\sqrt{x}|_r^9] = 3 - \lim_{r \to 0^+} \sqrt{r} = 3.$$

Let f be continuous for all values of x in the interval $[a, c]$, except at the value $x = b$ where $a < b < c$. The *Cauchy principal value* of f over $[a, c]$ is defined by

$$(3) \qquad \text{P.V.} \int_a^c f(x)\,dx = \lim_{r \to 0^+} \left[\int_a^{b-r} f(x)\,dx + \int_{b+r}^c f(x)\,dx \right]$$

provided that the limit exists.

EXAMPLE 7.17

$$\text{P.V.} \int_{-1}^8 \frac{dx}{x^{1/3}} = \lim_{r \to 0^+} \left[\int_{-1}^{-r} \frac{dx}{x^{1/3}} + \int_r^8 \frac{dx}{x^{1/3}} \right]$$
$$= \lim_{r \to 0^+} \left[\frac{3}{2} r^{2/3} - \frac{3}{2} + 6 - \frac{3}{2} r^{2/3} \right] = \frac{9}{2}.$$

In this section we extend the results of Sections 7.4 and 7.5 to include the case in which the integrand f has simple poles on the x axis. We now state how residues can be used to find the Cauchy principal value for the integral of f over $(-\infty, \infty)$.

Theorem 7.5 *Let $f(z) = P(z)/Q(z)$ where P and Q are polynomials of degree m and n, respectively, where $n \geqq m + 2$. If Q has simple zeros at the points $t_1, t_2, \ldots, t_l$ on the x axis, then*

$$(4) \qquad \text{P.V.} \int_{-\infty}^{\infty} \frac{P(x)\,dx}{Q(x)} = 2\pi i \sum_{j=1}^{k} \text{Res}[f, z_j] + \pi i \sum_{j=1}^{l} \text{Res}[f, t_j]$$

where $z_1, z_2, \ldots, z_k$ are the poles of f that lie in the upper half-plane.

Theorem 7.6 *Let P and Q be polynomials of degree m and n, respectively, where $n \geqq m + 1$, and let Q have simple zeros at the points $t_1, t_2, \ldots, t_l$ on the x axis. If α is a positive real number and if*

$$(5) \qquad f(z) = \frac{\exp(i\alpha z)P(z)}{Q(z)},$$

then

$$(6) \qquad \text{P.V.} \int_{-\infty}^{\infty} \frac{P(x)}{Q(x)} \cos \alpha x \, dx = -2\pi \sum_{j=1}^{k} \operatorname{Im} \operatorname{Res}[f, z_j] - \pi \sum_{j=1}^{l} \operatorname{Im} \operatorname{Res}[f, t_j]$$

and

$$(7) \qquad \text{P.V.} \int_{-\infty}^{\infty} \frac{P(x)}{Q(x)} \sin \alpha x \, dx = 2\pi \sum_{j=1}^{k} \operatorname{Re} \operatorname{Res}[f, z_j] + \pi \sum_{j=1}^{l} \operatorname{Re} \operatorname{Res}[f, t_j]$$

where $z_1, z_2, \ldots, z_k$ are the poles of f that lie in the upper half-plane.

Before we prove Theorems 7.5 and 7.6, let us make some observations and look at some examples. First, the formulas in equations (4), (6), and (7) give the Cauchy principal value in the integral. This answer is special because of the manner in which the limit in (3) is taken. Second, the formulas are similar to those in Sections 7.4 and 7.5, except that here we add one-half of the value of each residue at the points $t_1, t_2, \ldots, t_l$ on the x axis.

EXAMPLE 7.18 Show that P.V. $\displaystyle\int_{-\infty}^{\infty} \frac{x\,dx}{x^3 - 8} = \frac{\pi\sqrt{3}}{6}$.

Solution The complex integrand

$$f(z) = \frac{z}{z^3 - 8} = \frac{z}{(z-2)(z+1+i\sqrt{3})(z+1-i\sqrt{3})}$$

has simple poles at the points $t_1 = 2$ on the x axis and $z_1 = -1 + i\sqrt{3}$ in the upper half-plane. Now equation (4) gives

$$\text{P.V.} \int_{-\infty}^{\infty} \frac{x\,dx}{x^3 - 8} = 2\pi i \operatorname{Res}[f, z_1] + \pi i \operatorname{Res}[f, t_1]$$

$$= 2\pi i \,\frac{-1 - i\sqrt{3}}{12} + \pi i \,\frac{1}{6} = \frac{\pi\sqrt{3}}{6}.$$

EXAMPLE 7.19 Show that P.V. $\displaystyle\int_{-\infty}^{\infty} \frac{\sin x\,dx}{(x-1)(x^2+4)} = \frac{\pi}{5}\left(\cos 1 - \frac{1}{e^2}\right)$.

Solution The complex integrand $f(z) = \exp(iz)/(z-1)(z^2+4)$ has simple poles at the points $t_1 = 1$ on the x axis and $z_1 = 2i$ in the upper half-plane. Now equation (7) gives

$$\text{P.V.} \int_{-\infty}^{\infty} \frac{\sin x\,dx}{(x-1)(x^2+4)} = 2\pi \operatorname{Re} \operatorname{Res}[f, z_1] + \pi \operatorname{Re} \operatorname{Res}[f, t_1]$$

$$= 2\pi \operatorname{Re} \frac{-2+i}{20e^2} + \pi \operatorname{Re} \frac{\cos 1 + i \sin 1}{5}$$

$$= \frac{\pi}{5}\left(\cos 1 - \frac{1}{e^2}\right).$$

The proofs of Theorems 7.5 and 7.6 depend on the following result.

Lemma 7.4 *Let f have a simple pole at the point t_0 on the x axis. If C: $z = t_0 + re^{i\theta}$ for $0 \leqq \theta \leqq \pi$, then*

$$\lim_{r\to 0} \int_C f(z)\,dz = i\pi \operatorname{Res}[f, t_0]. \tag{8}$$

Proof The Laurent series for f at $z = t_0$ has the form

$$f(z) = \frac{\operatorname{Res}[f, t_0]}{z - t_0} + g(z) \tag{9}$$

where g is analytic at $z = t_0$. Using the parameterization of C and equation (9), we can write

$$\begin{aligned} \int_C f(z)\,dz &= \operatorname{Res}[f, t_0] \int_0^\pi \frac{ire^{i\theta}\,d\theta}{re^{i\theta}} + ir\int_0^\pi g(t_0 + re^{i\theta})e^{i\theta}\,d\theta \\ &= i\pi \operatorname{Res}[f, t_0] + ir\int_0^\pi g(t_0 + re^{i\theta})e^{i\theta}\,d\theta. \end{aligned} \tag{10}$$

Since g is continuous at t_0, there is an $M > 0$ so that $|g(t_0 + re^{i\theta})| \leqq M$. Hence

$$\left|\lim_{r\to 0} ir\int_0^\pi g(t_0 + re^{i\theta})e^{i\theta}\,d\theta\right| \leqq \lim_{r\to 0} r\int_0^\pi M\,d\theta = \lim_{r\to 0} r\pi M = 0. \tag{11}$$

When (11) is used in (10), the resulting limit is given by equation (8), and Lemma 7.4 is proven.

Proof of Theorems 7.5 and 7.6 Since f has only a finite number of poles, we can choose r small enough so that the semicircles

$$C_j\colon z = t_j + re^{i\theta} \qquad \text{for } 0 \leqq \theta \leqq \pi \text{ and } j = 1, 2, \ldots, l$$

are disjoint and the poles $z_1, z_2, \ldots, z_k$ of f in the upper half-plane lie above them as shown in Figure 7.4.

Let R be chosen large enough so that the poles of f in the upper half-plane lie under the semicircle C_R: $z = \mathrm{R}e^{i\theta}$ for $0 \leqq \theta \leqq \pi$ and the poles of f on the x axis lie in the interval $-R \leqq x \leqq R$. Let C be the simple closed positively oriented contour that consists of C_R and $-C_1, -C_2, \ldots, -C_l$ and the segments of the real axis that lie between the semicircles as shown in Figure 7.4. The Residue Theorem can be used to otain

$$\int_C f(z)\,dz = 2\pi i \sum_{j=1}^{k} \operatorname{Res}[f, z_j]. \tag{12}$$

Equation (12) can be written as

$$\int_{I_R} f(x)\,dx = 2\pi i \sum_{j=1}^{k} \operatorname{Res}[f, z_j] + \sum_{j=1}^{l} \int_{C_j} f(z)\,dz - \int_{C_R} f(z)\,dz \tag{13}$$

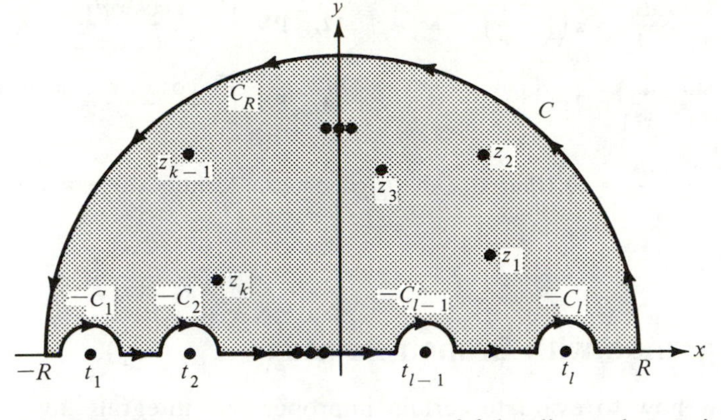

FIGURE 7.4 The poles $t_1, t_2, \ldots, t_l$ of f that lie on the x axis and the poles $z_1, z_2, \ldots, z_k$ that lie above the semicircles $C_1, C_2, \ldots, C_l$.

where I_R is the portion of the interval $-R \leqq x \leqq R$ that lies outside the intervals $(t_j - r, t_j + r)$ for $j = 1, 2, \ldots, l$. The proofs of Theorems 7.3 and 7.4 show that

$$(14) \qquad \lim_{R\to\infty} \int_{C_R} f(z)\, dz = 0.$$

If we let $R \to \infty$ and $r \to 0$ in (13) and use the result of (14) and Lemma 7.4, then we obtain

$$(15) \qquad \text{P.V.} \int_{-\infty}^{\infty} f(x)\, dx = 2\pi i \sum_{j=1}^{k} \text{Res}[f, z_j] + \pi i \sum_{j=1}^{l} \text{Res}[f, t_j].$$

If f is given in Theorem 7.5, then equation (15) becomes equation (4). If f is given in Theorem 7.6, then equating the real and imaginary parts of (15) results in equations (6) and (7), respectively, and the theorems are established.

EXERCISES FOR SECTION 7.6

Use residues to compute the integrals in Exercises 1–15.

1. $\text{P.V.} \int_{-\infty}^{\infty} \frac{dx}{x(x-1)(x-2)} = 0$

2. $\text{P.V.} \int_{-\infty}^{\infty} \frac{dx}{x^3 + x}$

3. $\text{P.V.} \int_{-\infty}^{\infty} \frac{x\, dx}{x^3 + 1} = \frac{\pi}{\sqrt{3}}$

4. $\text{P.V.} \int_{-\infty}^{\infty} \frac{dx}{x^3 + 1}$

5. $\text{P.V.} \int_{-\infty}^{\infty} \frac{x^2\, dx}{x^4 - 1} = \frac{\pi}{2}$

6. $\text{P.V.} \int_{-\infty}^{\infty} \frac{x^4\, dx}{x^6 - 1}$

7. $\text{P.V.} \int_{-\infty}^{\infty} \frac{\sin x\, dx}{x} = \pi$

8. $\text{P.V.} \int_{-\infty}^{\infty} \frac{\cos x\, dx}{x^2 - x}$

9. $\text{P.V.} \int_{-\infty}^{\infty} \frac{\sin x\, dx}{x(\pi^2 - x^2)} = \frac{2}{\pi}$

10. $\text{P.V.} \int_{-\infty}^{\infty} \frac{\cos x\, dx}{\pi^2 - 4x^2}$

11. P.V. $\int_{-\infty}^{\infty} \frac{\sin x\,dx}{x(x^2+1)} = \pi\left(1 - \frac{1}{e}\right)$

12. P.V. $\int_{-\infty}^{\infty} \frac{x\cos x\,dx}{x^2+3x+2}$

13. P.V. $\int_{-\infty}^{\infty} \frac{\sin x\,dx}{x(1-x^2)} = 2\pi$

14. P.V. $\int_{-\infty}^{\infty} \frac{\cos x\,dx}{a^2-x^2} = \frac{\pi \sin a}{a}$

15. P.V. $\int_{-\infty}^{\infty} \frac{\sin^2 x\,dx}{x^2} = \pi$. *Hint*: Use the trigonometric identity $\sin^2 x = \frac{1}{2} - \frac{1}{2}\cos 2x$.

7.7 Integrands with Branch Points

We now show how to evaluate certain improper real integrals involving the integrand $x^\alpha P(x)/Q(x)$. Since the complex function z^α is multivalued, we must first specify the branch that we will be using.

Let α be a real number with $0 < \alpha < 1$. Then in this section we will use the branch of z^α defined as follows:

(1) $$z^\alpha = e^{\alpha(\ln r + i\theta)} = r^\alpha(\cos \alpha\theta + i \sin \alpha\theta) \qquad \text{where } 0 \leqq \theta < 2\pi.$$

Using definition (1), we see that z^α is analytic in the domain $r > 0$, $0 < \theta < 2\pi$.

Theorem 7.7 *Let P and Q be polynomials of degree m and n, respectively, where $n \geqq m + 2$. If $Q(x) \neq 0$ for $x > 0$ and Q has a zero of order at most 1 at the origin and*

(2) $$f(z) = \frac{z^\alpha P(z)}{Q(z)} \qquad \textit{where } 0 < \alpha < 1,$$

then

(3) $$\text{P.V.} \int_0^\infty \frac{x^\alpha P(x)\,dx}{Q(x)} = \frac{2\pi i}{1 - e^{i\alpha 2\pi}} \sum_{j=1}^{k} \text{Res}[f, z_j]$$

where $z_1, z_2, \ldots, z_k$ are the nonzero poles of P/Q.

Proof Let C denote the simple, closed, positively oriented contour that consists of the portions of the circles $|z| = r$ and $|z| = R$ and the horizontal segments joining them as shown in Figure 7.5. A small value of r and a large value of R can be selected so that the nonzero poles $z_1, z_2, \ldots, z_k$ of P/Q lie inside C.

Using the Residue Theorem, we can write

(4) $$\int_C f(z)\,dz = 2\pi i \sum_{j=1}^{k} \text{Res}[f, z_j].$$

If we let $r \to 0$ in (4) and use property (17) of Section 5.2 to express the

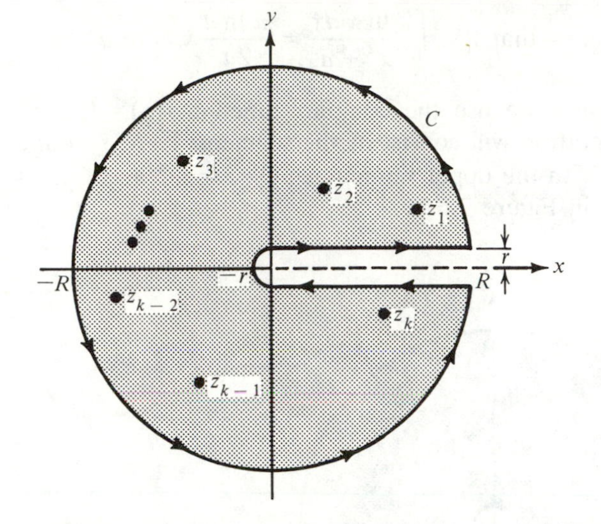

FIGURE 7.5 The contour C that encloses all the nonzero poles $z_1, z_2, \ldots, z_k$ of P/Q.

limiting value of the integral on the lower segment, we find that equation (4) becomes

$$\int_0^R \frac{x^\alpha P(x)\,dx}{Q(x)} - \int_0^R \frac{x^\alpha e^{i\alpha 2\pi} P(x)\,dx}{Q(x)} = 2\pi i \sum_{j=1}^{k} \operatorname{Res}[f, z_j] - \int_{C_R} f(z)\,dz,$$

which can be written as

$$(5) \qquad \int_0^R \frac{x^\alpha P(x)\,dx}{Q(x)} = \frac{2\pi i}{1 - e^{i\alpha 2\pi}} \sum_{j=1}^{k} \operatorname{Res}[f, z_j] - \frac{1}{1 - e^{i\alpha 2\pi}} \int_{C_R} f(z)\,dz.$$

Letting $R \to \infty$ in (5) results in (3), and Theorem 7.7 is established.

EXAMPLE 7.20 Show that P.V. $\displaystyle\int_0^\infty \frac{x^a\,dx}{x(x+1)} = \frac{\pi}{\sin a\pi}$ where $0 < a < 1$.

Solution The complex function $f(z) = (z^a/z(z+1))$ has a nonzero pole at the point $z_1 = -1$. Using equation (3) we find that

$$\int_0^\infty \frac{x^a\,dx}{x(x+1)} = \frac{2\pi i}{1 - e^{ia2\pi}} \operatorname{Res}[f, -1] = \frac{2\pi i}{1 - e^{ia2\pi}} \frac{e^{ia\pi}}{-1}$$

$$= \frac{\pi}{\dfrac{e^{ia\pi} - e^{-ia\pi}}{2i}} = \frac{\pi}{\sin a\pi}.$$

The above ideas can be applied to other multivalued functions.

EXAMPLE 7.21 Show that P.V. $\displaystyle\int_0^\infty \frac{\ln x\,dx}{x^2+a^2} = \frac{\pi \ln a}{2a}$ where $a > 0$.

Solution Here we use the complex function $f(z) = \operatorname{Log} z/(z^2+a^2)$. The path C of integration will consist of the segments $[-R, -r]$ and $[r, R]$ of the x axis together with the upper semicircles C_r: $z = re^{i\theta}$ and C_R: $z = Re^{i\theta}$ for $0 \leqq \theta \leqq \pi$ as shown in Figure 7.6.

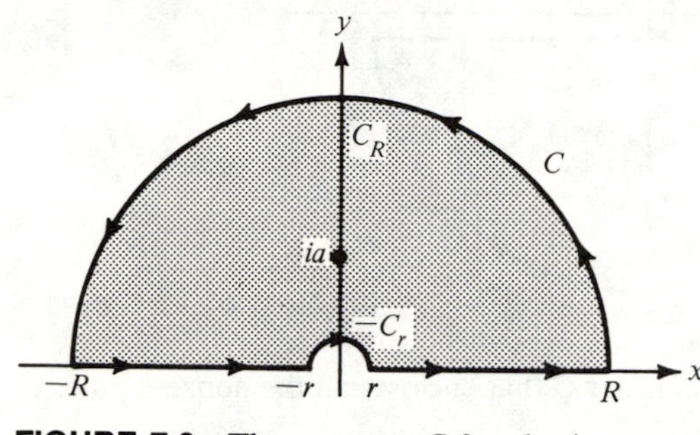

FIGURE 7.6 The contour C for the integrand $f(z) = (\operatorname{Log} z)/(z^2+a^2)$.

The Residue Theorem can be used to write

$$(6)\qquad \int_C f(z)\,dz = 2\pi i \operatorname{Res}[f, ai] = \frac{\pi \ln a}{a} + i\,\frac{\pi^2}{2a}.$$

The inequality

$$\left|\int_0^\pi \frac{\ln R + i\theta}{R^2 e^{i2\theta} + a^2}\, i\,Re^{i\theta}\,d\theta\right| \leqq \frac{R(\ln R + \pi)\pi}{R^2 - a^2}$$

and L'Hôpital's Rule can be used to show that

$$(7)\qquad \lim_{R\to\infty} \int_{C_R} f(z)\,dz = 0.$$

A similar computation will show that

$$(8)\qquad \lim_{r\to 0} \int_{C_r} f(z)\,dz = 0.$$

We can use the results of equations (7) and (8) in equation (6) to obtain

$$(9)\qquad \text{P.V.}\left[\int_{-\infty}^0 \frac{\ln|x| + i\pi}{x^2+a^2}\,dx + \int_0^\infty \frac{\ln x\,dx}{x^2+a^2}\right] = \frac{\pi \ln a}{a} + i\,\frac{\pi^2}{2a}.$$

Equating the real parts in equation (9), we obtain

$$\text{P.V.} \int_0^\infty \frac{2\ln x\,dx}{x^2+a^2} = \frac{\pi \ln a}{a},$$

and the result is established.

EXERCISES FOR SECTION 7.7

Use residues to compute the integrals in Exercises 1–11.

1. P.V. $\displaystyle\int_0^\infty \frac{dx}{x^{2/3}(1+x)} = \frac{2\pi}{\sqrt{3}}$

2. P.V. $\displaystyle\int_0^\infty \frac{dx}{x^{1/2}(1+x)}$

3. P.V. $\displaystyle\int_0^\infty \frac{x^{1/2}\,dx}{(1+x)^2} = \frac{\pi}{2}$

4. P.V. $\displaystyle\int_0^\infty \frac{x^{1/2}\,dx}{1+x^2}$

5. P.V. $\displaystyle\int_0^\infty \frac{\ln(x^2+1)\,dx}{x^2+1} = \pi \ln 2$. Use $f(z) = \dfrac{\text{Log}(z+i)}{z^2+1}$

6. P.V. $\displaystyle\int_0^\infty \frac{\ln x\,dx}{(1+x^2)^2}$

7. P.V. $\displaystyle\int_0^\infty \frac{\ln(1+x)}{x^{1+a}} = \frac{\pi}{a \sin a}$ where $0<a<1$

8. P.V. $\displaystyle\int_0^\infty \frac{\ln x\,dx}{(x+a)^2}$ where $a>0$

9. P.V. $\displaystyle\int_{-\infty}^\infty \frac{\sin x}{x}\,dx = \pi$

Hint: Use the integrand $f(z) = \exp(iz)/z$ and the contour C in Figure 7.6, and let $r \to 0$ and $R \to \infty$.

10. P.V. $\displaystyle\int_{-\infty}^\infty \frac{\sin^2 x}{x^2}\,dx = \pi$

Hint: Use the integrand $f(z) = (1 - \exp(i2z))/z^2$ and the contour C in Figure 7.6, and let $r \to 0$ and $R \to \infty$.

11. The Fresnel integrals

$$\text{P.V.} \int_0^\infty \cos(x^2)\,dx = \text{P.V.} \int_0^\infty \sin(x^2)\,dx = \frac{\sqrt{\pi}}{2\sqrt{2}}$$

are important in the study of optics. Use the integrand $f(z) = \exp(-z^2)$ and the contour C shown in Figure 7.7, and let $R \to \infty$; then establish these integrals. Also use the fact from calculus that P.V. $\int_0^\infty e^{-x^2}\,dx = \sqrt{\pi}/2$.

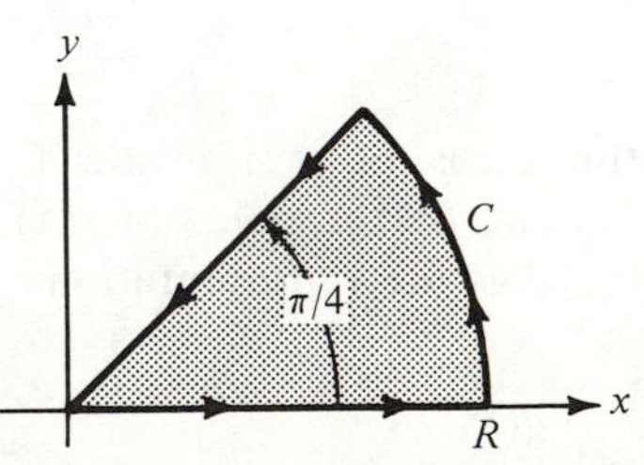

FIGURE 7.7 Accompanies Exercise 11.

7.8 The Argument Principle and Rouché's Theorem

We will now derive two results based on Cauchy's Residue Theorem. They have important practical applications and pertain only to functions all of whose isolated singularities are poles.

Definition 7.1 A function $f(z)$ *is said to be meromorphic in a domain D provided that the only singularities of* $f(z)$ *are isolated poles (and removable singularities). A meromorphic function does not have essential singularities!*

Observe that analytic functions are a special case of meromorphic functions. Rational functions $f(z) = P(z)/Q(z)$ where $P(z)$ and $Q(z)$ are polynomials are meromorphic in the entire complex plane.

Suppose that $f(z)$ is analytic at each point on a simple closed contour C and $f(z)$ is meromorphic in the domain that is the interior of C. An extension of Theorem 6.20 can be made that shows that $f(z)$ has at most finitely many zeros that lie inside C. Since the function $g(z) = 1/f(z)$ is also meromorphic, it can have only finitely many zeros inside C. Therefore $f(z)$ can have at most a finite number of poles that lie inside C.

An application of the Residue Theorem that is useful in determining the number of zeros and poles of a function is called the argument principle.

Theorem 7.8 (Argument Principle) *Let* $f(z)$ *be meromorphic in the simply connected domain D. Let C be a simple closed positively oriented contour in D along which* $f(z) \neq 0$ *and* $f(z) \neq \infty$. *Then*

$$(1) \qquad \frac{1}{2\pi i} \int_C \frac{f'(z)}{f(z)}\, dz = N - P$$

where N is the number of zeros of $f(z)$ *that lie inside C and P is the number of poles that lie inside C.*

Proof Let $a_1, a_2, \ldots, a_N$ be the zeros of $f(z)$ inside C counted according to multiplicity and let $b_1, b_2, \ldots, b_P$ be the poles of $f(z)$ inside C counted according to multiplicity. Then $f(z)$ has the representation

$$(2) \qquad f(z) = \frac{(z - a_1)(z - a_2)\cdots(z - a_N)}{(z - b_1)(z - b_2)\cdots(z - b_P)}\, g(z)$$

where $g(z)$ is analytic and nonzero on C and inside C. An elementary calculation shows that

$$\text{(3)} \qquad \frac{f'(z)}{f(z)} = \frac{1}{(z-a_1)} + \frac{1}{(z-a_2)} + \cdots + \frac{1}{(z-a_N)} - \frac{1}{(z-b_1)} - \frac{1}{(z-b_2)} - \cdots - \frac{1}{(z-b_P)} + \frac{g'(z)}{g(z)}.$$

According to Example 5.12, we have

$$\int_C \frac{dz}{(z-a_j)} = 2\pi i \qquad \text{for } j = 1, 2, \ldots, N$$

and

$$\int_C \frac{dz}{(z-b_k)} = 2\pi i \qquad \text{for } k = 1, 2, \ldots, P.$$

Since $g'(z)/g(z)$ is analytic inside and on C, it follows from the Cauchy-Goursat Theorem that

$$\int_C \frac{g'(z)\,dz}{g(z)} = 0.$$

These facts can be used to integrate both sides of (3) over C. The result is equation (1), and the theorem is proven.

Corollary 7.1 *Suppose that $f(z)$ is analytic in the simply connected domain D. Let C be a simple closed positively oriented contour in D along which $f(z) \neq 0$. Then*

$$\text{(4)} \qquad \frac{1}{2\pi i}\int_C \frac{f'(z)}{f(z)}\,dz = N$$

where N is the number of zeros of $f(z)$ that lie inside C.

Theorem 7.9 (Rouché's Theorem) *Let $f(z)$ and $g(z)$ be analytic functions defined in the simply connected domain D. Let C be a simply closed contour in D. If the strict inequality*

$$\text{(5)} \qquad |f(z) - g(z)| < |f(z)| \qquad \textit{holds for all } z \textit{ on } C,$$

then $f(z)$ and $g(z)$ have the same number of zeros inside C (counting multiplicity).

Proof The condition $|f(z) - g(z)| < |f(z)|$ precludes the possibility of $f(z)$ or $g(z)$ having zeros on the contour C. Therefore division by $f(z)$ is permitted, and we obtain

$$\text{(6)} \qquad \left|\frac{g(z)}{f(z)} - 1\right| < 1 \qquad \text{for all } z \text{ on } C.$$

Let $F(z) = g(z)/f(z)$. Then $F(C)$, the image of the curve C under the mapping $w = F(z)$, is contained in the disk $|w - 1| < 1$ in the w plane. Therefore $F(C)$ is a closed curve that does not wind around $w = 0$. Hence $1/w$ is analytic on the curve $f(C)$, and we obtain

$$(7) \qquad \int_{F(C)} \frac{dw}{w} = 0.$$

Using the change of variable $w = f(z)$ and $dw = F'(z)\,dz$, we see that the integral in (7) can be expressed as

$$(8) \qquad \int_C \frac{F'(z)}{F(z)}\,dz = 0.$$

Since $F'(z) = [g'(z)f(z) - f'(z)g(z)]/[f(z)]^2$, it follows that

$$(9) \qquad \frac{F'(z)}{F(z)} = \frac{g'(z)}{g(z)} - \frac{f'(z)}{f(z)}.$$

Hence equations (8) and (9) can be used to obtain

$$(10) \qquad \frac{1}{2\pi i}\int_C \frac{f'(z)}{f(z)}\,dz = \frac{1}{2\pi i}\int_C \frac{g'(z)}{g(z)}\,dz.$$

Corollary 7.1 and equation (10) imply that the number of zeros of $f(z)$ inside C equals the number of zeros of $g(z)$ inside C, and the theorem is proven.

One can use Rouché's Theorem to gain information about the location of the zeros of an analytic function.

EXAMPLE 7.22 Show that all four zeros of the polynomial

$$g(z) = z^4 - 7z - 1$$

lie in the disk $|z| < 2$.

Solution Let $f(z) = z^4$, then $f(z) - g(z) = 7z + 1$. At points on the circle $|z| = 2$ we have the relation

$$|f(z) - g(z)| = |7z + 1| \leqq |7z| + 1 = 7(2) + 1 = 15 < 16 = |f(z)|.$$

The function $f(z)$ has a zero of order 4 at the origin, and the hypothesis of Rouché's Theorem holds true for the circle $|z| = 2$. Therefore $g(z)$ has four zeros inside $|z| = 2$.

EXAMPLE 7.23 Show that the polynomial $g(z) = z^4 - 7z - 1$ has one zero in the disk $|z| < 1$.

Solution Let $f(z) = -7z - 1$, then $f(z) - g(z) = -z^4$. At points on the circle $|z| = 1$ we have the relation

$$|f(z) - g(z)| = |-z^4| = 1 < 6 = |7 - 1| = ||7z| - |-1||$$
$$\leqq |7z - 1| = |f(z)|.$$

The function $f(z)$ has one zero at $z = -1/7$ in the disk $|z| < 1$, and the hypothesis of Rouché's Theorem holds true on the circle $|z| = 1$. Therefore $g(z)$ has one zero inside $|z| = 1$.

Certain feedback control systems in engineering must be stable. A test for stability involves the function $G(z) = 1 + F(z)$ where $F(z)$ is a rational function. If $G(z)$ does not have any zeros for $\text{Re}\, z \geqq 0$, then the system is stable. The number of zeros of $G(z)$ can be determined by writing $F(z) = P(z)/Q(z)$ where $P(z)$ and $Q(z)$ are polynomials with no common zero. Then $G(z) = [Q(z) + P(z)]/Q(z)$. We can check for zeros of $Q(z) + P(z)$ using Theorem 7.8. A value R is selected so that $G(z) \neq 0$ for $|z| > R$. Contour integration is then performed along the contour consisting of the right half of the circle $|z| = R$ and the line segment between iR and $-iR$. The method is known as the Nyquist stability criterion.

The Winding Number

Suppose that C: $z(t) = x(t) + iy(t)$ for $a \leqq t \leqq b$ is a simple closed contour. Let $a = t_0 < t_1 < \cdots < t_n = b$ be a partition of the interval and let $z_k = z(t_k)$ (for $k = 0, 1, \ldots, n$) denote points on C where $z_0 = z_n$. If z^* lies inside C, then $z(t)$ winds around z^* once as t goes from a to b (see Figure 7.8).

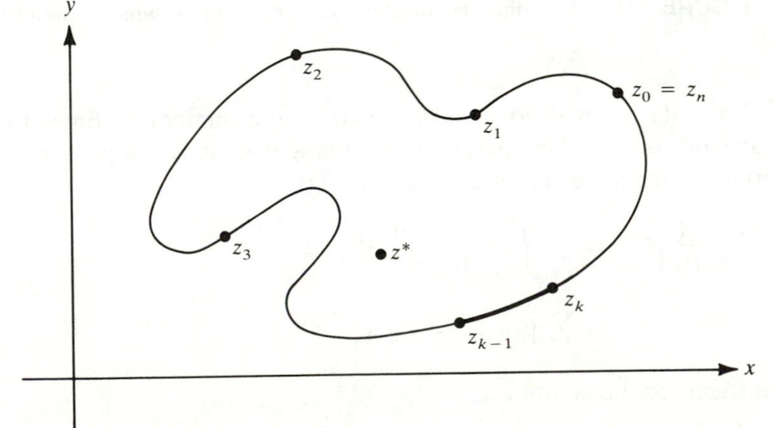

FIGURE 7.8 The points z_k on the contour C that winds around z^*.

Now suppose that $f(z)$ is analytic at each point on C and meromorphic inside C. Then $f(C)$ is a closed curve in the w plane that passes through $w_k = f(z_k)$ (for $k = 0, 1, \ldots, n$) where $w_0 = w_n$. The subintervals $[t_{k-1}, t_k]$ can be chosen small enough so that a continuous branch $\log w = \ln|w| + i \arg w = \ln \rho + i\phi$ can be defined on the portion of $f(C)$ between w_{k-1} and w_k (see Figure 7.9). Then

$$\text{(11)} \qquad \log f(z_k) - \log f(z_{k-1}) = \ln \rho_k - \ln \rho_{k-1} + i\Delta\phi_k$$

where $\Delta\phi_k = \phi_k - \phi_{k-1}$ measures the amount that the portion of the curve $f(C)$ between w_k and w_{k-1} winds around the origin $w = 0$.

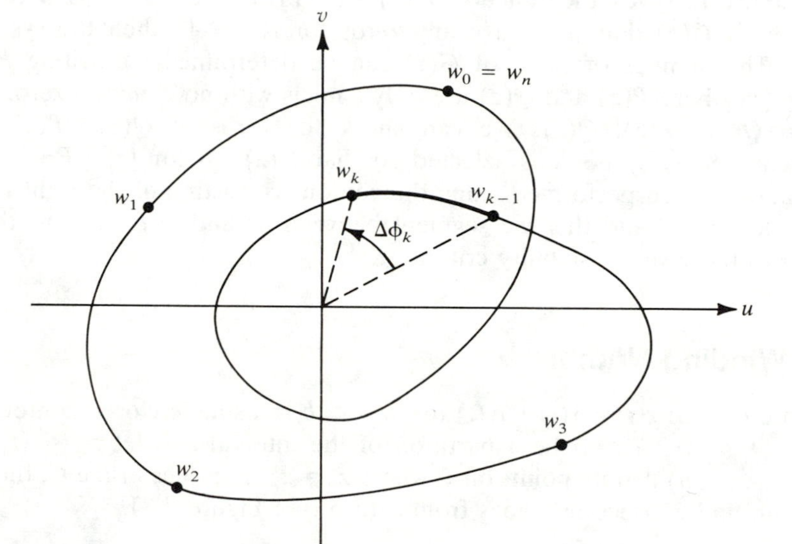

FIGURE 7.9 The points w_k on the contour $f(C)$ that winds around 0.

Formula (1) will now be shown to be the number of times that $f(C)$ winds around $w = 0$. The parameterization given above together with the appropriate branches of $\log w$ are used to write

$$\int_C \frac{f'(z)}{f(z)}\, dz = \sum_{k=1}^{n} \int_{t_{k-1}}^{t_k} \frac{f'(z(t))}{f(z(t))}\, z'(t)\, dt$$

$$= \sum_{k=1}^{n} [\log w_k - \log w_{k-1}],$$

which in turn can be written as

$$\text{(12)} \qquad \int_C \frac{f'(z)}{f(z)}\, dz = \sum_{k=1}^{n} [\ln \rho_k - \ln \rho_{k-1}] + i \sum_{k=1}^{n} \Delta\phi_k.$$

By using the fact that $\rho_0 = \rho_n$ the first summation in (12) vanishes. The summation of the quantities $\Delta\phi_k$ is the total amount that $f(C)$ winds around $w = 0$ in radians. When the quantities in (12) are divided by $2\pi i$, we are left with an integer that is the number of times $f(C)$ winds around $w = 0$. For example, the image of the circle $C: |z| = 2$ under the mapping $w = f(z) = z^2 + z$ is the curve $x = 4\cos 2t + 2\cos t$, $y = 4\sin 2t + 2\sin t$ for $0 < t \leqq 2\pi$ that is shown in Figure 7.10. Notice that the image curve $f(C)$ winds twice around the origin $w = 0$.

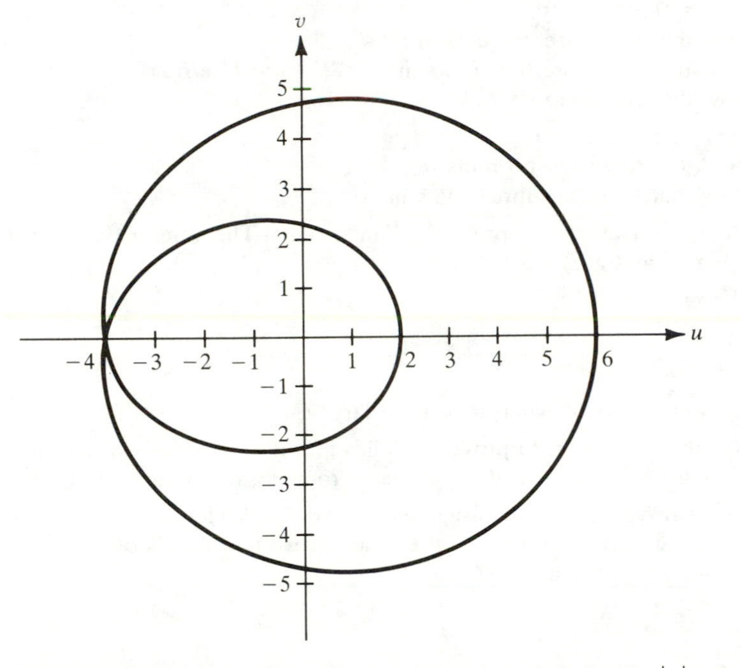

FIGURE 7.10 The image curve $f(C)$ of the circle C: $|z| = 2$ under the mapping $w = f(z) = z^2 + z$.

EXERCISES FOR SECTION 7.8

For Exercises 1–5, use Rouché's Theorem to show that the roots lie in the indicated region.

1. Let $P(z) = z^5 + 4z - 15$.
   (a) Show that there are no roots in $|z| < 1$. *Hint*: Use $f(z) = 15$.
   (b) Show that there are five roots in $|z| < 2$. *Hint*: $f(z) = z^5$.
   Remark: A factorization of the polynomial using numerical approximations for the coefficients is

$$(z - 1.546)(z^2 - 1.340z + 2.857)(z^2 + 2.885z + 3.397).$$

2. Let $P(z) = z^3 + 9z + 27$.
   (a) Show that there are no roots in $|z| < 2$. *Hint*: Use $f(z) = 27$.
   (b) Show that there are three roots in $|z| < 4$. *Hint*: $f(z) = z^3$.
   Remark: A factorization of the polynomial using numerical approximations for the coefficients is

$$(z + 2.047)(z^2 - 2.047z + 13.19).$$

3. Let $P(z) = z^5 + 6z^2 + 2z + 1$.
   (a) Show that there are two roots in $|z| < 1$. *Hint*: Use $f(z) = 6z^2$.
   (b) Show that there are five roots in $|z| < 2$.
4. Let $P(z) = z^6 - 5z^4 + 10$.
   (a) Show that there are no roots in $|z| < 1$.
   (b) Show that there are four roots in $|z| < 2$. *Hint*: Use $f(z) = 5z^4$.
   (c) Show that there are six roots in $|z| < 3$.
5. Let $P(z) = 3z^3 - 2iz^2 + iz - 7$.
   (a) Show that there are no roots in $|z| < 1$.
   (b) Show that there are three roots in $|z| < 2$.
6. Use Rouché's Theorem to prove the Fundamental Theorem of Calculus. *Hint*: Let $f(z) = -a_n z^n$ and $g(z) = a_0 + a_1 z + \cdots + a_{n-1} z^{n-1}$. Then show that for points z on the circle $|z| = R$ we have

$$\left| \frac{g(z)}{f(z)} \right| < \frac{|a_0| + |a_1| + \cdots + |a_{n-1}|}{|a_n| R},$$

   and see what happens when R is made large.
7. Use Rouché's Theorem to prove the following. If $h(z)$ is analytic and nonzero and $|h(z)| < 1$ for $|z| < 1$, then $h(z) - z^n$ has n roots inside the unit circle $|z| = 1$.
8. Suppose that $f(z)$ is analytic inside and on the simple closed contour C. If $f(z)$ is a one-to-one function at points z on C, then prove that $f(z)$ is one-to-one inside C. *Hint*: Consider the image of C.

8

Conformal Mapping

8.1 Basic Properties of Conformal Mappings

Let f be an analytic function in the domain D, and let z_0 be a point in D. If $f'(z_0) \neq 0$, then we can express f in the form

$$(1) \qquad f(z) = f(z_0) + f'(z_0)(z - z_0) + \eta(z)(z - z_0)$$

where $\eta(z) \to 0$ as $z \to z_0$.

If z is near z_0, then the transformation $w = f(z)$ has the *linear approximation*

$$(2) \qquad S(z) = A + B(z - z_0) \qquad \text{where } A = f(z_0) \text{ and } B = f'(z_0) .$$

Since $\eta(z) \to 0$ when $z \to z_0$, it is reasonable that near z_0 the transformation $w = f(z)$ has an effect much like the linear mapping $w = S(z)$. The effect of the linear mapping S is a rotation of the plane through the angle $\alpha = \arg f'(z_0)$, followed by a magnification by the factor $|f'(z_0)|$, followed by a rigid translation by the vector $A - Bz_0$. Consequently, the mapping $w = S(z)$ preserves angles at the point z_0. We now show that the mapping $w = f(z)$ preserves angles at z_0.

Let $C\colon z(t) = x(t) + iy(t)$, $-1 \leqq t \leqq 1$ denote a smooth curve that passes through the point $z(0) = z_0$. A vector $\mathbf{T}$ tangent to C at the point z_0 is given by

$$(3) \qquad \mathbf{T} = z'(0)$$

where the complex number $z'(0)$ has assumed its vector interpretation.

The angle of inclination of $\mathbf{T}$ with respect to the positive x axis is

$$(4) \qquad \beta = \arg z'(0) .$$

The image of C under the mapping $w = f(z)$ is the curve K given by K: $w(t) = u(x(t), y(t)) + iv(x(t), y(t))$. The chain rule can be used to show that a vector $\mathbf{T}^*$ tangent to K at the point $w_0 = f(z_0)$ is given by

$$(5) \qquad \mathbf{T}^* = w'(0) = f'(z_0)z'(0) .$$

The angle of inclination of $\mathbf{T}^*$ with respect to the positive u axis is

$$(6) \qquad \gamma = \arg f'(z_0) + \arg z'(0) = \alpha + \beta, \qquad \text{where } \alpha = \arg f'(z_0).$$

Therefore the effect of the transformation $w = f(z)$ is to rotate the angle of inclination of the tangent vector $\mathbf{T}$ at z_0 through the angle $\alpha = \arg f'(z_0)$ to obtain the angle of inclination of the tangent vector $\mathbf{T}^*$ at w_0. The situation is illustrated in Figure 8.1.

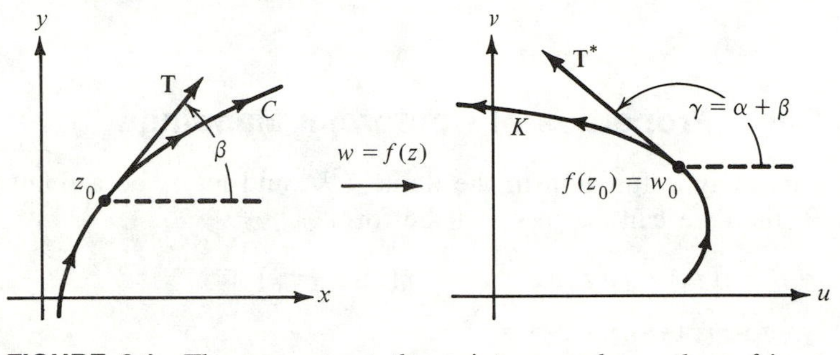

FIGURE 8.1 The tangents at the points z_0 and w_0 where f is an analytic function and $f'(z_0) \neq 0$.

A mapping $w = f(z)$ is said to be angle preserving, or *conformal at* z_0, if it preserves angles between oriented curves in magnitude as well as in sense. The following result shows where a mapping by an analytic function is conformal.

Theorem 8.1 *Let f be an analytic function in the domain D, and let z_0 be a point in D. If $f'(z_0) \neq 0$, then f is conformal at z_0.*

Proof Let C_1 and C_2 be two smooth curves passing through z_0 with tangents given by $\mathbf{T}_1$ and $\mathbf{T}_2$, respectively. Let β_1 and β_2 denote the angles of inclination of $\mathbf{T}_1$ and $\mathbf{T}_2$, respectively. The image curves K_1 and K_2 that pass through the point $w_0 = f(z_0)$ will have tangents denoted by $\mathbf{T}_1^*$ and $\mathbf{T}_2^*$, respectively. Using equation (6), we see that the angles of inclination γ_1 and γ_2 of $\mathbf{T}_1^*$ and $\mathbf{T}_2^*$ are related to β_1 and β_2 by the equations

$$(7) \qquad \gamma_1 = \alpha + \beta_1 \qquad \text{and} \qquad \gamma_2 = \alpha + \beta_2$$

where $\alpha = \arg f'(z_0)$. Hence from (7) we conclude that

$$(8) \qquad \gamma_2 - \gamma_1 = \beta_2 - \beta_1.$$

That is, the angle $\gamma_2 - \gamma_1$ from K_1 to K_2 is the same in magnitude and sense as the angle $\beta_2 - \beta_1$ from C_1 to C_2. Therefore the mapping $w = f(z)$ is conformal at z_0. The situation is shown in Figure 8.2.

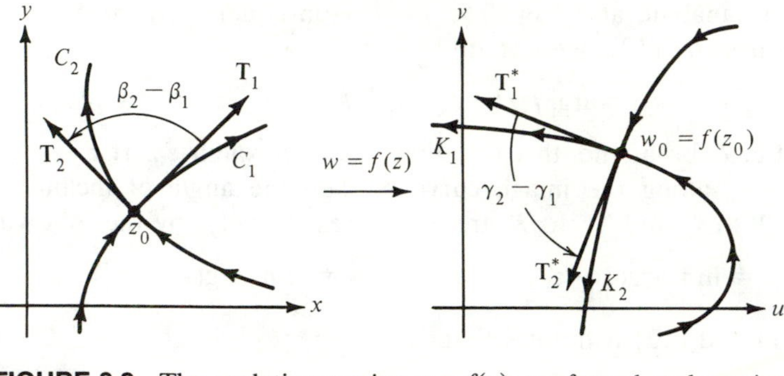

FIGURE 8.2 The analytic mapping $w = f(z)$, conformal at the point z_0 where $f'(z_0) \neq 0$.

EXAMPLE 8.1 Show that the mapping $w = f(z) = \cos z$ is conformal at the points $z_1 = i$, $z_2 = 1$, and $z_3 = \pi + i$, and determine the angle of rotation $\alpha = \arg f'(z)$ at the given points.

Solution Since $f'(z) = -\sin z$, we conclude that the mapping $w = \cos z$ is conformal at all points except $z = n\pi$ where n is an integer. Calculation reveals that

$$f'(i) = -i \sinh 1, \qquad f'(1) = -\sin 1, \qquad \text{and} \qquad f'(\pi + i) = i \sinh 1.$$

Therefore the angle of rotation is given by

$$\alpha_1 = \arg f'(i) = \frac{-\pi}{2}, \qquad \alpha_2 = \arg f'(1) = \pi, \qquad \text{and}$$

$$\alpha_3 = \arg f'(\pi + i) = \frac{\pi}{2}.$$

Let f be a nonconstant analytic function. If $f'(z_0) = 0$, then z_0 is called a *critical point* of f, and the mapping $w = f(z)$ is not conformal at z_0. The next result shows what happens at a critical point.

Theorem 8.2 *Let f be analytic at z_0. If $f'(z_0) = 0, \ldots, f^{(k-1)}(z_0) = 0$, and $f^{(k)}(z_0) \neq 0$, then the mapping $w = f(z)$ magnifies angles at the vertex z_0 by the factor k.*

Proof Since f is analytic at z_0, it has the representation

$$(9) \qquad f(z) = f(z_0) + a_k(z - z_0)^k + a_{k+1}(z - z_0)^{k+1} + \cdots.$$

From (9) we conclude that

$$(10) \qquad f(z) - f(z_0) = (z - z_0)^k g(z)$$

where g is analytic at z_0 and $g(z_0) \neq 0$. Consequently, if $w = f(z)$ and $w_0 = f(z_0)$, then using (10), we obtain

$$(11) \qquad \arg[w - w_0] = \arg[f(z) - f(z_0)] = k \arg(z - z_0) + \arg g(z) .$$

Let C be a smooth curve that passes through z_0. If $z \to z_0$, along C, then $w \to w_0$ along the image curve K, and the angle of inclination of the tangents **T** to C and **T*** to K are given, respectively, by the following limits:

$$(12) \qquad \beta = \lim_{z \to z_0} \arg(z - z_0) \qquad \text{and} \qquad \gamma = \lim_{w \to w_0} \arg(w - w_0) .$$

From (11) and (12) it follows that

$$(13) \qquad \gamma = \lim_{z \to z_0} [k \arg(z - z_0) + \arg g(z)] = k\beta + \delta$$

where $\delta = \arg g(z_0) = \arg a_k$.

Let C_1 and C_2 be two smooth curves that pass through z_0, and let K_1 and K_2 be their images. Then from (13) it follows that

$$(14) \qquad \Delta\gamma = \gamma_2 - \gamma_1 = k(\beta_2 - \beta_1) = k\Delta\beta.$$

That is, the angle $\Delta\gamma$ from K_1 to K_2 is k times as large as the angle $\Delta\beta$ from C_1 to C_2. Therefore angles at the vertex z_0 are magnified by the factor k. The situation is shown in Figure 8.3.

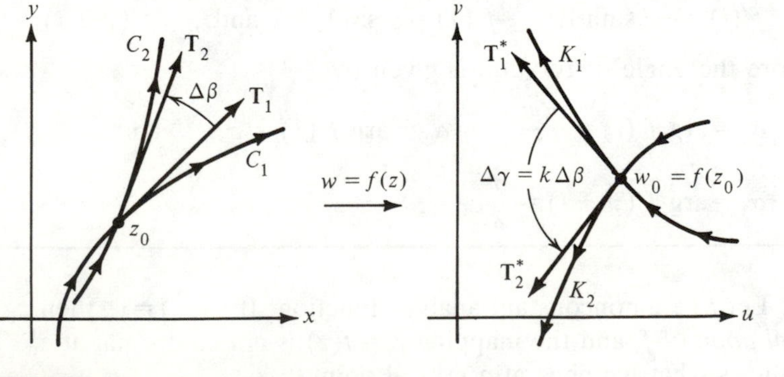

FIGURE 8.3 The analytic mapping $w = f(z)$ at a point z_0 where $f'(z_0) = 0, \ldots, f^{(k-1)}(z_0) = 0$ and $f^{(k)}(z_0) \neq 0$.

EXAMPLE 8.2 The mapping $w = f(z) = z^2$ maps the square $S = \{x + iy: 0 < x < 1, 0 < y < 1\}$ onto the region in the upper half-plane $\operatorname{Im} w > 0$, which lies under the parabolas

$$u = 1 - \tfrac{1}{4}v^2 \qquad \text{and} \qquad u = -1 + \tfrac{1}{4}v^2$$

as shown in Figure 8.4. The derivative is $f'(z) = 2z$, and we conclude that the mapping $w = z^2$ is conformal for all $z \neq 0$. It is worthwhile to observe that the

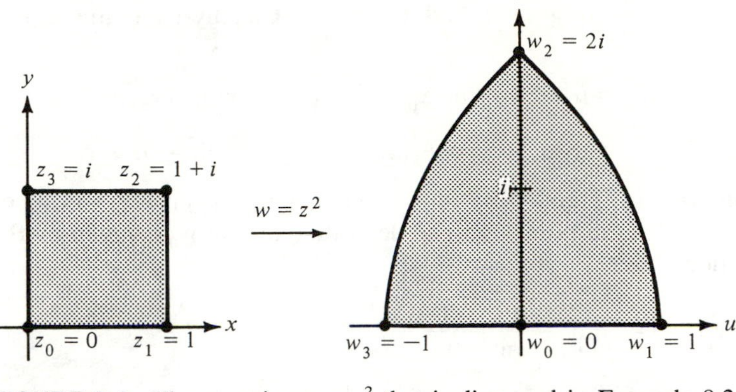

FIGURE 8.4 The mapping $w = z^2$ that is discussed in Example 8.2.

right angles at the vertices $z_1 = 1$, $z_2 = 1 + i$, and $z_3 = i$ are mapped onto right angles at the vertices $w_1 = 1$, $w_2 = 2i$, and $w_3 = -1$, respectively. At the point $z_0 = 0$ we have $f'(0) = 0$ and $f''(0) \neq 0$. Hence angles at the vertex $z_0 = 0$ are magnified by the factor $k = 2$. In particular, we see that the right angle at $z_0 = 0$ is mapped onto the straight angle at $w_0 = 0$.

Another property of a conformal mapping $w = f(z)$ is obtained by considering the modulus of $f'(z_0)$. If z_1 is near z_0, we can use equation (1) and neglect the term $\eta(z_1)(z_1 - z_0)$. We then have the approximation

$$(15) \qquad w_1 - w_0 = f(z_1) - f(z_0) \approx f'(z_0)(z_1 - z_0).$$

Using (15), we see that the distance $|w_1 - w_0|$ between the images of the points z_1 and z_0 is given approximately by $|f'(z_0)||z_1 - z_0|$. Therefore we say that the transformation $w = f(z)$ changes small distances near z_0 by the *scale factor* $|f'(z_0)|$. For example, the scale factor of the transformation $w = f(z) = z^2$ near the point $z_0 = 1 + i$ is $|f'(1 + i)| = |2(1 + i)| = 2\sqrt{2}$.

It is also necessary to say a few things about the inverse transformation $z = g(w)$ of a conformal mapping $w = f(z)$ near a point z_0 where $f'(z_0) \neq 0$. A complete justification of the following relies on theorems studied in advanced calculus*. Let the mapping $w = f(z)$ be expressed in the coordinate form

$$(16) \qquad u = u(x, y) \qquad \text{and} \qquad v = v(x, y).$$

The mapping in (16) represents a transformation from the xy plane into the uv plane, and the *Jacobian determinant* $J(x, y)$ is defined by

$$(17) \qquad J(x, y) = \begin{vmatrix} u_x(x, y) & u_y(x, y) \\ v_x(x, y) & v_y(x, y) \end{vmatrix}.$$

* See, for instance, R. Creighton Buck, *Advanced Calculus*, 3rd ed. (New York, McGraw-Hill Book Company) pp. 358–361, 1978.

It is known that the transformation in (16) has a local inverse provided that $J(x, y) \neq 0$. Expanding (17) and using the Cauchy-Riemann equations, we obtain

$$(18) \qquad J(x_0, y_0) = u_x(x_0, y_0)v_y(x_0, y_0) - v_x(x_0, y_0)u_y(x_0, y_0)$$
$$= u_x^2(x_0, y_0) + v_x^2(x_0, y_0) = |f'(z_0)|^2 \neq 0.$$

Consequently, (17) and (18) imply that a local inverse $z = g(w)$ exists in a neighborhood of the point w_0. The derivative of g at w_0 is given by the familiar computation

$$(19) \qquad g'(w_0) = \lim_{w \to w_0} \frac{g(w) - g(w_0)}{w - w_0}$$
$$= \lim_{z \to z_0} \frac{z - z_0}{f(z) - f(z_0)} = \frac{1}{f'(z_0)} = \frac{1}{f'(g(w_0))}.$$

EXERCISES FOR SECTION 8.1

1. State where the following mappings are conformal.
(a) $w = \exp z$ **(b)** $w = \sin z$ **(c)** $w = z^2 + 2z$
(d) $w = \exp(z^2 + 1)$ **(e)** $w = \dfrac{1}{z}$ **(f)** $w = \dfrac{z+1}{z-1}$

For Exercises 2–5, find the angle of rotation $\alpha = \arg f'(z)$ and the scale factor $|f'(z)|$ of the mapping $w = f(z)$ at the indicated points.

2. $w = 1/z$ at the points 1, $1 + i$, i.

3. $w = \ln r + i\theta$ where $-\pi/2 < \theta < 3\pi/2$ at the points 1, $1 + i$, i, -1.

4. $w = r^{1/2} \cos(\theta/2) + ir^{1/2} \sin(\theta/2)$, where $-\pi < \theta < \pi$, at the points i, 1, $-i$, $3 + 4i$.

5. $w = \sin z$ at the points $\pi/2 + i$, 0, $-\pi/2 + i$.

6. Consider the mapping $w = z^2$. If $a \neq 0$ and $b \neq 0$, show that the lines $x = a$ and $y = b$ are mapped onto orthogonal parabolas.

7. Consider the mapping $w = z^{1/2}$ where $z^{1/2}$ denotes the principal branch of the square root function. If $a > 0$ and $b > 0$, show that the lines $x = a$ and $y = b$ are mapped onto orthogonal curves.

8. Consider the mapping $w = \exp z$. Show that the lines $x = a$ and $y = b$ are mapped onto orthogonal curves.

9. Consider the mapping $w = \sin z$. Show that the line segment $-\pi/2 < x < \pi/2$, $y = 0$, and the vertical line $x = a$ where $|a| < \pi/2$ are mapped onto orthogonal curves.

10. Consider the mapping $w = \operatorname{Log} z$ where $\operatorname{Log} z$ denotes the principal branch of the logarithm function. Show that the positive x axis and the vertical line $x = 1$ are mapped onto orthogonal curves.

11. Let f be analytic at z_0 and $f'(z_0) \neq 0$. Show that the function $g(z) = \overline{f(z)}$ preserves the magnitude, but reverses the sense, of angles at z_0.

8.2 Bilinear Transformations

Another important class of elementary mappings was studied by Augustus Ferdinand Möbius (1790–1868). These mappings are conveniently expressed as the quotient of two linear expressions and are commonly known as linear fractional or bilinear transformations. They arise naturally in mapping problems involving arctan z. In this section we will show how they are used to map a disk one-to-one and onto a half-plane.

Let a, b, c, and d denote four complex constants with the restriction that $ad \neq bc$. Then the transformation

$$(1)\qquad w = S(z) = \frac{az + b}{cz + d}$$

is called a *bilinear transformation* or Möbius transformation. If the expression for S in (1) is multiplied through by the quantity $cz + d$, then the resulting expression has the bilinear form $cwz - az + dw - b = 0$. We can collect terms involving z and write $z(cw - a) = -dw + b$. For values of $w \neq a/c$ the inverse transformation is given by

$$(2)\qquad z = S^{-1}(w) = \frac{-dw + b}{cw - a}.$$

We can extend S and S^{-1} to mappings in the extended complex plane. The value $S(\infty)$ should be chosen so that $S(z)$ has a limit at ∞. Therefore we define

$$(3)\qquad S(\infty) = \lim_{z\to\infty} S(z) = \lim_{z\to\infty} \frac{a + (b/z)}{c + (d/z)} = \frac{a}{c},$$

and the inverse is $S^{-1}(a/c) = \infty$. Similarly, the value $S^{-1}(\infty)$ is obtained by

$$(4)\qquad S^{-1}(\infty) = \lim_{w\to\infty} S^{-1}(w) = \lim_{w\to\infty} \frac{-d + (b/w)}{c - (a/w)} = \frac{-d}{c},$$

and the inverse is $S(-d/c) = \infty$. With these extensions we conclude that the transformation $w = S(z)$ is a one-to-one mapping of the extended complex z plane onto the extended complex w plane.

We now show that a bilinear transformation carries the class of circles and lines onto itself. Let S be an arbitrary bilinear transformation given by equation (1). If $c = 0$, then S reduces to a linear transformation, which carries lines onto lines and circles onto circles. If $c \neq 0$, then we can write S in the form

$$(5)\qquad S(z) = \frac{a(cz + d) + bc - ad}{c(cz + d)} = \frac{a}{c} + \frac{bc - ad}{c} \frac{1}{cz + d}.$$

The condition $ad \neq bc$ precludes the possibility that S reduces to a constant. It is easy to see from equation (5) that S can be considered as a composition of functions. It is a linear mapping $\xi = cz + d$, followed by the reciprocal transformation $Z = 1/\xi$, followed by $w = (a/c) + ((bc - ad)/c)Z$. It was shown in

Chapter 2 that each function in the composition maps the class of circles and lines onto itself, it follows that the bilinear transformation S has this property. A half-plane can be considered a family of parallel lines and a disk as a family of circles. Therefore it is reasonable to conclude that a bilinear transformation maps the class of half-planes and disks onto itself. Example 8.3 illustrates this idea.

EXAMPLE 8.3 Show that $w = S(z) = i(1 - z)/(1 + z)$ maps the unit disk $|z| < 1$ one-to-one and onto the upper half-plane $\operatorname{Im} w > 0$.

Solution Let us first consider the unit circle C: $|z| = 1$, which forms the boundary of the disk, and find its image in the w plane. If we write $S(z) = (-iz + i)/(z + 1)$, then we see that $a = -i$, $b = i$, $c = 1$, and $d = 1$. Using equation (2), we find that the inverse is given by

$$(6) \qquad z = S^{-1}(w) = \frac{-dw + b}{cw - a} = \frac{-w + i}{w + i}.$$

If $|z| = 1$, then equation (6) implies that the images of points on the unit circle satisfy the equation

$$(7) \qquad |w + i| = |-w + i|.$$

Squaring both sides of equation (7), we obtain $u^2 + (1 + v)^2 = u^2 + (1 - v)^2$, which can be simplified to yield $v = 0$, which is the equation of the u axis in the w plane.

The circle C divides the z plane into two portions, and its image is the u axis, which divides the w plane into two portions. Since the image of the point $z = 0$ is $w = S(0) = i$, we expect that the interior of the circle C is mapped onto the portion of the w plane that lies above the u axis. To show that this is true, we let $|z| < 1$. Then (6) implies that the image values must satisfy the inequality $|-w + i| < |w + i|$, which can be written as

$$(8) \qquad d_1 = |w - i| < |w - (-i)| = d_2.$$

If we interpret d_1 as the distance from w to i and d_2 as the distance from w to $-i$, then a geometric argument shows that the image point w must lie in the upper half-plane $\operatorname{Im} w > 0$, as shown in Figure 8.5. Since S is one-to-one and onto in the extended complex plane, it follows that S maps the disk onto the half-plane.

The general formula (1) of a bilinear transformation appears to involve four independent coefficients a, b, c, d. But since either $a \neq 0$ or $c \neq 0$, the transformation can be expressed with three unknown coefficients and can be written as either

$$S(z) = \frac{z + b/a}{cz/a + d/a} \qquad \text{or} \qquad S(z) = \frac{az/c + b/c}{z + d/c},$$

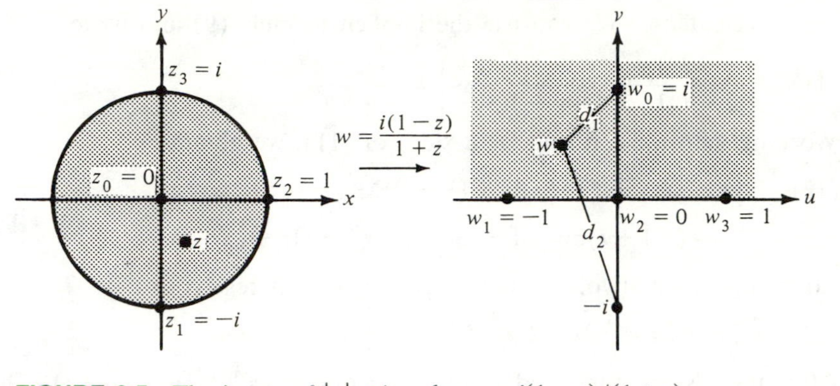

FIGURE 8.5 The image of $|z| < 1$ under $w = i(1 - z)/(1 + z)$.

respectively. This permits us to uniquely determine a bilinear transformation if three distinct image values $S(z_1) = w_1$, $S(z_2) = w_2$, and $S(z_3) = w_3$ are specified. To determine such a mapping, it is convenient to use an implicit formula involving z and w.

Theorem 8.3 (The Implicit Formula) *There exists a unique bilinear transformation that maps three distinct points z_1, z_2, and z_3 onto three distinct points w_1, w_2, and w_3, respectively. An implicit formula for the mapping is given by the equation*

$$\frac{z - z_1}{z - z_3} \frac{z_2 - z_3}{z_2 - z_1} = \frac{w - w_1}{w - w_3} \frac{w_2 - w_3}{w_2 - w_1}. \tag{9}$$

Proof Equation (9) can be algebraically manipulated, and we can solve for w in terms of z. The result will be an expression for w that has the form (1), where the coefficients a, b, c, and d involve the values z_1, z_2, z_3, w_1, w_2, and w_3. The details are left as an exercise.

If we set $z = z_1$ and $w = w_1$ in (9), then both sides of the equation are zero. This shows that w_1 is the image of z_1. If we set $z = z_2$ and $w = w_2$ in (9), then both sides of the equation take on the value 1. Hence w_2 is the image of z_2. Taking reciprocals, we can write equation (9) in the form

$$\frac{z - z_3}{z - z_1} \frac{z_2 - z_1}{z_2 - z_3} = \frac{w - w_3}{w - w_1} \frac{w_2 - w_1}{w_2 - w_3}. \tag{10}$$

If we set $z = z_3$ and $w = w_3$ in equation (10), then both sides of the equation are zero. Therefore w_3 is the image of z_3, and we have shown that the transformation has the required mapping properties.

EXAMPLE 8.4 Construct the bilinear transformation $w = S(z)$ that maps the points $z_1 = -i$, $z_2 = 1$, $z_3 = i$ onto the points $w_1 = -1$, $w_2 = 0$, $w_3 = 1$, respectively.

Solution We can use the implicit formula (9) and write

$$\frac{z-i}{z+i}\,\frac{1+i}{1-i}=\frac{w-1}{w+1}\,\frac{0+1}{0-1}=\frac{-w+1}{w+1}. \tag{11}$$

Working with the left and right sides of (11), we obtain

$$\begin{aligned}(1+i)zw+(1-i)w+(1+i)z+(1-i)\\ =(-1+i)zw+(-1-i)w+(1-i)z+(1+i).\end{aligned} \tag{12}$$

Collecting terms involving w and zw on the left results in

$$2w+2zw=2i-2iz. \tag{13}$$

After the 2's are cancelled in (13), we obtain $w(1+z)=i(1-z)$. Therefore the desired bilinear transformation is

$$w=S(z)=\frac{i(1-z)}{1+z}.$$

EXAMPLE 8.5 Find the bilinear transformation $w=S(z)$ that maps the points $z_1=-2$, $z_2=-1-i$, and $z_3=0$ onto $w_1=-1$, $w_2=0$, and $w_3=1$, respectively.

Solution We can use the implicit formula (9) and write

$$\frac{z-(-2)}{z-0}\,\frac{-1-i-0}{-1-i-(-2)}=\frac{w-(-1)}{w-1}\,\frac{0-1}{0-(-1)}. \tag{14}$$

From the fact that $(-1-i)/(1-i)=1/i$, equation (14) can be written as

$$\frac{z+2}{iz}=\frac{1+w}{1-w}. \tag{15}$$

Equation (15) is equivalent to $z+2-zw-2w=iz+izw$, which can be solved for w in terms of z, giving the solution

$$w=S(z)=\frac{(1-i)z+2}{(1+i)z+2}.$$

Let D be a region in the z plane that is bounded by either a circle or a straight line C. Let z_1, z_2, and z_3 be three distinct points that lie on C with the property that an observer moving along C from z_1 to z_3 through z_2 finds the region D on the left. In the case that C is a circle and D is the interior of C we say that C is positively oriented. Conversely, the ordered triple z_1, z_2, z_3 uniquely determines a region that lies on the left of C.

Let G be a region in the w plane that is bounded by either a circle or a straight line K. Let w_1, w_2, w_3 be three distinct points that lie on K such that an observer moving along K from w_1 to w_3 through w_2 finds the region G on the left. Since a bilinear transformation is a conformal mapping that maps the class of circles and straight lines onto itself, we can use the implicit formula (9) to construct a bilinear transformation $w = S(z)$ that is a one-to-one mapping of D onto G.

EXAMPLE 8.6 Show that the mapping

$$w = S(z) = \frac{(1-i)z+2}{(1+i)z+2}$$

maps the disk D: $|z+1|<1$ onto the upper half-plane $\operatorname{Im} w > 0$.

Solution For convenience we choose the ordered triple $z_1 = -2$, $z_2 = -1 - i$, $z_3 = 0$, which will give the circle C: $|z+1| = 1$ a positive orientation and the disk D a "left orientation." We saw in Example 8.5 that the corresponding image points are

$$w_1 = S(z_1) = -1, \qquad w_2 = S(z_2) = 0, \qquad \text{and} \qquad w_3 = S(z_3) = 1.$$

Since the ordered triple of points w_1, w_2, w_3 lie on the u axis, it follows that the image of the circle C is the u axis. The points w_1, w_2, and w_3 give the upper half-plane G: $\operatorname{Im} w > 0$ a "left orientation." Therefore $w = S(z)$ maps the disk D onto the upper half-plane G. To check our work, we choose a point z_0 that lies in D and find the half-plane where its image w_0 lies. The choice $z_0 = -1$ yields $w_0 = S(-1) = i$. Hence the upper half-plane is the correct image. The situation is illustrated in Figure 8.6.

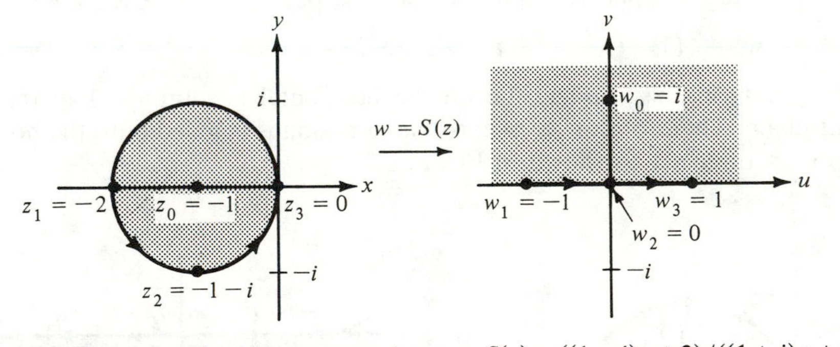

FIGURE 8.6 The bilinear mapping $w = S(z) = ((1-i)z+2)/((1+i)z+2)$.

In equation (9) the point at infinity can be introduced as one of the prescribed points in either the z plane or the w plane. For example, if $w_3 = \infty$, then we are permitted to write

$$\text{(16)} \qquad \frac{w_2 - w_3}{w - w_3} = \frac{w_2 - \infty}{w - \infty} = 1,$$

and substitution of (16) into (9) yields

$$(17)\qquad \frac{z-z_1}{z-z_3}\,\frac{z_2-z_3}{z_2-z_1}=\frac{w-w_1}{w_2-w_1}\qquad \text{where } w_3=\infty.$$

Equation (17) is sometimes used to map the crescent-shaped region that lies between tangent circles onto an infinite strip.

EXAMPLE 8.7 Find a bilinear transformation that maps the crescent-shaped region that lies inside the disk $|z-2|<2$ and outside the circle $|z-1|=1$ onto a horizontal strip.

Solution For convenience we choose $z_1=4$, $z_2=2+2i$, $z_3=0$ and the image values $w_1=0$, $w_2=1$, $w_3=\infty$, respectively. The ordered triple z_1, z_2, z_3 gives the circle $|z-2|=2$ a positive orientation and the disk $|z-2|<2$ a "left orientation." The image points w_1, w_2, w_3 all lie on the extended u axis, and they determine a left orientation for the upper half-plane $\operatorname{Im} w>0$. Therefore we can use the implicit formula (17) to write

$$(18)\qquad \frac{z-4}{z-0}\,\frac{2+2i-0}{2+2i-4}=\frac{w-0}{1-0},$$

which determines a mapping of the disk $|z-2|<2$ onto the upper half-plane $\operatorname{Im} w>0$. We can simplify (18) to obtain the desired solution

$$w=S(z)=\frac{-iz+4i}{z}.$$

A straightforward calculation shows that the points $z_4=1-i$, $z_5=2$, and $z_6=1+i$ are mapped respectively onto the points

$$w_4=S(1-i)=-2+i,\quad w_5=S(2)=i,\quad \text{and}\quad w_6=S(1+i)=2+i$$

The points w_4, w_5, and w_6 lie on the horizontal line $\operatorname{Im} w=1$ in the upper half-plane. Therefore the crescent-shaped region is mapped onto the horizontal strip $0<\operatorname{Im} w<1$ as shown in Figure 8.7.

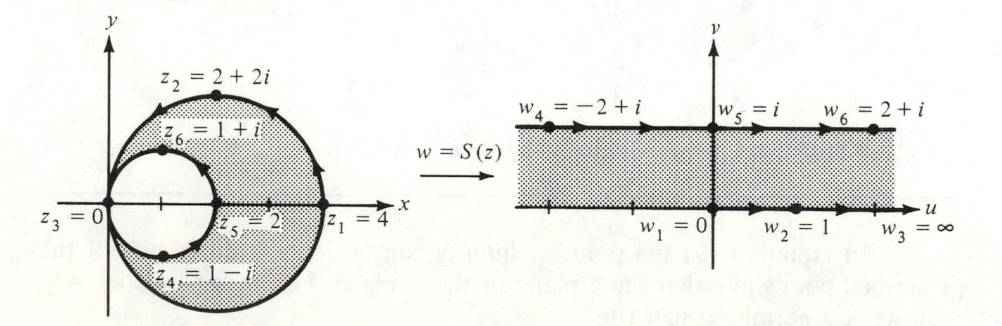

FIGURE 8.7 The mapping $w=S(z)=(-iz+4i)/z$.

EXERCISES FOR SECTION 8.2

1. Let $w = S(z) = ((1 - i)z + 2)/((1 + i)z + 2)$. Find $S^{-1}(w)$.

2. Let $w = S(z) = (i + z)/(i - z)$. Find $S^{-1}(w)$.

3. Find the image of the right half-plane $\text{Re } z > 0$ under the transformation $w = i(1 - z)/(1 + z)$.

4. Show that the bilinear transformation $w = i(1 - z)/(1 + z)$ maps the portion of the disk $|z| < 1$ that lies in the upper half-plane $\text{Im } z > 0$ onto the first quadrant $u > 0$, $v > 0$.

5. Find the image of the upper half-plane $\text{Im } z > 0$ under the transformation

$$w = \frac{(1 - i)z + 2}{(1 + i)z + 2}.$$

6. Find the bilinear transformation $w = S(z)$ that maps the points $z_1 = 0$, $z_2 = i$, $z_3 = -i$ onto $w_1 = -1$, $w_2 = 1$, $w_3 = 0$, respectively.

7. Find the bilinear transformation $w = S(z)$ that maps the points $z_1 = -i$, $z_2 = 0$, $z_3 = i$ onto $w_1 = -1$, $w_2 = i$, $w_3 = 1$, respectively.

8. Find the bilinear transformation $w = S(z)$ that maps the points $z_1 = 0$, $z_2 = 1$, $z_3 = 2$ onto $w_1 = 0$, $w_2 = 1$, $w_3 = \infty$, respectively.

9. Find the bilinear transformation $w = S(z)$ that maps the points $z_1 = 1$, $z_2 = i$, $z_3 = -1$ onto $w_1 = 0$, $w_2 = 1$, $w_3 = \infty$, respectively.

10. Show that the transformation $w = (i + z)/(i - z)$ maps the unit disk $|z| < 1$ onto the right half-plane $\text{Re } w > 0$.

11. Find the image of the lower half-plane $\text{Im } z < 0$ under the transformation $w = (i + z)/(i - z)$.

12. Let $S_1(z) = (z - 2)/(z + 1)$ and $S_2(z) = z/(z + 3)$. Find $S_1(S_2(z))$ and $S_2(S_1(z))$.

13. Find the image of the first quadrant $x > 0$, $y > 0$ under the mapping $w = (z - 1)/(z + 1)$.

14. Find the image of the horizontal strip $0 < y < 2$ under the mapping $w = z/(z - i)$.

15. Show that equation (9) can be written in the form of equation (1).

16. Show that the bilinear transformation $w = S(z) = (az + b)/(cz + d)$ is conformal at all points $z \neq -d/c$.

17. A *fixed point* of a mapping $w = f(z)$ is a point z_0 such that $f(z_0) = z_0$. Show that a bilinear transformation can have at most two fixed points.

18. **(a)** Find the fixed points of $w = (z - 1)/(z + 1)$.
(b) Find the fixed points of $w = (4z + 3)/(2z - 1)$.

8.3 Mappings Involving Elementary Functions

In Section 4.2 we saw that the function $w = f(z) = \exp z$ is a one-to-one mapping of the fundamental period strip $-\pi < y \leqq \pi$ in the z plane onto the w plane with the point $w = 0$ deleted. Since $f'(z) \neq 0$, the mapping $w = \exp z$ is a conformal mapping at each point z in the complex plane. The family of horizontal lines $y = c$, $-\pi < c \leqq \pi$, and the segments $x = a$, $-\pi < y \leqq \pi$ form an orthogonal grid in the fundamental period strip. Their images under the mapping $w = \exp z$ are the rays $\rho > 0$, $\phi = c$ and the circles $|w| = e^a$, respectively. These images form an orthogonal curvilinear grid in the w plane, as shown in Figure 8.8. If $-\pi < c < d \leqq \pi$, then the rectangle $R = \{x + iy: a < x < b,\ c < y < d\}$ is mapped one-to-one and onto the region $G = \{\rho e^{i\phi}: e^a < \rho < e^b,\ c < \phi < d\}$. The inverse mapping is the principal branch of the logarithm $z = \text{Log}\, w$.

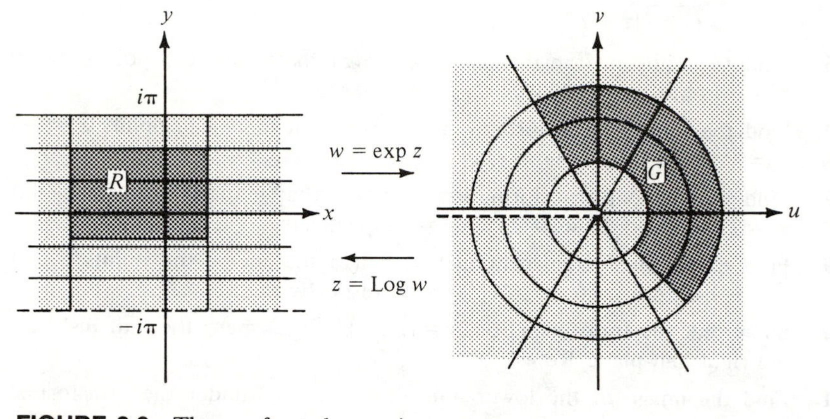

FIGURE 8.8 The conformal mapping $w = \exp z$.

In this section we will show how compositions of conformal transformations are used to construct mappings with specified characteristics.

EXAMPLE 8.8 The transformation $w = f(z) = (e^z - i)/(e^z + i)$ is a one-to-one conformal mapping of the horizontal strip $0 < y < \pi$ onto the disk $|w| < 1$. Furthermore, the x axis is mapped onto the lower semicircle bounding the disk, and the line $y = \pi$ is mapped onto the upper semicircle.

Solution The function $w = f(z)$ can be considered as a composition of the exponential mapping $Z = \exp z$ followed by the bilinear transformation $w = (Z - i)/(Z + i)$. The image of the horizontal strip $0 < y < \pi$ under the mapping $Z = \exp z$ is the upper half-plane $\text{Im}\, Z > 0$; the x axis is mapped onto the positive X axis; and the line $y = \pi$ is mapped onto the negative X axis. The bilinear transformation $w = (Z - i)/(Z + i)$ then maps the upper half-plane $\text{Im}\, Z > 0$ onto the disk $|w| < 1$; the positive X axis is mapped onto the lower semicircle; and the negative X axis onto the upper semicircle. Figure 8.9 illustrates the composite mapping.

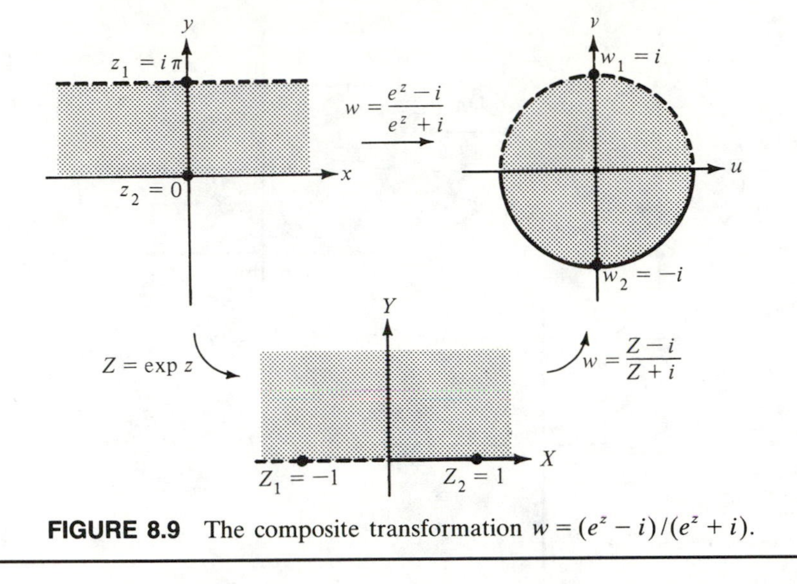

FIGURE 8.9 The composite transformation $w = (e^z - i)/(e^z + i)$.

EXAMPLE 8.9 The transformation $w = f(z) = \text{Log}(1 + z)/(1 - z)$ is a one-to-one conformal mapping of the unit disk $|z| < 1$ onto the horizontal strip $|v| < \pi/2$. Furthermore, the upper semicircle of the disk is mapped onto the line $v = \pi/2$ and the lower semicircle onto $v = -\pi/2$.

Solution The function $w = f(z)$ is the composition of the bilinear transformation $Z = (1 + z)/(1 - z)$ followed by the logarithmic mapping $w = \text{Log}\, Z$. The image of the disk $|z| < 1$ under the bilinear transformation $Z = (1 + z)/(1 - z)$ is the right half-plane $\text{Re}\, Z > 0$; the upper semicircle is mapped onto the positive Y axis; and the lower semicircle is mapped onto the negative Y axis. The logarithmic function $w = \text{Log}\, Z$ then maps the right half-plane onto the horizontal strip; the image of the positive Y axis is the line $v = \pi/2$; and the image of the negative Y axis is the line $v = -\pi/2$. Figure 8.10 shows the composite mapping.

EXAMPLE 8.10 The transformation $w = f(z) = (1 + z)^2/(1 - z)^2$ is a one-to-one conformal mapping of the portion of the disk $|z| < 1$ that lies in the upper half-plane $\text{Im}\, z > 0$ onto the upper half-plane $\text{Im}\, w > 0$. Furthermore, the image of the semicircular portion of the boundary is mapped onto the negative u axis, and the segment $-1 < x < 1$, $y = 0$ is mapped onto the positive u axis.

Solution The function $w = f(z)$ is the composition of the bilinear transformation $Z = (1 + z)/(1 - z)$ followed by the mapping $w = Z^2$. The image of the half-disk under the bilinear mapping $Z = (1 + z)/(1 - z)$ is the first quadrant $X > 0$, $Y > 0$; the image of the segment $y = 0$, $-1 < x < 1$ is the

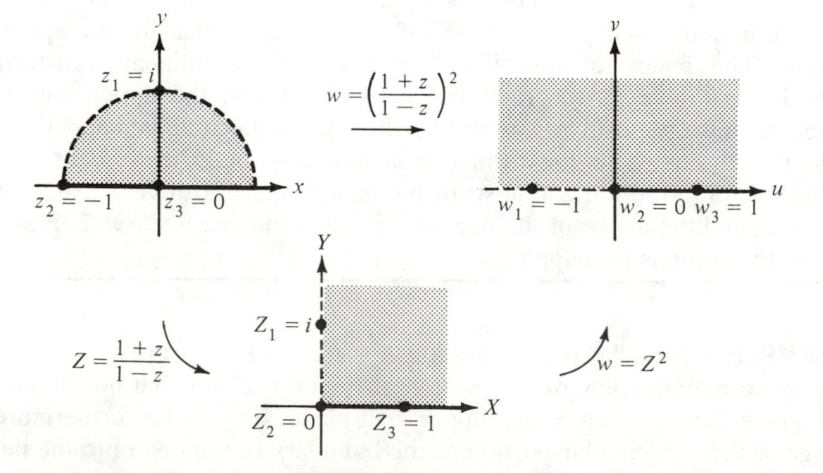

FIGURE 8.10 The composite transformation $w = \text{Log}((1+z)/(1-z))$.

positive X axis; and the image of the semicircle is the positive Y axis. The mapping $w = Z^2$ then maps the first quadrant in the Z plane onto the upper half-plane $\text{Im}\, w > 0$, as shown in Figure 8.11.

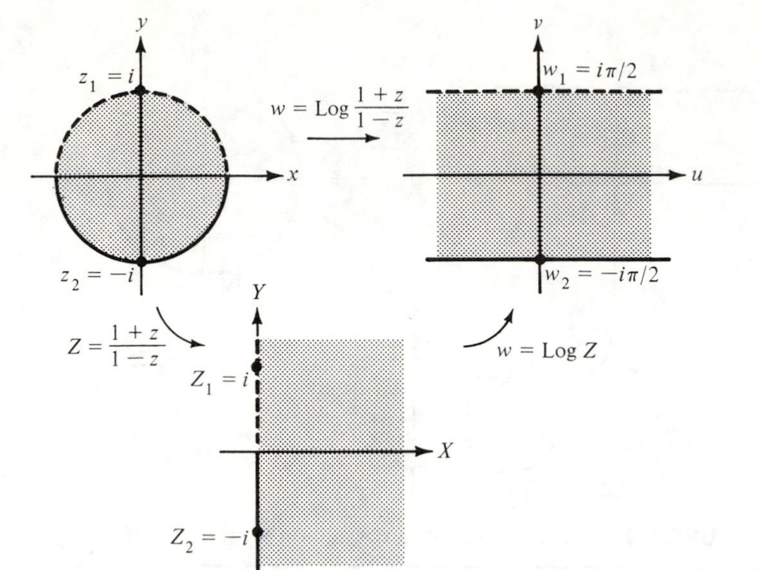

FIGURE 8.11 The composite transformation $w = ((1+z)/(1-z))^2$.

EXAMPLE 8.11 Consider the function $w = f(z) = (z^2 - 1)^{1/2}$, which is the composition of the functions $Z = z^2 - 1$ and $w = Z^{1/2}$ where the branch of the square root is $Z^{1/2} = R^{1/2}[\cos(\theta/2) + i\sin(\theta/2)]$, ($\theta = \arg Z$ and $0 \leqq \theta < 2\pi$).

Then the transformation $w = f(z)$ maps the upper half-plane $\text{Im}\, z > 0$ one-to-one and onto the upper half-plane $\text{Im}\, w > 0$ slit along the segment $u = 0$, $0 < v \leqq 1$.

Solution The function $Z = z^2 - 1$ maps the upper half-plane $\text{Im}\, z > 0$ one-to-one and onto the Z-plane slit along the ray $Y = 0$, $X \geqq -1$. Then the function $w = Z^{1/2}$ maps the slit plane onto the slit half-plane as shown in Figure 8.12.

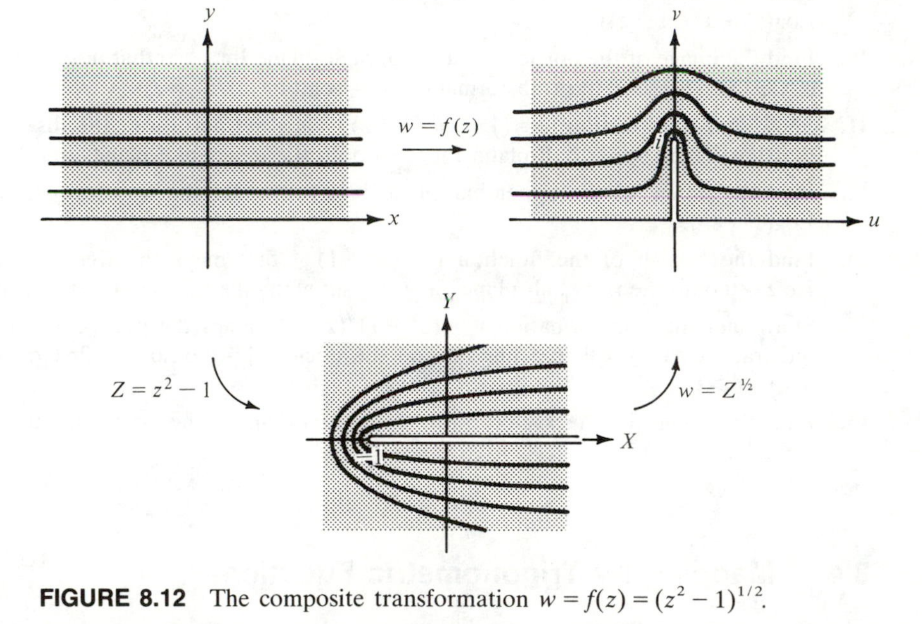

FIGURE 8.12 The composite transformation $w = f(z) = (z^2 - 1)^{1/2}$.

Remark The images of the horizontal lines $y = b$ are curves in the w plane that bend around the segment from 0 to i. The curves represent the streamlines of a fluid flowing across the w plane. We will study fluid flows in more detail in Chapter 9.

EXERCISES FOR SECTION 8.3

1. Find the image of the semi-infinite strip $0 < x < \pi/2$, $y > 0$ under the transformation $w = \exp(iz)$.
2. Find the image of the rectangle $0 < x < \ln 2$, $0 < y < \pi/2$ under the transformation $w = \exp z$.
3. Find the image of the first quadrant $x > 0$, $y > 0$ under the transformation $w = (2/\pi)\,\text{Log}\, z$.
4. Find the image of the annulus $1 < |z| < e$ under the transformation $w = \text{Log}\, z$.

5. Show that the multivalued function $w = \log z$ maps the annulus $1 < |z| < e$ onto the vertical strip $0 < \text{Re } w < 1$.
6. Show that the transformation $w = (2 - z^2)/z^2$ maps the portion of the right half-plane $\text{Re } z > 0$ that lies to the right of the hyperbola $x^2 - y^2 = 1$ onto the unit disk $|w| < 1$.
7. Show that the function $w = (e^z - i)/(e^z + i)$ maps the horizontal strip $-\pi < \text{Im } z < 0$ onto the region $1 < |w|$.
8. Show that the transformation $w = (e^z - 1)/(e^z + 1)$ maps the horizontal strip $|y| < \pi/2$ onto the unit disk $|w| < 1$.
9. Find the image of the upper half-plane $\text{Im } z > 0$ under the transformation $w = \text{Log}((1 + z)/(1 - z))$.
10. Find the image of the portion of the upper half-plane $\text{Im } z > 0$ that lies outside the circle $|z| = 1$ under the transformation $w = \text{Log}((1 + z)/(1 - z))$.
11. Show that the function $w = (1 + z)^2/(1 - z)^2$ maps the portion of the disk $|z| < 1$ that lies in the upper half-plane $\text{Im } z > 0$ onto the upper half-plane $\text{Im } w > 0$.
12. Find the image of the upper half-plane $\text{Im } z > 0$ under the transformation $w = \text{Log}(1 - z^2)$.
13. Find the branch of the function $w = (z^2 + 1)^{1/2}$ that maps the right half-plane $\text{Re } z > 0$ onto the right half-plane $\text{Re } w > 0$ slit along the segment $0 < u \leqq 1$, $v = 0$.
14. Show that the transformation $w = (z^2 - 1)/(z^2 + 1)$ maps the portion of the first quadrant $x > 0$, $y > 0$ that lies outside the circle $|z| = 1$ onto the first quadrant $u > 0$, $v > 0$.
15. Find the image of the sector $r > 0$, $0 < \theta < \pi/4$ under the transformation $w = (i - z^4)/(i + z^4)$.

8.4 Mapping by Trigonometric Functions

The trigonometric functions can be expressed with compositions that involve the exponential function followed by a bilinear function. We will be able to find images of certain regions by following the shapes of successive images in the composite mapping.

EXAMPLE 8.12 The transformation $w = \tan z$ is a one-to-one conformal mapping of the vertical strip $|x| < \pi/4$ onto the unit disk $|w| < 1$.

Solution Using identities (3) and (4) in Section 4.3, we write

$$(1) \qquad w = \tan z = \frac{1}{i}\frac{e^{iz} - e^{-iz}}{e^{iz} + e^{-iz}} = \frac{-ie^{i2z} + i}{e^{i2z} + 1}.$$

From (1) it is easy to see that the mapping $w = \tan z$ can be considered as the composition

$$(2) \qquad w = \frac{-iZ + i}{Z + 1} \quad \text{and} \quad Z = e^{i2z}.$$

The function $Z = \exp(i2z)$ maps the vertical strip $|x| < \pi/4$ one-to-one and onto the right half-plane $\operatorname{Re} Z > 0$. Then the bilinear transformation $w = (-iZ + i)/(Z + 1)$ maps the half-plane one-to-one and onto the disk as shown in Figure 8.13.

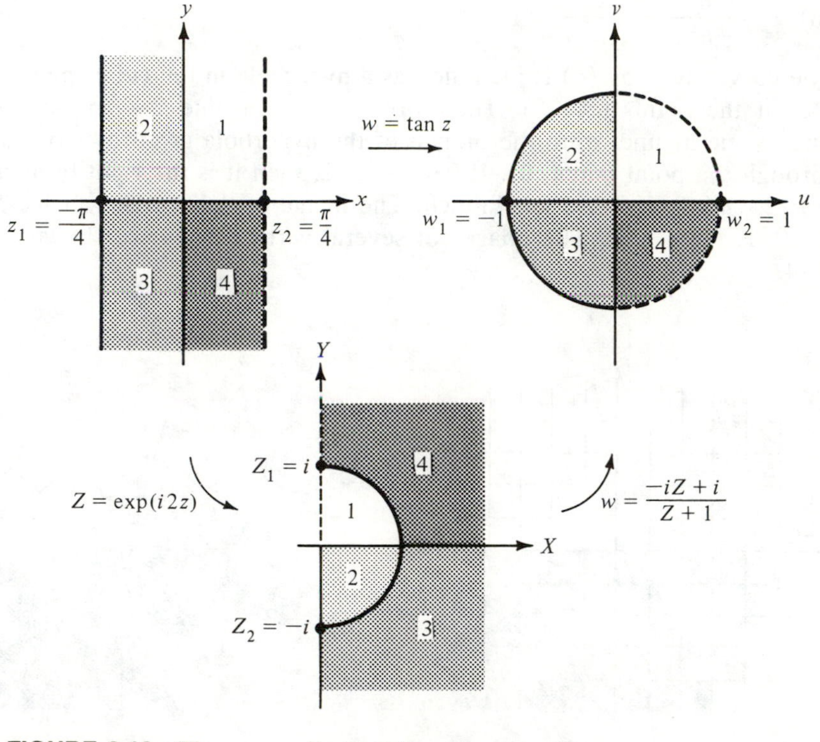

FIGURE 8.13 The composite transformation $w = \tan z$.

EXAMPLE 8.13 The transformation $w = f(z) = \sin z$ is a one-to-one conformal mapping of the vertical strip $|x| < \pi/2$ onto the w plane slit along the rays $u \leqq -1$, $v = 0$ and $u \geqq 1$, $v = 0$.

Solution Since $f'(z) = \cos z \neq 0$ for values of z satisfying $-\pi/2 < \operatorname{Re} z < \pi/2$, it follows that $w = \sin z$ is a conformal mapping. Using equation (15) in Section 4.3, we write

(3) $\qquad u + iv = \sin z = \sin x \cosh y + i \cos x \sinh y.$

If $|a| < \pi/2$, then the image of the vertical line $x = a$ is the curve in the w plane given by the parametric equations

(4) $\qquad u = \sin a \cosh y \qquad \text{and} \qquad v = \cos a \sinh y$

for $-\infty < y < \infty$. We can rewrite (4) in the form

$$\text{(5)} \qquad \cosh y = \frac{u}{\sin a} \qquad \text{and} \qquad \sinh y = \frac{v}{\cos a}.$$

We can eliminate y in (5) by squaring and using the hyperbolic identity $\cosh^2 y - \sinh^2 y = 1$, and the result is the single equation

$$\text{(6)} \qquad \frac{u^2}{\sin^2 a} - \frac{v^2}{\cos^2 a} = 1.$$

The curve given by (6) is identified as a hyperbola in the (u, v) plane that has foci at the points $(\pm 1, 0)$. Therefore the vertical line $x = a$ is mapped in a one-to-one manner onto the branch of the hyperbola given by (6) that passes through the point $(\sin a, 0)$. If $0 < a < \pi/2$, then it is the right branch; and if $-\pi/2 < a < 0$, it is the left branch. The image of the y axis, which is the line $x = 0$, is the v axis. The images of several vertical lines are shown in Figure 8.14.

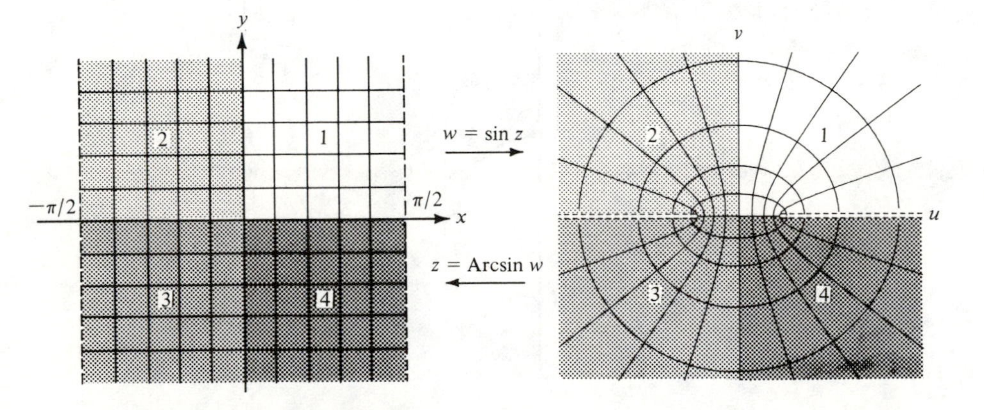

FIGURE 8.14 The transformation $w = \sin z$.

The image of the horizontal segment $-\pi/2 < x < \pi/2$, $y = b$ is the curve in the w plane given by the parametric equations

$$\text{(7)} \qquad u = \sin x \cosh b \qquad \text{and} \qquad v = \cos x \sinh b$$

for $-\pi/2 < x < \pi/2$. We can rewrite (7) in the form

$$\text{(8)} \qquad \sin x = \frac{u}{\cosh b} \qquad \text{and} \qquad \cos x = \frac{v}{\sinh b}.$$

We can eliminate x in (8) by squaring and using the trigonometric identity $\sin^2 x + \cos^2 x = 1$, and the result is the single equation

$$\text{(9)} \qquad \frac{u^2}{\cosh^2 b} + \frac{v^2}{\sinh^2 b} = 1.$$

The curve given by equation (9) is identified as an ellipse in the (u, v) plane that passes through the points $(\pm\cosh b, 0)$ and $(0, \pm\sinh b)$ and has foci at the points $(\pm 1, 0)$. Therefore if $b > 0$, then $v = \cos x \sinh b > 0$, and the image of the horizontal segment is the portion of the ellipse given by equation (9) that lies in the upper half-plane $\operatorname{Im} w > 0$. If $b < 0$, then it is the portion that lies in the lower half-plane. The images of several segments are shown in Figure 8.14.

We now develop explicit formulas for the real and imaginary parts of the principal value of the arcsine function $w = f(z) = \operatorname{Arcsin} z$. This mapping will be used to solve certain problems involving steady temperatures and ideal fluid flow in Chapter 9. The mapping is found by solving the equation

$$(10) \qquad x + iy = \sin w = \sin u \cosh v + i \cos u \sinh v$$

for u and v expressed as functions of x and y. To solve for u, we first equate the real and imaginary parts of equation (10) and obtain the system of equations

$$(11) \qquad \cosh v = \frac{x}{\sin u} \quad \text{and} \quad \sinh v = \frac{y}{\cos u}.$$

Then we eliminate v in (11) and obtain the single equation

$$(12) \qquad \frac{x^2}{\sin^2 u} - \frac{y^2}{\cos^2 u} = 1.$$

If we treat u as a constant, then equation (12) represents a hyperbola in the (x, y) plane, the foci occur at the points $(\pm 1, 0)$, and the traverse axis is given by $2 \sin u$. Therefore a point (x, y) on the hyperbola must satisfy the equation

$$(13) \qquad 2 \sin u = \sqrt{(x+1)^2 + y^2} - \sqrt{(x-1)^2 + y^2}.$$

The quantity on the right side of equation (13) represents the difference of the distances from (x, y) to $(-1, 0)$ and from (x, y) to $(1, 0)$.

Solving equation (13) for u yields the real part

$$(14) \qquad u(x, y) = \arcsin\left(\frac{\sqrt{(x+1)^2 + y^2} - \sqrt{(x-1)^2 + y^2}}{2}\right).$$

The principal branch of the real function $\arcsin t$ is used in equation (14) where the range values satisfy the inequality $-\pi/2 < \arcsin t < \pi/2$.

Similarly, we can start with equation (10) and obtain the system of equations

$$(15) \qquad \sin u = \frac{x}{\cosh v} \quad \text{and} \quad \cos u = \frac{y}{\sinh v}.$$

Then we eliminate u in (15) and obtain the single equation

$$(16) \qquad \frac{x^2}{\cosh^2 v} + \frac{y^2}{\sinh^2 v} = 1.$$

If we treat v as a constant, then equation (16) represents an ellipse in the (x, y)

plane, the foci occur at the points $(\pm 1, 0)$, and the major axis has length $2\cosh v$. Therefore a point (x, y) on this ellipse must satisfy the equation

$$(17)\qquad 2\cosh v = \sqrt{(x+1)^2 + y^2} + \sqrt{(x-1)^2 + y^2}.$$

The quantity on the right side of equation (17) represents the sum of the distances from (x, y) to $(-1, 0)$ and from (x, y) to $(1, 0)$.

The function $z = \sin w$ maps points in the upper-half (lower-half) of the vertical strip $-\pi/2 < u < \pi/2$ onto the upper half-plane (lower half-plane), respectively. Hence we can solve equation (17) to obtain v as a function of x and y:

$$(18)\qquad v(x, y) = (\text{sign } y)\operatorname{arccosh}\left(\frac{\sqrt{(x+1)^2 + y^2} + \sqrt{(x-1)^2 + y^2}}{2}\right)$$

where $\text{sign } y = +1$ if $y \geqq 0$ and $\text{sign } y = -1$ if $y < 0$. The real function $\operatorname{arccosh} t = \ln(t + \sqrt{t^2 - 1})$ with $t \geqq 1$ is used in equation (18).

Therefore the mapping $w = \operatorname{Arcsin} z$ is a one-to-one conformal mapping of the z plane cut along the rays $x \leqq -1$, $y = 0$ and $x \geqq 1$, $y = 0$ onto the vertical strip $-\pi/2 < u < \pi/2$ in the w plane. The Arcsine transformation is indicated in Figure 8.14. The formulas in equations (14) and (18) are also convenient for evaluating $\operatorname{Arcsin} z$ as shown in the following example.

EXAMPLE 8.14 Find the principal value $\operatorname{Arcsin}(1 + i)$.

Solution Using formulas (14) and (18), we find that

$$u(1, 1) = \arcsin \frac{\sqrt{5} - 1}{2} \approx 0.666239432 \qquad \text{and}$$

$$v(1, 1) = \operatorname{arccosh} \frac{\sqrt{5} + 1}{2} \approx 1.061275062.$$

Therefore we obtain

$$\arcsin(1 + i) \approx 0.666239432 + i\, 1.061275062.$$

Is there any reason to assume that there exists a conformal mapping for some specified domain D onto another domain G? The theorem concerning the existence of conformal mappings is attributed to Riemann and can be found in Lars V. Ahlfors, *Complex Analysis* (New York: McGraw-Hill Book Co.) Chapter 6, 1966.

Theorem 8.4 (Riemann Mapping Theorem) *If D is any simply connected domain in the plane (other than the entire plane itself), then there exists a one-to-one conformal mapping $w = f(z)$ that maps D onto the unit disk $|w| < 1$.*

EXERCISES FOR SECTION 8.4

1. Find the image of the semi-infinite strip $-\pi/4 < x < 0$, $y > 0$ under the mapping $w = \tan z$.
2. Find the image of the vertical strip $0 < \text{Re } z < \pi/2$ under the mapping $w = \tan z$.
3. Find the image of the vertical line $x = \pi/4$ under the transformation $w = \sin z$.
4. Find the image of the horizontal line $y = 1$ under the transformation $w = \sin z$.
5. Find the image of the rectangle $R = \{x + iy\colon 0 < x < \pi/4,\ 0 < y < 1\}$ under the transformation $w = \sin z$.
6. Find the image of the semi-infinite strip $-\pi/2 < x < 0$, $y > 0$ under the mapping $w = \sin z$.
7. **(a)** Find $\lim_{y \to +\infty} \text{Arg}[\sin((\pi/6) + iy)]$.

   (b) Find $\lim_{y \to +\infty} \text{Arg}[\sin((-2\pi/3) + iy)]$.
8. Use formulas (14) and (18) to find the following:
   (a) $\text{Arcsin}(2 + 2i)$ **(b)** $\text{Arcsin}(-2 + i)$
   (c) $\text{Arcsin}(1 - 3i)$ **(d)** $\text{Arcsin}(-4 - i)$
9. Show that the function $w = \sin z$ maps the rectangle $R = \{x + iy\colon -\pi/2 < x < \pi/2,\ 0 < y < b\}$ one-to-one and onto the portion of the upper half-plane $\text{Im } w > 0$ that lies inside the ellipse

$$\frac{u^2}{\cosh^2 b} + \frac{v^2}{\sinh^2 b} = 1.$$

10. Find the image of the vertical strip $-\pi/2 < x < 0$ under the mapping $w = \cos z$.
11. Find the image of the horizontal strip $0 < \text{Im } z < \pi/2$ under the mapping $w = \sinh z$.
12. Find the image of the right half-plane $\text{Re } z > 0$ under the mapping

$$w = \arctan z = \frac{i}{2} \text{Log} \frac{i + z}{i - z}.$$

13. Find the image of the first quadrant $x > 0$, $y > 0$ under the mapping $w = \text{Arcsin } z$.
14. Find the image of the first quadrant $x > 0$, $y > 0$ under the mapping $w = \text{Arcsin}(z^2)$.
15. Show that the transformation $w = \sin^2 z$ is a one-to-one conformal mapping of the semi-infinite strip $0 < x < \pi/2$, $y > 0$ onto the upper half-plane $\text{Im } w > 0$.
16. Find the image of the semi-infinite strip $|x| < \pi/2$, $y > 0$ under the mapping $w = \text{Log}(\sin z)$.

9

Applications of Harmonic Functions

9.1 Preliminaries

In most applications involving harmonic functions it is required to find a harmonic function that takes on prescribed values along certain contours. We will assume that the reader is familiar with the material in Sections 3.3 and 4.4.

EXAMPLE 9.1 Find the function $u(x, y)$ that is harmonic in the vertical strip $a \leqq \operatorname{Re} z \leqq b$ and takes on the boundary values

$$(1) \qquad u(a, y) = U_1 \quad \text{and} \quad u(b, y) = U_2$$

along the vertical lines $x = a$ and $x = b$, respectively.

Solution Intuition suggests that we should seek a solution that takes on constant values along the vertical lines $x = x_0$ and that $u(x, y)$ should be a function of x alone; that is,

$$(2) \qquad u(x, y) = P(x) \quad \text{for } a \leqq x \leqq b \text{ and for all } y.$$

Laplace's equation, $u_{xx}(x, y) + u_{yy}(x, y) = 0$, implies that $P''(x) = 0$, so $P(x) = mx + c$ where m and c are constants. The boundary conditions $u(a, y) = P(a) = U_1$ and $u(b, y) = P(b) = U_2$ lead to the solution

$$(3) \qquad u(x, y) = U_1 + \frac{U_2 - U_1}{b - a}(x - a).$$

The level curves $u(x, y) =$ constant are vertical lines as indicated in Figure 9.1.

EXAMPLE 9.2 Find the function $\Psi(x, y)$ that is harmonic in the sector $0 < \operatorname{Arg} z < \alpha$ $(\alpha \leqq \pi)$ and takes on the boundary values

$$(4) \qquad \begin{aligned} &\Psi(x, 0) = C_1 && \text{for } x > 0 \quad \text{and} \\ &\Psi(x, y) = C_2 && \text{at points on the ray } r > 0,\ \theta = \alpha. \end{aligned}$$

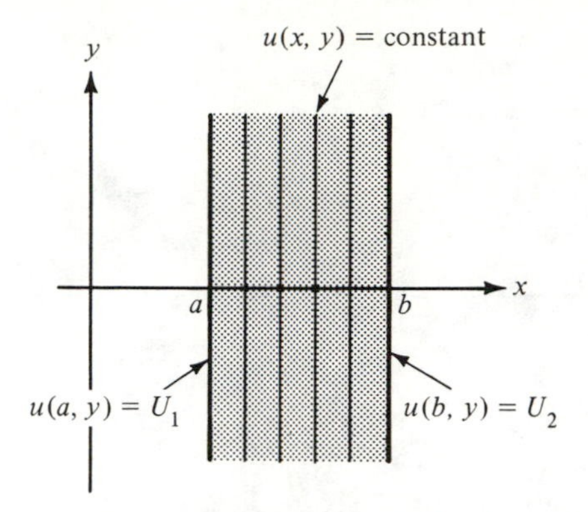

FIGURE 9.1 The harmonic function $u(x, y) = U_1 + (U_2 - U_1)(x - a)/(b - a)$.

Solution If we recall that the function Arg z is harmonic and takes on constant values along rays emanating from the origin, then we see that a solution has the form

$$\text{(5)} \qquad \Psi(x, y) = a + b \operatorname{Arg} z$$

where a and b are constants. The boundary conditions (4) lead to

$$\text{(6)} \qquad \Psi(x, y) = C_1 + \frac{C_2 - C_1}{\alpha} \operatorname{Arg} z.$$

The situation is shown in Figure 9.2.

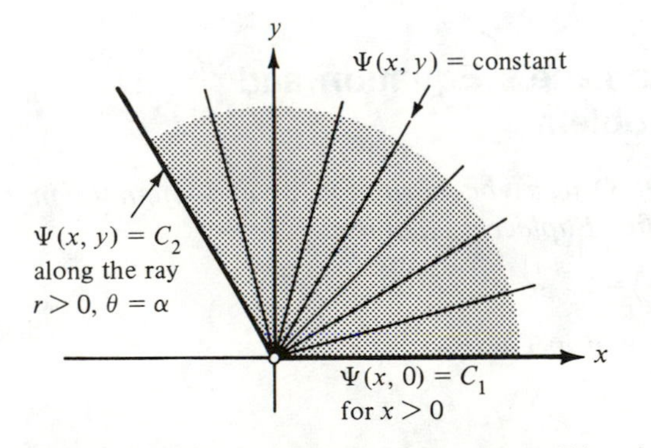

FIGURE 9.2 The harmonic function $\Psi(x, y) = C_1 + (C_2 - C_1)(1/\alpha) \operatorname{Arg} z$.

EXAMPLE 9.3 Find the function $\Phi(x, y)$ that is harmonic in the annulus $1 < |z| < R$ and takes on the boundary values

$$\text{(7)} \qquad \begin{aligned} &\Phi(x, y) = K_1 \quad \text{when } |z| = 1 \quad \text{and} \\ &\Phi(x, y) = K_2 \quad \text{when } |z| = R. \end{aligned}$$

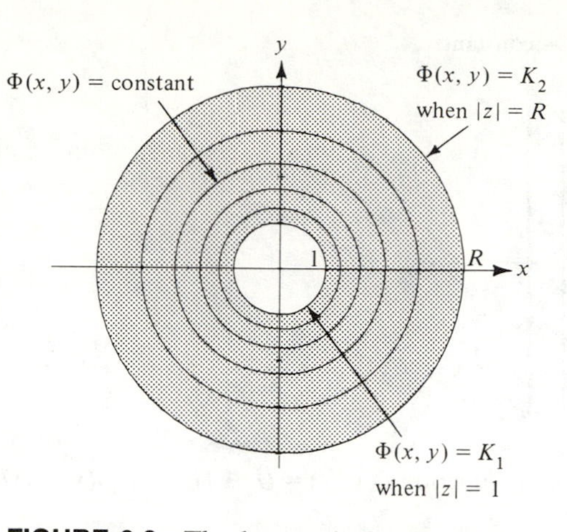

FIGURE 9.3 The harmonic function $\Phi(x, y) = K_1 + \ln|z|(K_2 - K_1)/\ln R$.

Solution This is a companion problem to Example 9.2. Here we use the fact that $\ln|z|$ is a harmonic function for all $z \neq 0$. Let us announce that the solution is

$$\Phi(x, y) = K_1 + \frac{K_2 - K_1}{\ln R} \ln|z| \tag{8}$$

and that the level curves $\Phi(x, y) =$ constant are concentric circles, as illustrated in Figure 9.3.

9.2 Invariance of Laplace's Equation and the Dirichlet Problem

Theorem 9.1 *Let $\Phi(u, v)$ be harmonic in a domain G in the w plane. Then Φ satisfies Laplace's equation*

$$\Phi_{uu}(u, v) + \Phi_{vv}(u, v) = 0 \tag{1}$$

at each point $w = u + iv$ in G. If

$$w = f(z) = u(x, y) + iv(x, y) \tag{2}$$

is a conformal mapping from a domain D in the z plane onto G, then the composition

$$\phi(x, y) = \Phi(u(x, y), v(x, y)) \tag{3}$$

is harmonic in D, and ϕ satisfies Laplace's equation

$$\phi_{xx}(x, y) + \phi_{yy}(x, y) = 0 \tag{4}$$

at each point $z = x + iy$ in D.

Proof Equations (1) and (4) are facts about the harmonic functions Φ and ϕ that were studied in Section 3.3. A direct proof that the function ϕ in (3) is harmonic would involve a tedious calculation of the partial derivatives ϕ_{xx} and ϕ_{yy}. An easier proof uses a complex variable technique. Let us assume that there is a harmonic conjugate $\Psi(u, v)$ so that the function

$$(5) \qquad g(w) = \Phi(u, v) + i\Psi(u, v)$$

is analytic in a neighborhood of the point $w_0 = f(z_0)$. Then the composition $h(z) = g(f(z))$ is analytic in a neighborhood of z_0 and can be written as

$$(6) \qquad h(z) = \Phi(u(x, y), v(x, y)) + i\Psi(u(x, y), v(x, y)).$$

If we use Theorem 3.6, it follows that $\Phi(u(x, y), v(x, y))$ is harmonic in a neighborhood of z_0, and Theorem 9.1 is established.

EXAMPLE 9.4 Show that $\phi(x, y) = \arctan(2x/(x^2 + y^2 - 1))$ is harmonic in the disk $|z| < 1$ where $-\pi/2 < \arctan t < \pi/2$.

Solution The results of Exercise 10 of Section 8.2 show that the function

$$(7) \qquad f(z) = \frac{i + z}{i - z} = \frac{1 - x^2 - y^2}{x^2 + (y - 1)^2} - \frac{i2x}{x^2 + (y - 1)^2}$$

is a conformal mapping of the unit disk $|z| < 1$ onto the right half-plane $\text{Re}\, w > 0$. The results from Exercise 12 in Section 4.4 show that the function

$$(8) \qquad \Phi(u, v) = \arctan \frac{v}{u} = \text{Arg}(u + iv)$$

is harmonic in the right half-plane $\text{Re}\, w > 0$. We can use (7) to write

$$(9) \qquad u(x, y) = \frac{1 - x^2 - y^2}{x^2 + (y - 1)^2} \quad \text{and} \quad v(x, y) = \frac{-2x}{x^2 + (y - 1)^2}.$$

Substituting (9) into (8) and using (3), we see that $\phi(x, y) = \arctan(v(x, y)/u(x, y)) = \arctan(2x/(x^2 + y^2 - 1))$ is harmonic for $|z| < 1$.

Let D be a domain whose boundary is made up of piecewise smooth contours joined end to end. The *Dirichlet problem* is to find a function ϕ that is harmonic in D such that ϕ takes on prescribed values at points on the boundary. Let us first study this problem in the upper half-plane.

EXAMPLE 9.5 The function

$$(10) \qquad \Phi(u, v) = \frac{1}{\pi} \text{Arctan} \frac{v}{u - u_0} = \frac{1}{\pi} \text{Arg}(w - u_0)$$

is harmonic in the upper half-plane $\operatorname{Im} w > 0$ and takes on the boundary values

$$(11)\quad \begin{aligned} &\Phi(u, 0) = 0 \quad \text{for } u > u_0 \quad \text{and} \\ &\Phi(u, 0) = 1 \quad \text{for } u < u_0. \end{aligned}$$

Solution The function

$$(12)\quad g(w) = \frac{1}{\pi} \operatorname{Log}(w - u_0) = \frac{1}{\pi} \ln|w - u_0| + \frac{i}{\pi} \operatorname{Arg}(w - u_0)$$

is analytic in the upper half-plane $\operatorname{Im} w > 0$, and its imaginary part is the harmonic function $(1/\pi) \operatorname{Arg}(w - u_0)$.

Remark Let t be a real number. We shall use the convention $\operatorname{Arctan}(\pm\infty) = \pi/2$ so that the function $\operatorname{Arctan} t$ denotes the branch of the inverse tangent that lies in the range $0 < \operatorname{Arctan} t < \pi$. This will permit us to write $\Phi(u, v) = (1/\pi) \operatorname{Arctan}(v/(u - u_0))$ in (10).

Theorem 9.2 (*N*-Value Dirichlet Problem for the Upper Half-Plane) *Let $u_1 < u_2 < \cdots < u_{N-1}$ denote $N - 1$ real constants. The function*

$$(13)\quad \begin{aligned} \Phi(u, v) &= a_{N-1} + \frac{1}{\pi} \sum_{k=1}^{N-1} (a_{k-1} - a_k) \operatorname{Arg}(w - u_k) \\ &= a_{N-1} + \frac{1}{\pi} \sum_{k=1}^{N-1} (a_{k-1} - a_k) \operatorname{Arctan} \frac{v}{u - u_k} \end{aligned}$$

is harmonic in the upper half-plane $\operatorname{Im} w > 0$ *and takes on the boundary values*

$$(14)\quad \begin{aligned} &\Phi(u, 0) = a_0 && \textit{for } u < u_1, \\ &\Phi(u, 0) = a_k && \textit{for } u_k < u < u_{k+1} \qquad \text{for } k = 1, 2, \ldots, N-2, \\ &\Phi(u, 0) = a_{N-1} && \textit{for } u > u_{N-1}. \end{aligned}$$

The situation is illustrated in Figure 9.4.

Proof Since each term in the sum in (13) is harmonic, it follows that Φ is harmonic for $\operatorname{Im} w > 0$. To show that Φ has the prescribed boundary conditions, we fix j and let $u_j < u < u_{j+1}$. Using Example 9.5, we see that

$$(15)\quad \frac{1}{\pi} \operatorname{Arg}(u - u_k) = 0 \ \text{ if } k \leqq j \quad \text{and} \quad \frac{1}{\pi} \operatorname{Arg}(u - u_k) = 1 \ \text{ if } k > j.$$

Using (15) in (13) results in

$$(16)\quad \begin{aligned} \Phi(u, 0) &= a_{N-1} + \sum_{k=1}^{j} (a_{k-1} - a_k)(0) + \sum_{k=j+1}^{N-1} (a_{k-1} - a_k)(1) \\ &= a_{N-1} + (a_{N-2} - a_{N-1}) + \cdots + (a_{j+1} - a_{j+2}) + (a_j - a_{j+1}) \\ &= a_j \qquad \text{for } u_j < u < u_{j+1}. \end{aligned}$$

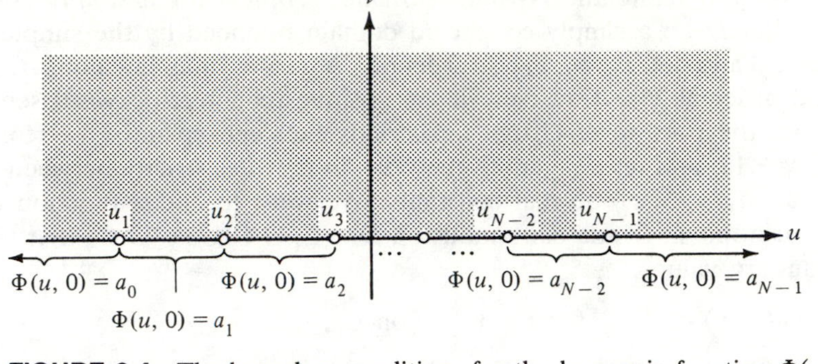

FIGURE 9.4 The boundary conditions for the harmonic function $\Phi(u, v)$ in the statement of Theorem 9.2.

The reader can verify that the boundary conditions are correct for $u < u_1$ and $u > u_{N-1}$, and the result will be established.

EXAMPLE 9.6 Find the function $\phi(x, y)$ that is harmonic in the upper half-plane Re $z > 0$, which takes on the boundary values indicated in Figure 9.5.

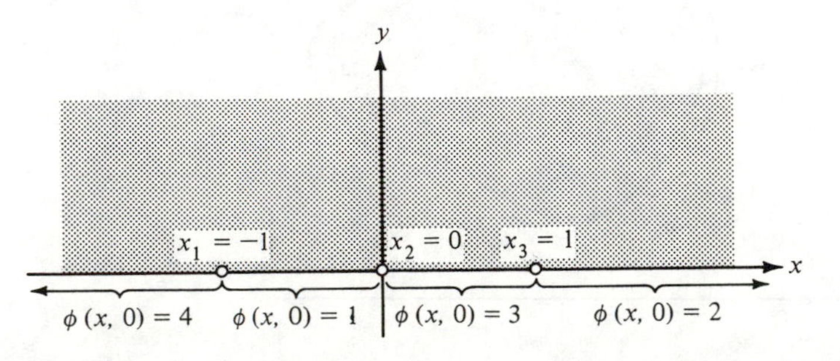

FIGURE 9.5 The boundary values for the Dirichlet problem in Example 9.6.

Solution This is a four-value Dirichlet problem in the upper half-plane Im $z > 0$. For the z plane the solution in (13) becomes

$$\phi(x, y) = a_3 + \frac{1}{\pi} \sum_{k=1}^{3} (a_{k-1} - a_k) \operatorname{Arg}(z - x_k). \tag{17}$$

Here we have $a_0 = 4$, $a_1 = 1$, $a_2 = 3$, $a_3 = 2$ and $x_1 = -1$, $x_2 = 0$, $x_3 = 1$, which can be substituted into (17) to obtain

$$\begin{aligned}\phi(x, y) &= 2 + \frac{4-1}{\pi} \operatorname{Arg}(z+1) + \frac{1-3}{\pi} \operatorname{Arg}(z-0) + \frac{3-2}{\pi} \operatorname{Arg}(z-1) \\ &= 2 + \frac{3}{\pi} \operatorname{Arctan} \frac{y}{x+1} - \frac{2}{\pi} \operatorname{Arctan} \frac{y}{x} + \frac{1}{\pi} \operatorname{Arctan} \frac{y}{x-1}.\end{aligned}$$

We now state the *N-value Dirichlet problem* for a *simply connected domain*. Let D be a simply connected domain bounded by the simple closed contour C, and let $z_1, z_2, \ldots, z_N$ denote N points that lie along C in this specified order as C is traversed in the positive (counterclockwise) sense. Let C_k denote the portion of C that lies strictly between z_k and z_{k+1} (for $k = 1, 2, \ldots, N-1$), and let C_N denote the portion that lies strictly between z_N and z_1. Let $a_1, a_2, \ldots, a_N$ be real constants. We want to find a function $\phi(x, y)$ that is harmonic in D and continuous on $D \cup C_1 \cup C_2 \cup \cdots \cup C_N$ that takes on the boundary values:

$$(18) \qquad \begin{aligned} \phi(x, y) &= a_1 \qquad \text{for } z = x + iy \text{ on } C_1, \\ \phi(x, y) &= a_2 \qquad \text{for } z = x + iy \text{ on } C_2, \\ &\vdots \\ \phi(x, y) &= a_N \qquad \text{for } z = x + iy \text{ on } C_N. \end{aligned}$$

The situation is illustrated in Figure 9.6.

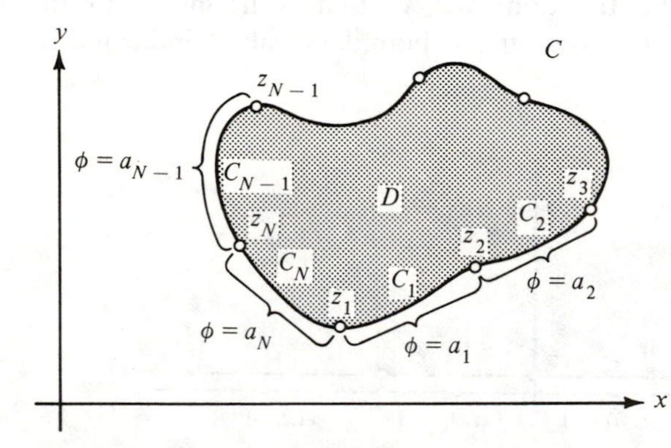

FIGURE 9.6 The boundary values for $\phi(x, y)$ for the Dirichlet problem in the simply connected domain D.

One method for finding ϕ is to find a conformal mapping

$$(19) \qquad w = f(z) = u(x, y) + iv(x, y)$$

of D onto the upper half-plane $\text{Im}\, w > 0$, such that the N points $z_1, z_2, \ldots, z_N$ are mapped onto the points $u_k = f(z_k)$ for $k = 1, 2, \ldots, N-1$ and z_N is mapped onto $u_N = +\infty$ along the u axis in the w plane.

Using Theorem 9.1, we see that the mapping in (19) gives rise to a new N-value Dirichlet problem in the upper half-plane $\text{Im}\, w > 0$ for which the solution is given by Theorem 9.2. If we set $a_0 = a_N$, then the solution to the

Dirichlet problem in D with boundary values (18) is

$$\phi(x, y) = a_{N-1} + \frac{1}{\pi}\sum_{k=1}^{N-1}(a_{k-1} - a_k)\operatorname{Arg}(f(z) - u_k) = a_{N-1} + \frac{1}{\pi}\sum_{k=1}^{N-1}(a_{k-1} - a_k)\operatorname{Arctan}\frac{v(x, y)}{u(x, y) - u_k}. \tag{20}$$

This method relies on our ability to construct a conformal mapping from D onto the upper half-plane $\operatorname{Im} w > 0$. Theorem 8.4 guarantees the existence of such a conformal mapping.

EXAMPLE 9.7 Find a function $\phi(x, y)$ that is harmonic in the unit disk $|z| < 1$ and takes on the boundary values

$$\begin{aligned} \phi(x, y) &= 0 \quad \text{for } x + iy = e^{i\theta},\ 0 < \theta < \pi, \\ \phi(x, y) &= 1 \quad \text{for } x + iy = e^{i\theta},\ \pi < \theta < 2\pi. \end{aligned} \tag{21}$$

Solution Example 8.3 showed that the function

$$u + iv = \frac{i(1 - z)}{1 + z} = \frac{2y}{(x+1)^2 + y^2} + i\,\frac{1 - x^2 - y^2}{(x+1)^2 + y^2} \tag{22}$$

is a one-to-one conformal mapping of the unit disk $|z| < 1$ onto the upper half-plane $\operatorname{Im} w > 0$. Using (22), we see that the points $z = x + iy$ that lie on the upper semicircle $y > 0$, $1 - x^2 - y^2 = 0$ are mapped onto the positive u axis. Similarly, the lower semicircle is mapped onto the negative u axis as shown in Figure 9.7.

The mapping (22) gives rise to a new Dirichlet problem of finding a

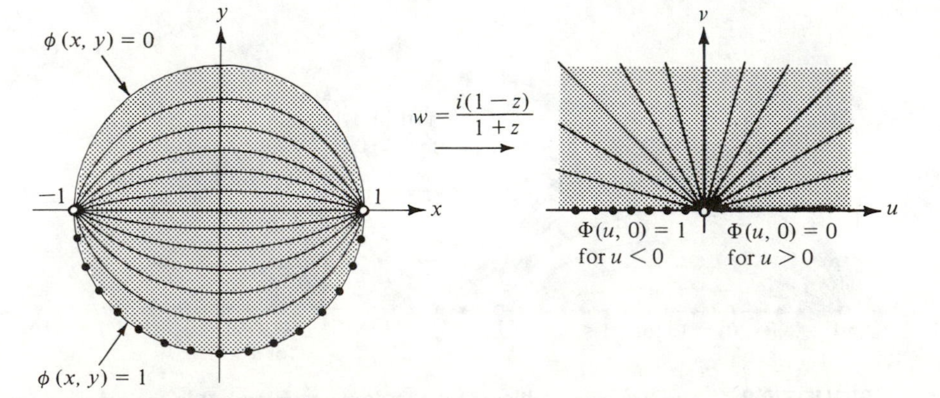

FIGURE 9.7 The Dirichlet problems for $|z| < 1$ and $\operatorname{Im} w > 0$ in the solution of Example 9.7.

harmonic function $\Phi(u, v)$ that has the boundary values

$$(23) \qquad \Phi(u, 0) = 0 \quad \text{for } u > 0 \qquad \text{and} \qquad \Phi(u, 0) = 1 \quad \text{for } u < 0$$

as shown in Figure 9.7. Using the result of Example 9.5 and the functions u and v in the mapping (22), we find that the solution to (21) is

$$\phi(x, y) = \frac{1}{\pi} \text{ Arctan } \frac{v(x, y)}{u(x, y)} = \frac{1}{\pi} \text{ Arctan } \frac{1 - x^2 - y^2}{2y}.$$

EXAMPLE 9.8 Find a function $\phi(x, y)$ that is harmonic in the upper half-disk H: $y > 0$, $|z| < 1$ and takes on the boundary values

$$(24) \qquad \begin{aligned} \phi(x, y) &= 0 \qquad \text{for } x + iy = e^{i\theta},\ 0 < \theta < \pi, \\ \phi(x, 0) &= 1 \qquad \text{for } -1 < x < 1. \end{aligned}$$

Solution By using the result of Exercise 4 in Section 8.2 the function in (22) is seen to map the upper half-disk H onto the first quadrant Q: $u > 0$, $v > 0$. Using (22), we see that points $z = x + iy$ that lie on the segment $y = 0$, $-1 < x < 1$ are mapped onto the positive v axis.

The mapping (22) gives rise to a new Dirichlet problem of finding a harmonic function $\Phi(u, v)$ in Q that has the boundary values

$$(25) \qquad \Phi(u, 0) = 0 \quad \text{for } u > 0 \qquad \text{and} \qquad \Phi(0, v) = 1 \quad \text{for } v > 0$$

as shown in Figure 9.8. In this case the method in Example 9.2 can be used to see that $\Phi(u, v)$ is given by

$$(26) \qquad \Phi(u, v) = 0 + \frac{1 - 0}{\pi/2} \text{ Arg } w = \frac{2}{\pi} \text{ Arg } w = \frac{2}{\pi} \text{ Arctan } \frac{v}{u}.$$

Using the functions u and v in (22) in equation (26), we find that the solution

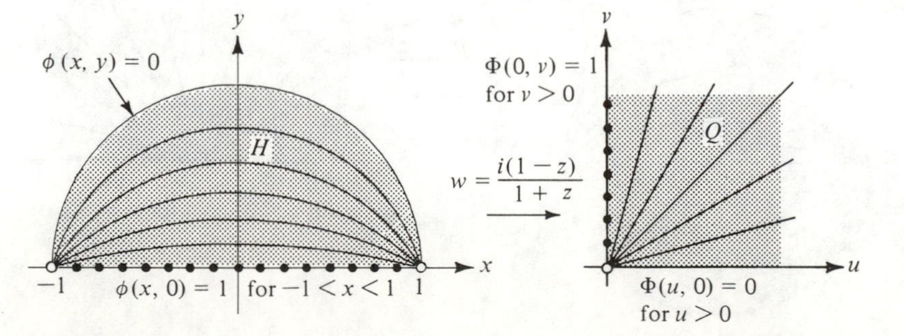

FIGURE 9.8 The Dirichlet problems for the domains H and Q in the solution of Example 9.8.

of the Dirichlet problem in H is

$$\phi(x, y) = \frac{2}{\pi} \operatorname{Arctan} \frac{v(x, y)}{u(x, y)} = \frac{2}{\pi} \operatorname{Arctan} \frac{1 - x^2 - y^2}{2y}.$$

EXAMPLE 9.9 Find a function $\phi(x, y)$ that is harmonic in the quarter disk G: $x > 0$, $y > 0$, $|z| < 1$ and takes on the boundary values

$$\phi(x, y) = 0 \quad \text{for } z = e^{i\theta},\ 0 < \theta < \pi/2,$$

$$\phi(x, 0) = 1 \quad \text{for } 0 < x < 1, \tag{27}$$

$$\phi(0, y) = 1 \quad \text{for } 0 \leqq y < 1.$$

Solution The function

$$u + iv = z^2 = x^2 - y^2 + i2xy \tag{28}$$

maps the quarter disk onto the upper half-disk H: $v > 0$, $|w| < 1$. The new Dirichlet problem in H is shown in Figure 9.9. From the result of Example 9.8 the solution $\Phi(u, v)$ in H is

$$\Phi(u, v) = \frac{2}{\pi} \operatorname{Arctan} \frac{1 - u^2 - v^2}{2v}. \tag{29}$$

Using (28), one can show that $u^2 + v^2 = (x^2 + y^2)^2$ and $2v = 4xy$, which can be used in (29) to show that the solution ϕ in G is

$$\phi(x, y) = \frac{2}{\pi} \operatorname{Arctan} \frac{1 - (x^2 + y^2)^2}{4xy}.$$

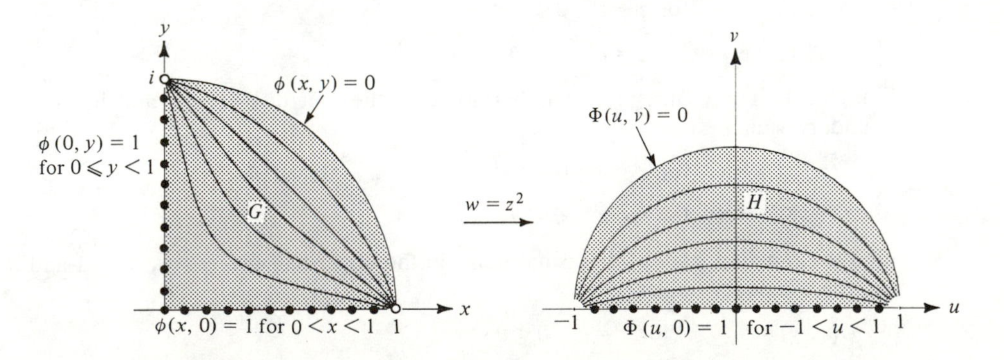

FIGURE 9.9 The Dirichlet problems for the domains G and H in the solution of Example 9.9.

EXERCISES FOR SECTION 9.2

For all of the following exercises, find a solution $\phi(x, y)$ of the Dirichlet problem in the domain indicated that takes on the prescribed boundary values.

1. Find the function $\phi(x, y)$ that is harmonic in the horizontal strip $1 < \text{Im } z < 2$ and has the boundary values

$$\phi(x, 1) = 6 \quad \text{for all } x, \qquad \phi(x, 2) = -3 \quad \text{for all } x.$$

2. Find the function $\phi(x, y)$ that is harmonic in the sector $0 < \text{Arg } z < \pi/3$ and has the boundary values

$$\phi(x, y) = 2 \quad \text{for Arg } z = \pi/3, \qquad \phi(x, 0) = 1 \quad \text{for } x > 0.$$

3. Find the function $\phi(x, y)$ that is harmonic in the annulus $1 < |z| < 2$ and has the boundary values

$$\phi(x, y) = 5 \quad \text{when } |z| = 1, \qquad \phi(x, y) = 8 \quad \text{when } |z| = 2.$$

4. Find the function $\phi(x, y)$ that is harmonic in the upper half-plane $\text{Im } z > 0$ and has the boundary values

$$\phi(x, 0) = 1 \quad \text{for } -1 < x < 1, \qquad \phi(x, 0) = 0 \quad \text{for } |x| > 1.$$

5. Find the function $\phi(x, y)$ that is harmonic in the upper half-plane $\text{Im } z > 0$ and has the boundary values

$$\phi(x, 0) = 3 \quad \text{for } x < -3, \qquad \phi(x, 0) = 7 \quad \text{for } -3 < x < -1,$$

$$\phi(x, 0) = 1 \quad \text{for } -1 < x < 2, \qquad \phi(x, 0) = 4 \quad \text{for } x > 2.$$

6. Find the function $\phi(x, y)$ that is harmonic in the first quadrant $x > 0$, $y > 0$ and has the boundary values

$$\phi(0, y) = 0 \quad \text{for } y > 1, \qquad \phi(0, y) = 1 \quad \text{for } 0 < y < 1,$$

$$\phi(x, 0) = 1 \quad \text{for } 0 \leqq x < 1, \qquad \phi(x, 0) = 0 \quad \text{for } x > 1.$$

7. Find the function $\phi(x, y)$ that is harmonic in the unit disk $|z| < 1$ and has the boundary values

$$\phi(x, y) = 0 \qquad \text{for } z = e^{i\theta},\ 0 < \theta < \pi,$$

$$\phi(x, y) = 5 \qquad \text{for } z = e^{i\theta},\ \pi < \theta < 2\pi.$$

8. Find the function $\phi(x, y)$ that is harmonic in the unit disk $|z| < 1$ and has the boundary values

$$\phi(x, y) = 8 \qquad \text{for } z = e^{i\theta},\ 0 < \theta < \pi,$$

$$\phi(x, y) = 4 \qquad \text{for } z = e^{i\theta},\ \pi < \theta < 2\pi.$$

9. Find the function $\phi(x, y)$ that is harmonic in the upper half-disk $y > 0$, $|z| < 1$ and has the boundary values

$$\phi(x, y) = 5 \qquad \text{for } z = e^{i\theta},\ 0 < \theta < \pi,$$

$$\phi(x, 0) = -5 \qquad \text{for } -1 < x < 1.$$

10. Find the function $\phi(x, y)$ that is harmonic in the portion of the upper half-plane

Im $z > 0$ that lies outside the circle $|z| = 1$ and has the boundary values

$$\phi(x, y) = 1 \qquad \text{for } z = e^{i\theta},\ 0 < \theta < \pi,$$

$$\phi(x, 0) = 0 \qquad \text{for } |x| > 1.$$

Hint: Use the mapping $w = -1/z$ and the result of Example 9.8.

11. Find the function $\phi(x, y)$ that is harmonic in the quarter disk $x > 0$, $y > 0$, $|z| < 1$ and has the boundary values

$$\phi(x, y) = 3 \qquad \text{for } z = e^{i\theta},\ 0 < \theta < \pi/2,$$

$$\phi(x, 0) = -3 \qquad \text{for } 0 \leqq x < 1,$$

$$\phi(0, y) = -3 \qquad \text{for } 0 < y < 1.$$

12. Find the function $\phi(x, y)$ that is harmonic in the unit disk $|z| < 1$ and has the boundary values

$$\phi(x, y) = 1 \qquad \text{for } z = e^{i\theta},\ -\pi/2 < \theta < \pi/2,$$

$$\phi(x, y) = 0 \qquad \text{for } z = e^{i\theta},\ \pi/2 < \theta < 3\pi/2.$$

9.3 Poisson's Integral Formula for the Upper Half-Plane

The Dirichlet problem for the upper half-plane Im $z > 0$ is to find a function $\phi(x, y)$ that is harmonic in the upper half-plane and has the boundary values $\phi(x, 0) = U(x)$ where $U(x)$ is a real-valued function of the real variable x.

Theorem 9.3 (Poisson's Integral Formula) *Let $U(t)$ be a real-valued function that is piecewise continuous and bounded for all real t. The function*

$$\phi(x, y) = \frac{y}{\pi}\int_{-\infty}^{\infty} \frac{U(t)\,dt}{(x-t)^2 + y^2} \tag{1}$$

is harmonic in the upper half-plane Im $z > 0$ *and has the boundary values*

$$\phi(x, 0) = U(x) \qquad \textit{at points of continuity of } U. \tag{2}$$

Proof The integral formula (1) is easy to motivate from the results of Theorem 9.2 regarding the Dirichlet problem. Let $t_1 < t_2 < \cdots < t_N$ denote N points that lie along the x axis. Let $t_0^* < t_1^* < \cdots < t_N^*$ be $N + 1$ points that are chosen so that $t_0^* < t_1$, $t_k < t_k^* < t_{k+1}$ (for $k = 1, 2, \ldots, N-1$), $t_N^* > t_N$, and $U(t)$ is continuous at each value t_k^*. Then according to Theorem 9.2, the function

$$\Phi(x, y) = U(t_N^*) + \frac{1}{\pi}\sum_{k=1}^{N} [U(t_{k-1}^*) - U(t_k^*)]\,\mathrm{Arg}(z - t_k) \tag{3}$$

is harmonic in the upper half-plane and takes on the boundary values

$$\Phi(x, 0) = U(t_0^*) \qquad \text{for } x < t_1,$$

$$(4) \qquad \Phi(x, 0) = U(t_k^*) \qquad \text{for } t_k < x < t_{k+1},$$

$$\Phi(x, 0) = U(t_N^*) \qquad \text{for } x > t_N,$$

as shown in Figure 9.10.

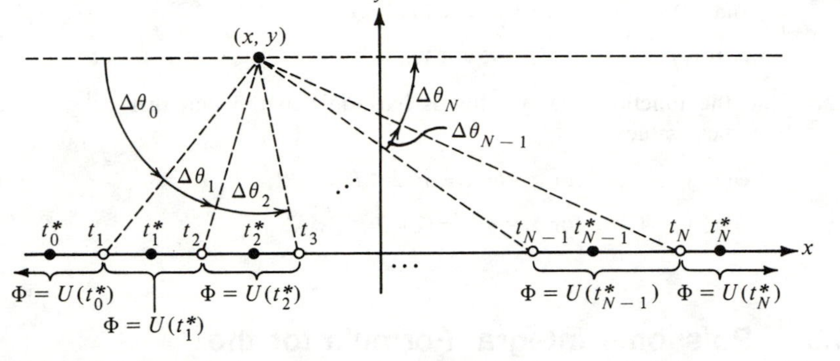

FIGURE 9.10 The boundary values for Φ in the proof of Theorem 9.3.

We can use properties of the argument of a complex number in Section 1.4 to write (3) in the form

$$(5) \qquad \Phi(x, y) = \frac{1}{\pi} U(t_0^*) \operatorname{Arg}(z - t_1) + \frac{1}{\pi} \sum_{k=1}^{N-1} U(t_k^*) \operatorname{Arg}\left(\frac{z - t_{k+1}}{z - t_k}\right)$$
$$+ \frac{1}{\pi} U(t_N^*)[\pi - \operatorname{Arg}(z - t_N)].$$

The value $\Phi(x, y)$ in (5) is given by the weighted mean

$$(6) \qquad \Phi(x, y) = \frac{1}{\pi} \sum_{k=0}^{N} U(t_k^*)\, \Delta\theta_k$$

where the angles $\Delta\theta_k$ $(k = 0, 1, \ldots, N)$ sum up to π and are shown in Figure 9.10.

Using the substitutions

$$(7) \qquad \theta = \operatorname{Arg}(z - t) = \operatorname{Arctan}\left(\frac{y}{x - t}\right) \qquad \text{and} \qquad d\theta = \frac{y\, dt}{(x - t)^2 + y^2},$$

we can write (6) as

$$(8) \qquad \Phi(x, y) = \frac{y}{\pi} \sum_{k=0}^{N} \frac{U(t_k^*)\, \Delta t_k}{(x - t_k^*)^2 + y^2}.$$

The limit of the Riemann sum (8) becomes an improper integral

$$\phi(x, y) = \frac{y}{\pi} \int_{-\infty}^{\infty} \frac{U(t)\, dt}{(x-t)^2 + y^2}.$$

and the result is established.

EXAMPLE 9.10 Find the function $\phi(x, y)$ that is harmonic in the upper half-plane $\operatorname{Im} z > 0$ and has the boundary values

$$(9) \qquad \phi(x, 0) = 1 \quad \text{for } -1 < x < 1, \qquad \phi(x, 0) = 0 \quad \text{for } |x| > 1.$$

Solution Using formula (1), we obtain

$$(10) \qquad \phi(x, y) = \frac{1}{\pi} \int_{-1}^{1} \frac{y\, dt}{(x-t)^2 + y^2}.$$

Using the antiderivative in (7), we can write the solution in (10) as

$$\phi(x, y) = \frac{1}{\pi} \operatorname{Arctan}\left(\frac{y}{x-t}\right)\Bigg|_{t=-1}^{t=1}$$

$$= \frac{1}{\pi} \operatorname{Arctan}\left(\frac{y}{x-1}\right) - \frac{1}{\pi} \operatorname{Arctan}\left(\frac{y}{x+1}\right).$$

EXAMPLE 9.11 Find the function $\phi(x, y)$ that is harmonic in the upper half-plane $\operatorname{Im} z > 0$ and has the boundary values

$$(11) \qquad \phi(x, 0) = x \quad \text{for } -1 < x < 1, \qquad \phi(x, 0) = 0 \quad \text{for } |x| > 1.$$

Solution Using formula (1), we obtain

$$(12) \qquad \phi(x, y) = \frac{y}{\pi} \int_{-1}^{1} \frac{t\, dt}{(x-t)^2 + y^2}$$

$$= \frac{y}{\pi} \int_{-1}^{1} \frac{(x-t)(-1)\, dt}{(x-t)^2 + y^2} + \frac{x}{\pi} \int_{-1}^{1} \frac{y\, dt}{(x-t)^2 + y^2}.$$

Using techniques of calculus and equations (7), we find that the solution in (12) is

$$\phi(x, y) = \frac{y}{2\pi} \ln \frac{(x-1)^2 + y^2}{(x+1)^2 + y^2} + \frac{x}{\pi} \operatorname{Arctan} \frac{y}{x-1} - \frac{x}{\pi} \operatorname{Arctan} \frac{y}{x+1}.$$

The *Poisson Integral Formula* for the *unit disk* $|z| < 1$ is

$$\phi(r\cos\theta, r\sin\theta) = \frac{1}{2\pi} \int_{-\pi}^{\pi} \frac{(1-r^2)U(t)\, dt}{1 + r^2 - 2r\cos(t-\theta)}$$

and is a representation of the harmonic function $\phi(x, y)$ that has the boundary values $\phi(\cos\theta, \sin\theta) = U(\theta)$ for $-\pi < \theta \leqq \pi$.

EXERCISES FOR SECTION 9.3

1. Use Poisson's Integral Formula to find the harmonic function $\phi(x, y)$ in the upper half-plane that takes on the boundary values

$$\begin{aligned} \phi(t,0) &= U(t) = -1 && \text{for } t < -1, \\ \phi(t,0) &= U(t) = t && \text{for } -1 < t < 1, \\ \phi(t,0) &= U(t) = 1 && \text{for } 1 < t. \end{aligned}$$

2. Use Poisson's Integral Formula to find the harmonic function $\phi(x, y)$ in the upper half-plane that takes on the boundary values

$$\begin{aligned} \phi(t,0) &= U(t) = 0 && \text{for } t < 0, \\ \phi(t,0) &= U(t) = t && \text{for } 0 < t < 1, \\ \phi(t,0) &= U(t) = 0 && \text{for } 1 < t. \end{aligned}$$

3. Use Poisson's Integral Formula for the upper half-plane to conclude that

$$\phi(x, y) = e^{-y}\cos x = \frac{y}{\pi}\int_{-\infty}^{\infty} \frac{\cos t\, dt}{(x-t)^2 + y^2}.$$

4. Use Poisson's Integral Formula for the upper half-plane to conclude that

$$\phi(x, y) = e^{-y}\sin x = \frac{y}{\pi}\int_{-\infty}^{\infty} \frac{\sin t\, dt}{(x-t)^2 + y^2}.$$

5. Show that the function $\phi(x, y)$ given by Poisson's Integral Formula is harmonic by applying Leibniz's Rule, which permits us to write

$$\left(\frac{\partial^2}{\partial x^2} + \frac{\partial^2}{\partial y^2}\right)\phi(x, y) = \frac{1}{\pi}\int_{-\infty}^{\infty} U(t)\left[\left(\frac{\partial^2}{\partial x^2} + \frac{\partial^2}{\partial y^2}\right)\frac{y}{(x-t)^2 + y^2}\right] dt.$$

6. Let $U(t)$ be a real-valued function that satisfies the conditions for Poisson's Integral Formula for the upper half-plane. If $U(t)$ is even function, that is, $U(-t) = U(t)$, then show that the harmonic function $\phi(x, y)$ has the property $\phi(-x, y) = \phi(x, y)$.

7. Let $U(t)$ be a real-valued function that satisfies the conditions for Poisson's Integral Formula for the upper half-plane. If $U(t)$ is an odd function, that is, $U(-t) = -U(t)$, then show that the harmonic function $\phi(x, y)$ has the property $\phi(-x, y) = -\phi(x, y)$.

9.4 Two-Dimensional Mathematical Models

We now turn our attention to problems involving steady state heat flow, electrostatics, and ideal fluid flow that can be solved by conformal mapping techniques. The method uses conformal mapping to carry a region in which the problem is posed to one in which the solution is easy to obtain. Since our solutions will involve only two independent variables, x and y, we first mention a basic assumption needed for the validity of the model.

The above-mentioned physical problems are real-world applications and involve solutions in three-dimensional Cartesian space. Such problems generally would involve the Laplacian in three variables and the divergence and curl of three-dimensional vector functions. Since complex analysis involves only x and y, we consider the special case in which the solution does not vary with the coordinate along the axis perpendicular to the xy plane. For steady state heat flow and electrostatics this assumption will mean that the temperature T, or the potential V, varies only with x and y. For the flow of ideal fluids this means that the fluid motion is the same in any plane that is parallel to the z plane. Curves drawn in the z plane are to be interpreted as cross sections that correspond to infinite cylinders perpendicular to the z plane. Since an infinite cylinder is the limiting case of a "long" physical cylinder, the mathematical model that we present is valid provided that the three-dimensional problem involves a physical cylinder long enough that the effects at the ends can be reasonably neglected.

In Sections 9.1 and 9.2 we learned how to obtain solutions $\phi(x, y)$ for harmonic functions. For applications it is important to consider the family of level curves

(1) $\{\phi(x, y) = K_1 \colon K_1 \text{ is a real constant}\}$

and the conjugate harmonic function $\psi(x, y)$ and its family of level curves

(2) $\{\psi(x, y) = K_2 \colon K_2 \text{ is a real constant}\}$.

It is convenient to introduce the terminology *complex potential* for the analytic function

(3) $F(z) = \phi(x, y) + i\psi(x, y)$.

The following result regarding the orthogonality of the above mentioned families of level curves will be used in developing ideas concerning the physical applications.

Theorem 9.4 (Orthogonal Families of Level Curves) *Let $\phi(x, y)$ be harmonic in a domain D. Let $\psi(x, y)$ be the harmonic conjugate, and let $F(z) = \phi(x, y) + i\psi(x, y)$ be the complex potential. Then the two families of level curves given in* (1) *and* (2), *respectively, are orthogonal in the sense that if (a, b) is a point common to two specific curves $\phi(x, y) = K_1$ and $\psi(x, y) = K_2$, and if $F'(a + ib) \neq 0$, then these two curves intersect orthogonally.*

Proof Since $\phi(x, y) = K_1$ is an implicit equation of a plane curve, the gradient vector grad ϕ, evaluated at (a, b), is perpendicular to the curve at

(a, b). This vector is given by

(4) $\mathbf{N}_1 = \phi_x(a, b) + i\phi_y(a, b).$

Similarly, the vector $\mathbf{N}_2$ defined by

(5) $\mathbf{N}_2 = \psi_x(a, b) + i\psi_y(a, b)$

is orthogonal to the curve $\psi(x, y) = K_2$ at (a, b). Using the Cauchy-Riemann equations, $\phi_x = \psi_y$ and $\phi_y = -\psi_x$, we have

(6) $$\begin{aligned}\mathbf{N}_1 \cdot \mathbf{N}_2 &= \phi_x(a, b)[\psi_x(a, b)] + \phi_y(a, b)[\psi_y(a, b)] \\ &= \phi_x(a, b)[-\phi_y(a, b)] + \phi_y(a, b)[\phi_x(a, b)] = 0.\end{aligned}$$

In addition, since $F'(a + ib) \neq 0$, we have

(7) $\phi_x(a, b) + i\psi_x(a, b) \neq 0.$

The Cauchy-Riemann equations and (7) imply that both $\mathbf{N}_1$ and $\mathbf{N}_2$ are nonzero. Therefore (6) implies that $\mathbf{N}_1$ is perpendicular to $\mathbf{N}_2$, and hence the curves are orthogonal.

The same complex potential $F(z) = \phi(x, y) + i\psi(x, y)$ has many physical interpretations. Suppose, for example, that we have solved a problem in steady state temperatures; then a similar problem with the same boundary conditions in electrostatics is obtained by interpreting the isothermals as equipotential curves and the heat flow lines as flux lines. This implies that heat flow and electrostatics correspond directly.

Or suppose we have solved a fluid flow problem; then an analogous problem in heat flow is obtained by interpreting the equipotentials as isothermals and streamlines as heat flow lines. Various interpretations of the families of level curves given in (1) and (2) and correspondences between families are summarized in Table 9.1.

TABLE 9.1 Interpretations for Level Curves

Physical Phenomenom	$\phi(x, y) = \text{constant}$	$\psi(x, y) = \text{constant}$
Heat flow	Isothermals	Heat flow lines
Electrostatics	Equipotential curves	Flux lines
Fluid flow	Equipotentials	Streamlines
Gravitational field	Gravitational potential	Lines of force
Magnetism	Potential	Lines of force
Diffusion	Concentration	Lines of flow
Elasticity	Strain function	Stress lines
Current flow	Potential	Lines of flow

9.5 Steady State Temperatures

In the theory of heat conduction the assumption is made that heat flows in the direction of decreasing temperature. We also assume that the time rate at which heat flows across a surface area is proportional to the component of the temperature gradient in the direction perpendicular to the surface area. If the temperature $T(x, y)$ does not depend on time, then the heat flow at the point (x, y) is given by the vector

$$(1) \qquad \mathbf{V}(x, y) = -K \operatorname{grad} T(x, y) = -K[T_x(x, y) + iT_y(x, y)]$$

where K is the thermal conductivity of the medium and is assumed to be constant. If Δz denotes a straight line segment of length Δs, then the amount of heat flowing across the segment per unit of time is

$$(2) \qquad \mathbf{V} \cdot \mathbf{N}\, \Delta s$$

where $\mathbf{N}$ is a unit vector perpendicular to the segment.

If we assume that no thermal energy is created or destroyed within the region, then the net amount of heat flowing through any small rectangle with sides of length Δx and Δy is identically zero (see Figure 9.11(a)). This leads to the conclusion that $T(x, y)$ is a harmonic function. The following heuristic argument is often used to suggest that $T(x, y)$ satisfies Laplace's equation. Using expression (2), we find that the amount of heat flowing out of the right edge of the rectangle in Figure 9.11(a) is approximately

$$(3) \qquad \mathbf{V} \cdot \mathbf{N}_1\, \Delta s_1 = -K[T_x(x + \Delta x, y) + iT_y(x + \Delta x, y)] \cdot [1 + 0i]\, \Delta y$$
$$= -KT_x(x + \Delta x, y)\, \Delta y,$$

and the amount of heat flowing out of the left edge is

$$(4) \qquad \mathbf{V} \cdot \mathbf{N}_2\, \Delta s_2 = -K[T_x(x, y) + iT_y(x, y)] \cdot [-1 + 0i]\, \Delta y = KT_x(x, y)\, \Delta y.$$

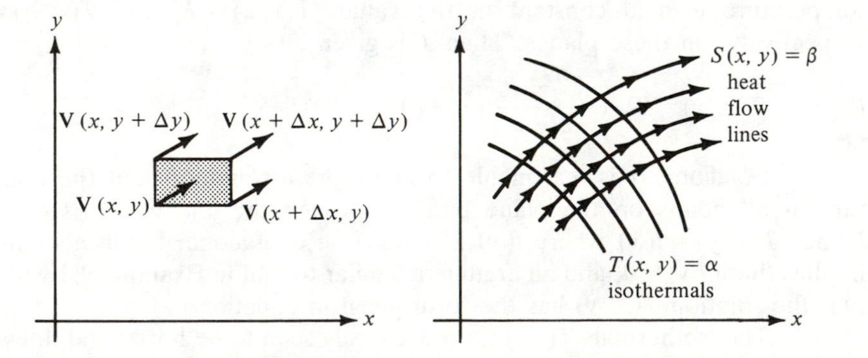

(a) The direction of heat flow. (b) Heat flow lines and isothermals.

FIGURE 9.11 Steady state temperatures.

If we add the contributions in (3) and (4), the result is

$$(5)\qquad -K\left[\frac{T_x(x+\Delta x, y) - T_x(x, y)}{\Delta x}\right]\Delta x\,\Delta y \approx -KT_{xx}(x, y)\,\Delta x\,\Delta y.$$

In a similar fashion it is found that the contribution for the amount of heat flowing out of the top and bottom edges is

$$(6)\qquad -K\left[\frac{T_y(x, y+\Delta y) - T_y(x, y)}{\Delta y}\right]\Delta x\,\Delta y \approx -KT_{yy}(x, y)\,\Delta x\,\Delta y.$$

Adding the quantities in (5) and (6), we find that the net heat flowing out of the rectangle is approximated by the equation

$$(7)\qquad -K[T_{xx}(x, y) + T_{yy}(x, y)]\,\Delta x\,\Delta y = 0,$$

which implies that $T(x, y)$ satisfies Laplace's equation and is a harmonic function.

If the domain in which $T(x, y)$ is defined is simply connected, then a conjugate harmonic function $S(x, y)$ exists, and

$$(8)\qquad F(z) = T(x, y) + iS(x, y)$$

is an analytic function. The curves $T(x, y) =$ constant are called *isothermals* and are lines connecting points of the same temperature. The curves $S(x, y) =$ constant are called the *heat flow lines*, and one can visualize the heat flowing along these curves from points of higher temperature to points of lower temperature. The situation is illustrated in Figure 9.11(b).

Boundary value problems for steady state temperatures are realizations of the Dirichlet problem where the value of the harmonic function $T(x, y)$ is interpreted as the temperature at the point (x, y).

EXAMPLE 9.12 Suppose that two parallel planes are perpendicular to the z plane and pass through the horizontal lines $y = a$ and $y = b$ and that the temperature is held constant at the values $T(x, a) = T_1$ and $T(x, b) = T_2$, respectively, on these planes. Then T is given by

$$(9)\qquad T(x, y) = T_1 + \frac{T_2 - T_1}{b - a}(y - a).$$

Solution It is reasonable to make the assumption that the temperature at all points on the plane passing through the line $y = y_0$ is constant. Hence $T(x, y) = t(y)$ where $t(y)$ is a function of y alone. Laplace's equation implies that $t''(y) = 0$, and an argument similar to that in Example 9.1 will show that the solution $T(x, y)$ has the form given in equation (9).

The isothermals $T(x, y) = \alpha$ are easily seen to be horizontal lines. The conjugate harmonic function is

$$S(x, y) = \frac{T_1 - T_2}{b - a}x,$$

and the heat flow lines $S(x, y) = \beta$ are vertical segments between the horizontal lines. If $T_1 > T_2$, then the heat flows along these segments from the plane through $y = a$ to the plane through $y = b$ as illustrated in Figure 9.12.

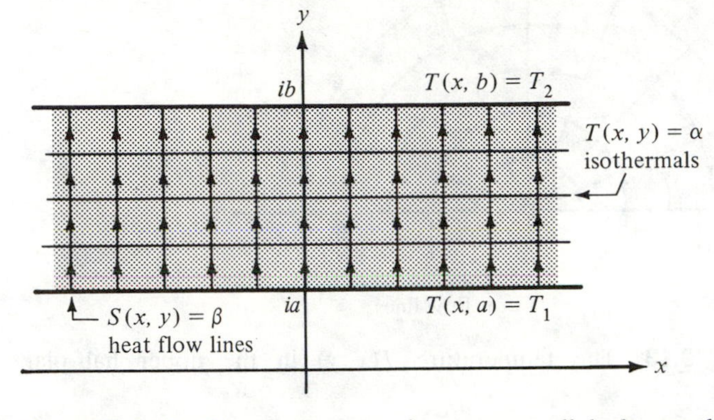

FIGURE 9.12 The temperature between parallel planes where $T_1 > T_2$.

EXAMPLE 9.13 Find the temperature $T(x, y)$ at each point in the upper half-plane $\text{Im } z > 0$ if the temperature along the x axis satisfies

$$T(x, 0) = T_1 \quad \text{for } x > 0 \qquad \text{and} \qquad T(x, 0) = T_2 \quad \text{for } x < 0. \tag{10}$$

Solution Since $T(x, y)$ is a harmonic function, this is an example of a Dirichlet problem. From Example 9.2 it follows that the solution is

$$T(x, y) = T_1 + \frac{T_2 - T_1}{\pi} \text{Arg } z. \tag{11}$$

The isothermals $T(x, y) = \alpha$ are rays emanating from the origin. The conjugate harmonic function is $S(x, y) = (1/\pi)(T_1 - T_2) \ln |z|$, and the heat flow lines $S(x, y) = \beta$ are semicircles centered at the origin. If $T_1 > T_2$, then the heat flows counterclockwise along the semicircles as shown in Figure 9.13.

EXAMPLE 9.14 Find the temperature $T(x, y)$ at each point in the upper half-disk H: $\text{Im } z > 0$, $|z| < 1$ if the temperature at points on the boundary satisfies

$$\begin{aligned} T(x, y) &= 100 \qquad \text{for } z = e^{i\theta},\ 0 < \theta < \pi, \\ T(x, 0) &= 50 \qquad \text{for } -1 < x < 1. \end{aligned} \tag{12}$$

Solution As discussed in Example 9.8, the function

$$u + iv = \frac{i(1 - z)}{1 + z} = \frac{2y}{(x + 1)^2 + y^2} + i\,\frac{1 - x^2 - y^2}{(x + 1)^2 + y^2} \tag{13}$$

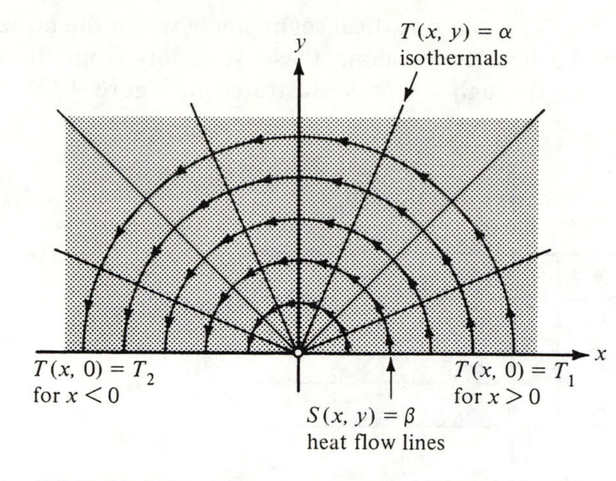

FIGURE 9.13 The temperature $T(x, y)$ in the upper half-plane where $T_1 > T_2$.

is a one-to-one conformal mapping of the half-disk H onto the first quadrant Q: $u > 0$, $v > 0$. The mapping (13) gives rise to a new problem of finding the temperature $T^*(u, v)$ that satisfies the boundary conditions

(14) $\quad T^*(u, 0) = 100 \quad \text{for } u > 0 \qquad \text{and} \qquad T^*(0, v) = 50 \quad \text{for } v > 0.$

If we use Example 9.2, the harmonic function $T^*(u, v)$ is given by

$$(15) \qquad T^*(u, v) = 100 + \frac{50 - 100}{\pi/2} \operatorname{Arg} w = 100 - \frac{100}{\pi} \operatorname{Arctan} \frac{v}{u}.$$

Substituting the expressions for u and v in (13) into equation (15) yields the desired solution

$$T(x, y) = 100 - \frac{100}{\pi} \operatorname{Arctan} \frac{1 - x^2 - y^2}{2y}.$$

The isothermals $T(x, y) = \text{constant}$ are circles that pass through the points ± 1 as shown in Figure 9.14.

We now turn our attention to the problem of finding the steady state temperature function $T(x, y)$ inside the simply connected domain D whose boundary consists of three adjacent curves C_1, C_2, and C_3 where $T(x, y) = T_1$ along C_1, $T(x, y) = T_2$ along C_2, and the region is insulated along C_3. Zero heat flowing across C_3 implies that

$$(16) \qquad \mathbf{V}(x, y) \cdot \mathbf{N}(x, y) = -K\mathbf{N}(x, y) \cdot \operatorname{grad} T(x, y) = 0$$

where $\mathbf{N}(x, y)$ is perpendicular to C_3. This means that the direction of heat flow must be parallel to this portion of the boundary. In other words, C_3 must be part of a heat flow line $S(x, y) = \text{constant}$ and the isothermals $T(x, y) = \text{constant}$ intersect C_3 orthogonally.

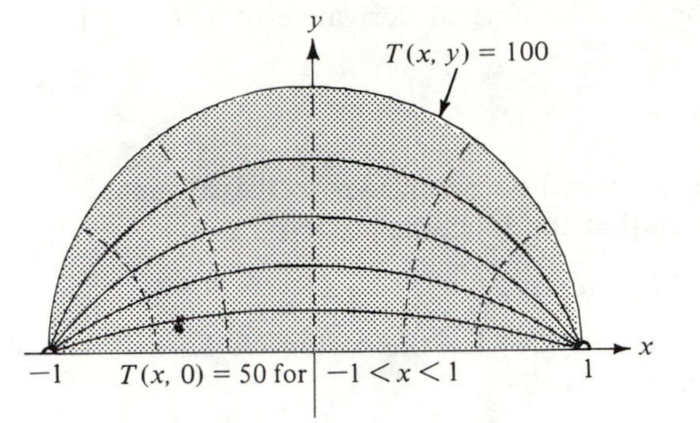

FIGURE 9.14 The temperature $T(x, y)$ in a half-disk.

This problem can be solved if we can find a conformal mapping

$$w = f(z) = u(x, y) + iv(x, y) \tag{17}$$

from D onto the semi-infinite strip G: $0 < u < 1$, $v > 0$ so that the image of the curve C_1 is the ray $u = 0$, $v > 0$; the image of the curve C_2 is the ray $u = 1$, $v > 0$; and the thermally insulated curve C_3 is mapped onto the segment $0 < u < 1$ of the u axis, as shown in Figure 9.15.

The new problem in G is to find the steady state temperature function $T^*(u, v)$ so that along the rays we have the boundary values

$$T^*(0, v) = T_1 \quad \text{for } v > 0 \qquad \text{and} \qquad T^*(1, v) = T_2 \qquad \text{for } v > 0. \tag{18}$$

The condition that a segment of the boundary is insulated can be expressed

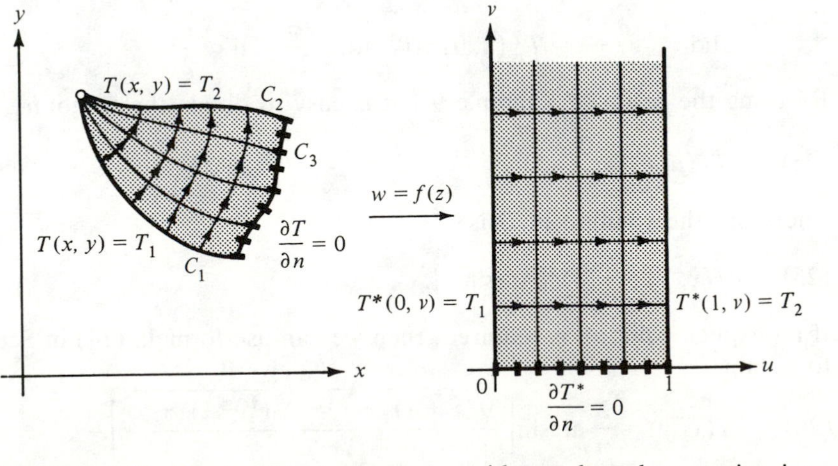

FIGURE 9.15 Steady state temperatures with one boundary portion insulated.

mathematically by saying that the normal derivative of $T^*(u, v)$ is zero. That is,

$$(19) \qquad \frac{\partial T^*}{\partial n} = T_v^*(u, 0) = 0$$

where n is a coordinate measured perpendicular to the segment.

It is easy to verify that the function

$$(20) \qquad T^*(u, v) = T_1 + (T_2 - T_1)u$$

satisfies the conditions (19) and (20) for the region G. Therefore using (17), we find that the solution in D is

$$(21) \qquad T(x, y) = T_1 + (T_2 - T_1)u(x, y).$$

The isothermals $T(x, y) = \text{constant}$, and their images under $w = f(z)$ are illustrated in Figure 9.15.

EXAMPLE 9.15 Find the steady state temperature $T(x, y)$ for the domain D consisting of the upper half-plane $\operatorname{Im} z > 0$ where $T(x, y)$ has the boundary conditions

$$(22) \qquad T(x, 0) = 1 \quad \text{for } x > 1 \qquad \text{and} \qquad T(x, 0) = -1 \quad \text{for } x < -1 \qquad \text{and}$$
$$\frac{\partial T}{\partial n} = T_y(x, 0) = 0 \quad \text{for } -1 < x < 1.$$

Solution The mapping $w = \operatorname{Arcsin} z$ conformally maps D onto the semi-infinite strip $v > 0$, $-\pi/2 < u < \pi/2$ where the new problem is to find the steady state temperature $T^*(u, v)$ that has the boundary conditions

$$(23) \qquad \begin{aligned} &T^*\left(\frac{\pi}{2}, v\right) = 1 \quad \text{for } v > 0 \qquad \text{and} \qquad T^*\left(\frac{-\pi}{2}, v\right) = -1 \quad \text{for } v > 0 \\ &\text{and} \qquad \frac{\partial T^*}{\partial n} = T_v^*(u, 0) = 0 \quad \text{for } \frac{-\pi}{2} < u < \frac{\pi}{2}. \end{aligned}$$

By using the result of Example 9.1 it is easy to obtain the solution

$$(24) \qquad T^*(u, v) = \frac{2}{\pi} u.$$

Therefore the solution in D is

$$(25) \qquad T(x, y) = \frac{2}{\pi} \operatorname{Re}[\operatorname{Arcsin} z].$$

If an explicit solution is required, then we can use formula (14) in Section 8.5 to obtain

$$(26) \qquad T(x, y) = \frac{2}{\pi} \arcsin\left[\frac{\sqrt{(x+1)^2 + y^2} - \sqrt{(x-1)^2 + y^2}}{2}\right]$$

where the real function $\arcsin t$ has range values satisfying $-\pi/2 < \arcsin t < \pi/2$.

EXERCISES FOR SECTION 9.5

1. Show that $H(x, y, z) = 1/\sqrt{x^2 + y^2 + z^2}$ satisfies Laplace's equation $H_{xx} + H_{yy} + H_{zz} = 0$ in three-dimensional Cartesian space but that $h(x, y) = 1/\sqrt{x^2 + y^2}$ *does not* satisfy equation $h_{xx} + h_{yy} = 0$ in two-dimensional Cartesian space.

2. Find the temperature function $T(x, y)$ in the infinite strip bounded by the lines $y = -x$ and $y = 1 - x$ that satisfies the boundary values in Figure 9.16.

$$T(x, -x) = 25 \qquad \text{for all } x$$

$$T(x, 1 - x) = 75 \qquad \text{for all } x$$

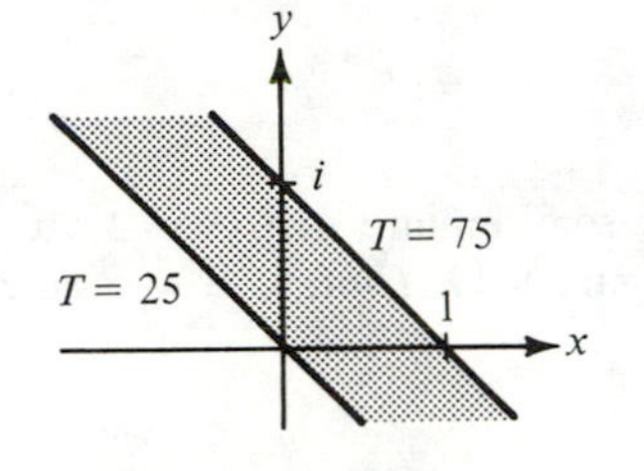

FIGURE 9.16 Accompanies Exercise 2.

3. Find the temperature function $T(x, y)$ in the first quadrant $x > 0$, $y > 0$ that satisfies the boundary values in Figure 9.17. *Hint*: Use $w = z^2$.

$$T(x, 0) = 10 \qquad \text{for } x > 1$$

$$T(x, 0) = 20 \qquad \text{for } 0 < x < 1$$

$$T(0, y) = 20 \qquad \text{for } 0 < y < 1$$

$$T(0, y) = 10 \qquad \text{for } y > 1$$

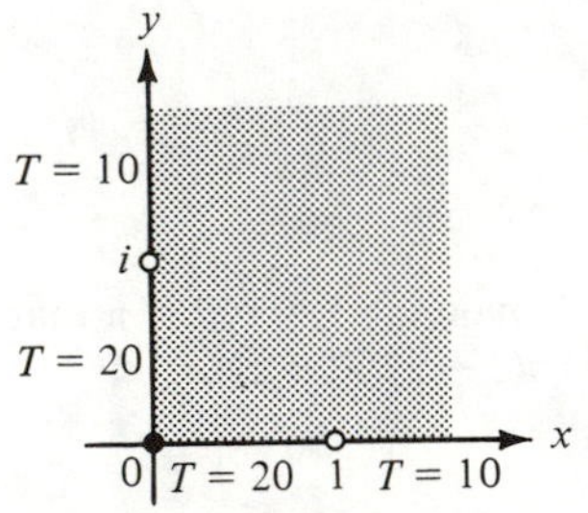

FIGURE 9.17 Accompanies Exercise 3.

4. Find the temperature function $T(x, y)$ inside the unit disk $|z| < 1$ that satisfies the boundary values in Figure 9.18. *Hint*: Use $w = i(1 - z)/(1 + z)$.

$$T(x, y) = 20 \qquad \text{for } z = e^{i\theta},\ 0 < \theta < \frac{\pi}{2}$$

$$T(x, y) = 60 \qquad \text{for } z = e^{i\theta}, \frac{\pi}{2} < \theta < 2\pi$$

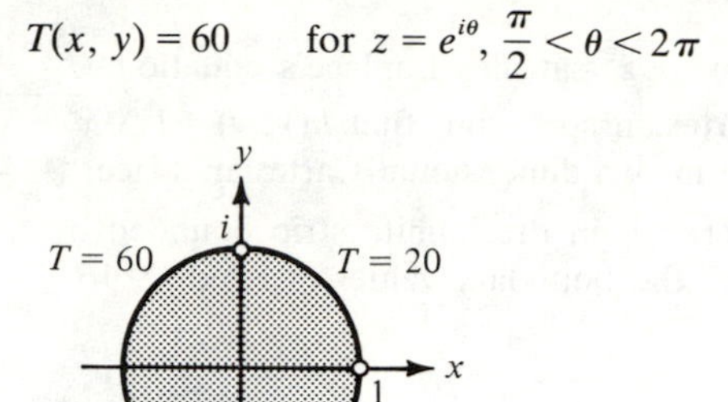

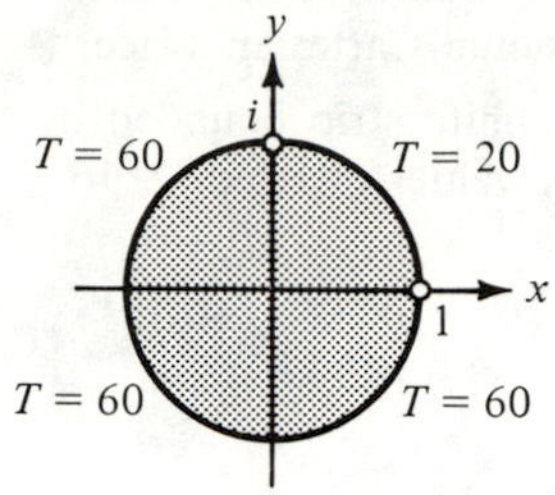

FIGURE 9.18 Accompanies Exercise 4.

5. Find the temperature function $T(x, y)$ in the semi-infinite strip $-\pi/2 < x < \pi/2$, $y > 0$ that satisfies the boundary values in Figure 9.19. *Hint*: Use $w = \sin z$.

$$T\left(\frac{\pi}{2}, y\right) = 100 \qquad \text{for } y > 0$$

$$T(x, y) = 0 \qquad \text{for } \frac{-\pi}{2} < x < \frac{\pi}{2}$$

$$T\left(\frac{-\pi}{2}, y\right) = 100 \qquad \text{for } y > 0$$

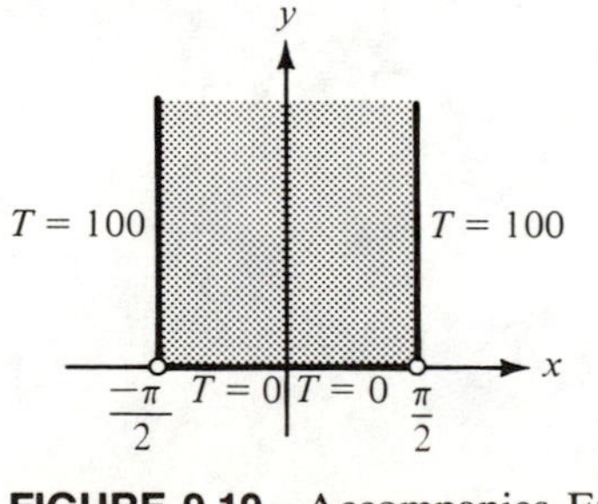

y

T = 100 T = 100

x

$\frac{-\pi}{2}$ T = 0 T = 0 $\frac{\pi}{2}$

FIGURE 9.19 Accompanies Exercise 5.

6. Find the temperature function $T(x, y)$ in the domain $r > 1$, $0 < \theta < \pi$ that satisfies the boundary values in Figure 9.20. *Hint*: $w = i(1 - z)/(1 + z)$.

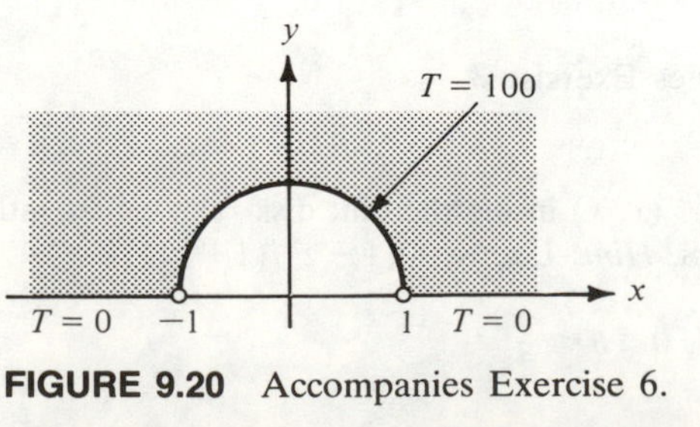

FIGURE 9.20 Accompanies Exercise 6.

$$T(x, 0) = 0 \qquad \text{for } x > 1$$

$$T(x, 0) = 0 \qquad \text{for } x < -1$$

$$T(x, y) = 100 \qquad \text{if } z = e^{i\theta},\ 0 < \theta < \pi$$

7. Find the temperature function $T(x, y)$ in the domain $1 < r < 2$, $0 < \theta < \pi/2$ that satisfies the boundary conditions in Figure 9.21. *Hint*: Use $w = \text{Log } z$.

$$T(x, y) = 0 \qquad \text{for } r = e^{i\theta},\ 0 < \theta < \frac{\pi}{2}$$

$$T(x, y) = 50 \qquad \text{for } r = 2e^{i\theta},\ 0 < \theta < \frac{\pi}{2}$$

$$\frac{\partial T}{\partial n} = T_y(x, 0) = 0 \qquad \text{for } 1 < x < 2$$

$$\frac{\partial T}{\partial n} = T_x(0, y) = 0 \qquad \text{for } 1 < y < 2$$

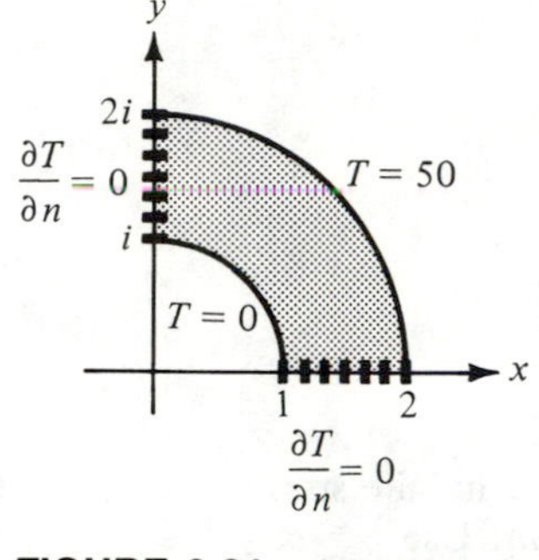

FIGURE 9.21 Accompanies Exercise 7.

8. Find the temperature function $T(x, y)$ in the domain $0 < r < 1$, $0 < \text{Arg } z < \alpha$ that satisfies the boundary conditions in Figure 9.22. *Hint*: Use $w = \text{Log } z$.

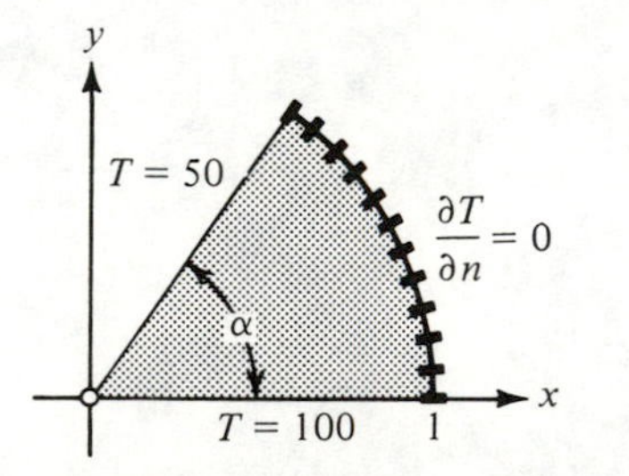

FIGURE 9.22 Accompanies Exercise 8.

$$T(x, 0) = 100 \qquad \text{for } 0 < x < 1$$

$$T(x, y) = 50 \qquad \text{for } z = re^{i\alpha},\ 0 < r < 1$$

$$\frac{\partial T}{\partial n} = 0 \qquad \text{for } z = e^{i\theta},\ 0 < \theta < \alpha$$

9. Find the temperature function $T(x, y)$ in the first quadrant $x>0$, $y>0$ that satisfies the boundary conditions in Figure 9.23. *Hint*: Use $w = \text{Arcsin } z^2$.

$$T(x, 0) = 100 \qquad \text{for } x > 1$$
$$T(0, y) = -50 \qquad \text{for } y > 1$$
$$\frac{\partial T}{\partial n} = T_y(x, 0) = 0 \qquad \text{for } 0 < x < 1$$
$$\frac{\partial T}{\partial n} = T_x(0, y) = 0 \qquad \text{for } 0 < y < 1$$

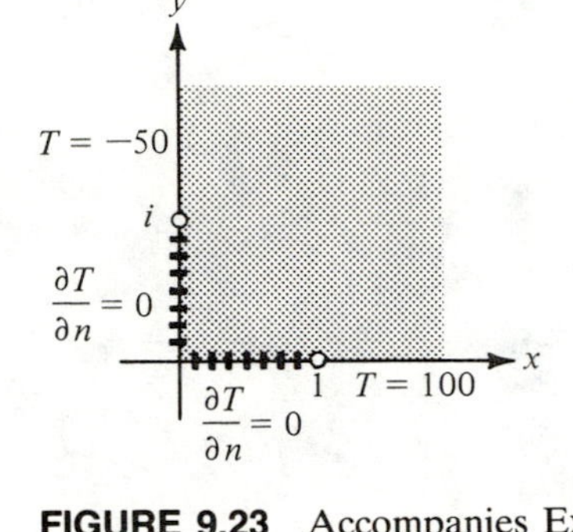

FIGURE 9.23 Accompanies Exercise 9.

10. Find the temperature function $T(x, y)$ in the infinite strip $0 < y < \pi$ that satisfies the boundary conditions in Figure 9.24. *Hint*: Use $w = e^z$.

$$T(x, 0) = 50 \qquad \text{for } x > 0$$
$$T(x, \pi) = -50 \qquad \text{for } x > 0$$
$$\frac{\partial T}{\partial n} = T_y(x, 0) = 0 \qquad \text{for } x < 0$$
$$\frac{\partial T}{\partial n} = T_y(x, \pi) = 0 \qquad \text{for } x < 0$$

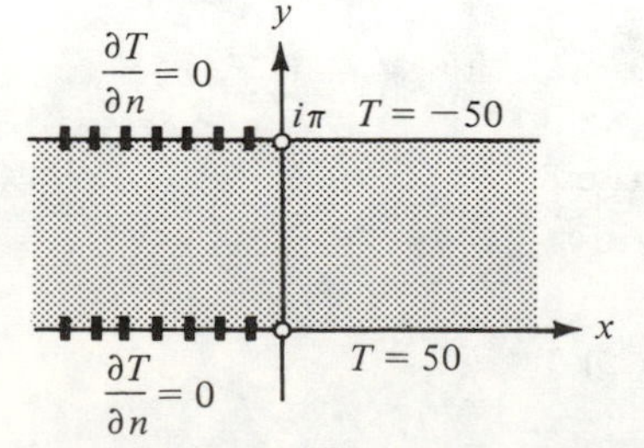

FIGURE 9.24 Accompanies Exercise 10.

11. Find the temperature function $T(x, y)$ in the upper half-plane $\operatorname{Im} z > 0$ that satisfies the boundary conditions in Figure 9.25. *Hint*: Use $w = 1/z$.

$$T(x, 0) = 100 \qquad \text{for } 0 < x < 1$$

$$T(x, 0) = -100 \qquad \text{for } -1 < x < 0$$

$$\frac{\partial T}{\partial n} = T_y(x, 0) = 0 \qquad \text{for } x > 1$$

$$\frac{\partial T}{\partial n} = T_y(x, 0) = 0 \qquad \text{for } x < -1$$

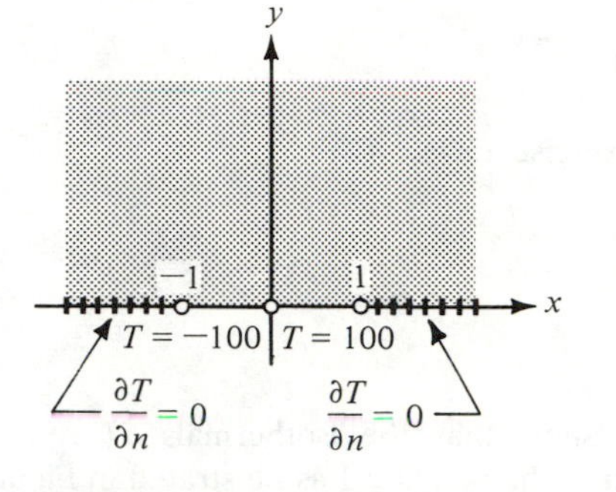

FIGURE 9.25 Accompanies Exercise 11.

12. Find the temperature function $T(x, y)$ in the first quadrant $x > 0$, $y > 0$ that satisfies the boundary conditions in Figure 9.26.

$$T(x, 0) = 50 \qquad \text{for } x > 0$$

$$T(0, y) = -50 \qquad \text{for } y > 1$$

$$\frac{\partial T}{\partial n} = T_x(0, y) = 0 \qquad \text{for } 0 < y < 1$$

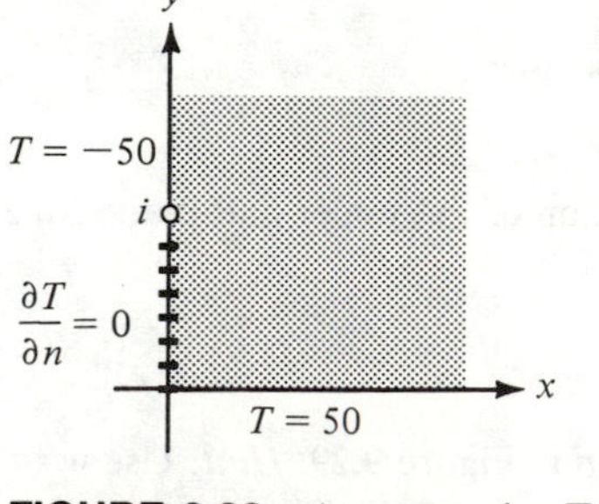

FIGURE 9.26 Accompanies Exercise 12.

13. For the temperature function

$$T(x, y) = 100 - \frac{100}{\pi} \arctan \frac{1 - x^2 - y^2}{2y}$$

in the upper half-disk $|z|<1$, $\operatorname{Im} z>0$, show that the isothermals $T(x, y)=\alpha$ are portions of circles that pass through the points $+1$ and -1 as illustrated in Figure 9.27.

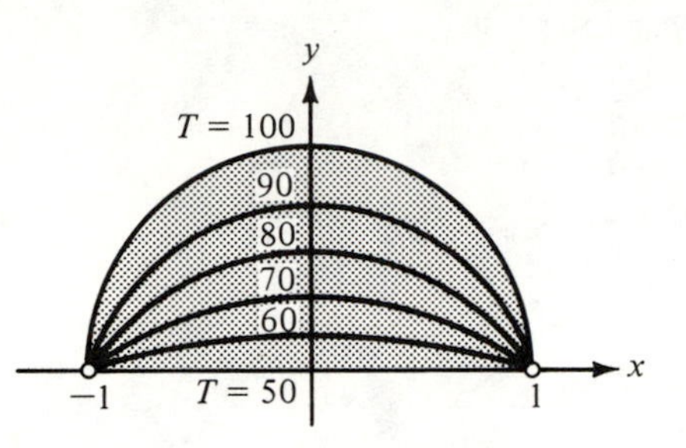

FIGURE 9.27 Accompanies Exercise 13.

14. For the temperature function

$$T(x, y)=\frac{300}{\pi} \operatorname{Re}[\operatorname{Arcsin} z]$$

in the upper half-plane $\operatorname{Im} z>0$, show that the isothermals $T(x, y)=\alpha$ are portions of hyperbolas that have foci at the points ± 1 as illustrated in Figure 9.28.

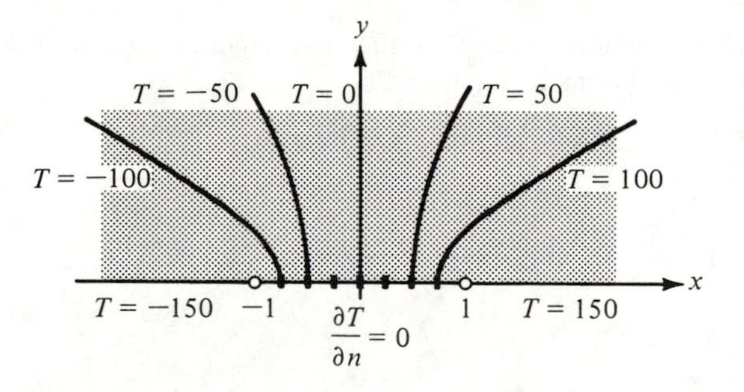

FIGURE 9.28 Accompanies Exercise 14.

15. Find the temperature function in the portion of the upper half-plane $\operatorname{Im} z>0$ that lies inside the ellipse

$$\frac{x^2}{\cosh^2(2)}+\frac{y^2}{\sinh^2(2)}=1$$

and satisfies the boundary conditions given in Figure 9.29. *Hint*: Use $w=\operatorname{Arcsin} z$.

$$T(x, y)=80 \quad \text{for } (x, y) \text{ on the ellipse}$$

$$T(x, 0)=40 \quad \text{for } -1<x<1$$

$$\frac{\partial T}{\partial n}=T_y(x, 0)=0 \quad \text{when } 1<|x|<\cosh(2)$$

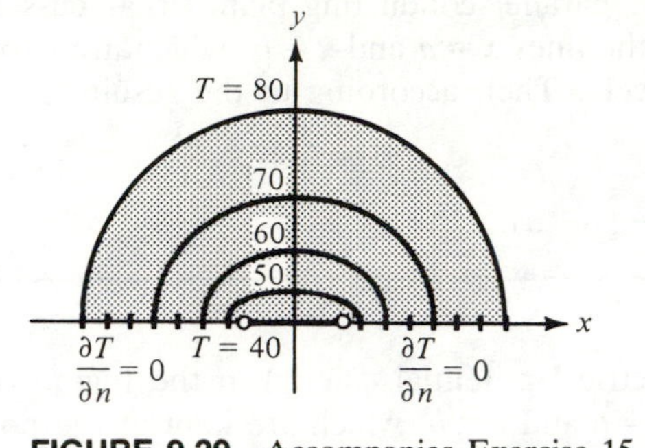

FIGURE 9.29 Accompanies Exercise 15.

9.6 Two-Dimensional Electrostatics

A two-dimensional electrostatic field is produced by a system of charged wires, plates, and cylindrical conductors that are perpendicular to the z plane. The wires, plates, and cylinders are assumed to be so long that the effects at the ends can be neglected as mentioned in Section 9.4. This sets up an electric field $\mathbf{E}(x, y)$ and can be interpreted as the force acting on a unit positive charge placed at the point (x, y). In the study of electrostatics the vector field $\mathbf{E}(x, y)$ is shown to be *conservative* and is derivable from a function $\phi(x, y)$ called the *electrostatic potential* as expressed by the equation

$$(1) \qquad \mathbf{E}(x, y) = -\operatorname{grad} \phi(x, y) = -\phi_x(x, y) - i\phi_y(x, y).$$

If we make the additional assumption that there are no charges within the domain D, then Gauss's Law for electrostatic fields implies that the line integral of the outward normal component of $\mathbf{E}(x, y)$ taken around any small rectangle lying inside D is identically zero. A heuristic argument similar to the one for steady state temperatures with $T(x, y)$ replaced by $\phi(x, y)$ will show that the value of the line integral is

$$(2) \qquad -[\phi_{xx}(x, y) + \phi_{yy}(x, y)]\,\Delta x\,\Delta y.$$

Since the quantity in (2) is zero, we conclude that $\phi(x, y)$ is a harmonic function. We let $\psi(x, y)$ denote the harmonic conjugate, and

$$(3) \qquad F(z) = \phi(x, y) + i\psi(x, y)$$

is the complex potential (not to be confused with the electrostatic potential).

The curves $\phi(x, y) = \text{constant}$ are called the *equipotential curves*, and the curves $\psi(x, y) = \text{constant}$ are called the *lines of flux*. If a small test charge is allowed to move under the influence of the field $\mathbf{E}(x, y)$, then it will travel along a line of flux. Boundary value problems for the potential function $\phi(x, y)$ are mathematically the same as those for steady state heat flow, and they are realizations of the Dirichlet problem where the harmonic function is $\phi(x, y)$.

EXAMPLE 9.16 Consider two parallel conducting planes that pass perpendicular to the z plane through the lines $x = a$ and $x = b$, which are kept at the potentials U_1 and U_2, respectively. Then according to the result of Example 9.1, the electrical potential is

$$(4) \qquad \phi(x, y) = U_1 + \frac{U_2 - U_1}{b - a}(x - a).$$

EXAMPLE 9.17 Find the electrical potential $\phi(x, y)$ in the region between two infinite coaxial cylinders $r = a$ and $r = b$, which are kept at the potentials U_1 and U_2, respectively.

Solution The function $w = \log z = \ln|z| + i \arg z$ maps the annular region between the circles $r = a$ and $r = b$ onto the infinite strip $\ln a < u < \ln b$ in the w plane as shown in Figure 9.30. The potential $\Phi(u, v)$ in the infinite strip will have the boundary values

$$(5) \qquad \Phi(\ln a, v) = U_1 \quad \text{and} \quad \Phi(\ln b, v) = U_2 \quad \text{for all } v.$$

If we use the result of Example 9.16, the electrical potential $\Phi(u, v)$ is

$$(6) \qquad \Phi(u, v) = U_1 + \frac{U_2 - U_1}{\ln b - \ln a}(u - \ln a).$$

Since $u = \ln|z|$, we can use equation (6) to conclude that the potential $\phi(x, y)$ is

$$\phi(x, y) = U_1 + \frac{U_2 - U_1}{\ln b - \ln a}(\ln|z| - \ln a).$$

The equipotentials $\phi(x, y) = \text{constant}$ are concentric circles centered at the origin, and the lines of flux are portions of rays emanating from the origin. If $U_2 < U_1$, then the situation is illustrated in Figure 9.30.

EXAMPLE 9.18 Find the electrical potential $\phi(x, y)$ produced by two charged half-planes that are perpendicular to the z plane and pass through the rays $x < -1$, $y = 0$ and $x > 1$, $y = 0$ where the planes are kept at the fixed potentials

$$(7) \qquad \phi(x, 0) = -300 \quad \text{for } x < -1 \quad \text{and} \quad \phi(x, 0) = 300 \quad \text{for } x > 1.$$

Solution The result of Example 8.13 shows that the function $w = \text{Arcsin}\, z$ is a conformal mapping of the z plane slit along the two rays $x < -1$, $y = 0$ and $x > 1$, $y = 0$ onto the vertical strip $-\pi/2 < u < \pi/2$ where the new problem is to find the potential $\Phi(u, v)$ that satisfies the boundary values

$$(8) \qquad \Phi\left(\frac{-\pi}{2}, v\right) = -300 \quad \text{and} \quad \Phi\left(\frac{\pi}{2}, v\right) = 300 \quad \text{for all } v.$$

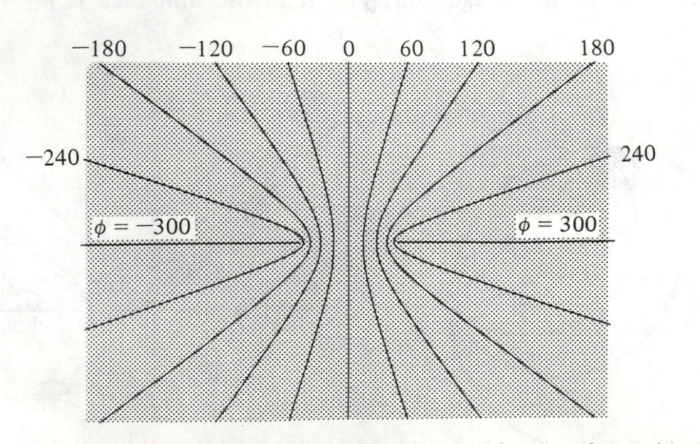

FIGURE 9.30 The electric field in a coaxial cylinder where $U_2 < U_1$.

Using the result of Example 9.1, we see that $\Phi(u, v)$ is

$$(9) \qquad \Phi(u, v) = \frac{600}{\pi} u.$$

As in the discussion of Example 9.15, the solution in the z plane is

$$(10) \qquad \phi(x, y) = \frac{600}{\pi} \operatorname{Re} \operatorname{Arcsin} z$$

$$= \frac{600}{\pi} \arcsin\left[\frac{\sqrt{(x+1)^2 + y^2} - \sqrt{(x-1)^2 + y^2}}{2}\right].$$

Several equipotential curves are shown in Figure 9.31.

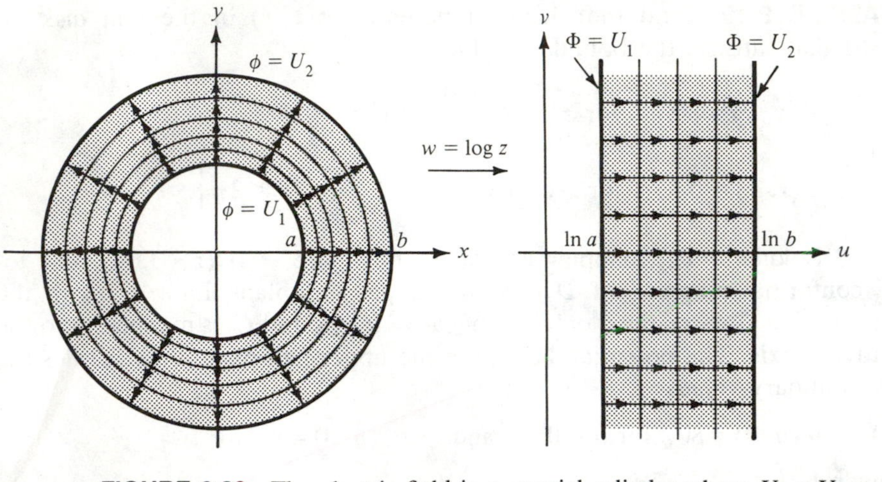

FIGURE 9.31 The electric field produced by two charged half-planes that are perpendicular to the complex plane.

EXAMPLE 9.19 Find the electrical potential $\phi(x, y)$ in the unit disk D: $|z| < 1$ that satisfies the boundary values

$$(11)\qquad \begin{aligned} \phi(x, y) &= 80 \quad \text{for } z \text{ on } C_1 = \left\{z = e^{i\theta}: 0 < \theta < \frac{\pi}{2}\right\} \quad \text{and} \\ \phi(x, y) &= 0 \quad \text{for } z \text{ on } C_2 = \left\{z = e^{i\theta}: \frac{\pi}{2} < \theta < 2\pi\right\}. \end{aligned}$$

Solution The mapping $w = S(z) = ((1 - i)(z - i))/(z - 1)$ is a one-to-one conformal mapping of D onto the upper half-plane $\text{Im}\, w > 0$ with the property that C_1 is mapped onto the negative u axis and C_2 is mapped onto the positive u axis. The potential $\Phi(u, v)$ in the upper half-plane that satisfies the new boundary values

$$(12)\qquad \Phi(u, 0) = 80 \quad \text{for } u < 0 \qquad \text{and} \qquad \Phi(u, 0) = 0 \quad \text{for } u > 0$$

is given by

$$(13)\qquad \Phi(u, v) = \frac{80}{\pi} \text{Arg}\, w = \frac{80}{\pi} \text{Arctan}\, \frac{v}{u}.$$

A straightforward calculation shows that

$$(14)\qquad u + iv = S(z) = \frac{(x - 1)^2 + (y - 1)^2 - 1 + i[1 - x^2 - y^2]}{(x - 1)^2 + y^2}.$$

The functions u and v in equation (14) can be substituted into equation (13) to obtain

$$\phi(x, y) = \frac{80}{\pi} \text{Arctan}\, \frac{1 - x^2 - y^2}{(x - 1)^2 + (y - 1)^2 - 1}.$$

The level curve $\Phi(u, v) = \alpha$ in the upper half-plane is a ray emanating from the origin, and the preimage $\phi(x, y) = \alpha$ in the unit disk is an arc of a

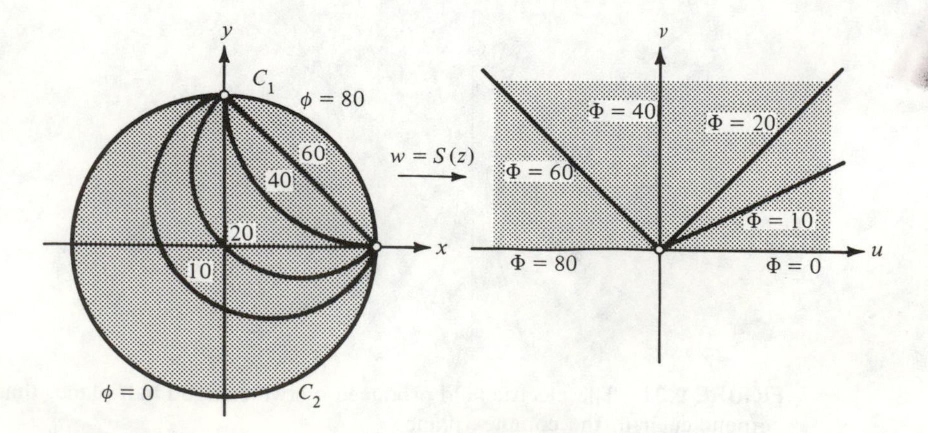

FIGURE 9.32 The potentials ϕ and Φ that are discussed in Example 9.19.

circle that passes through the points 1 and i. Several level curves are illustrated in Figure 9.32.

EXERCISES FOR SECTION 9.6

1. Find the electrostatic potential $\phi(x, y)$ between the two coaxial cylinders $r = 1$ and $r = 2$ that has the boundary values as shown in Figure 9.33.

$$\phi(x, y) = 100 \quad \text{when } |z| = 1$$

$$\phi(x, y) = 200 \quad \text{when } |z| = 2$$

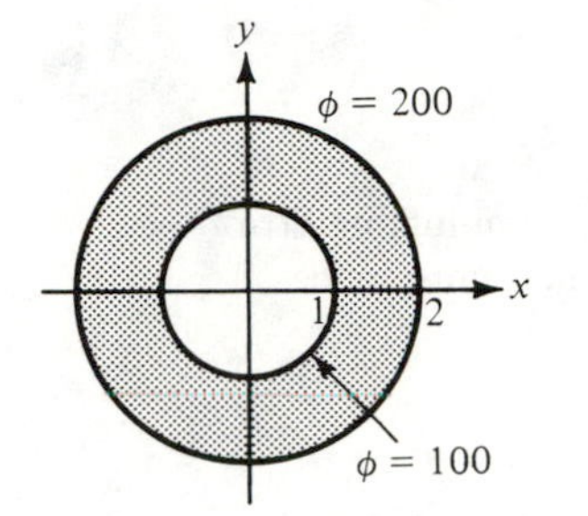

FIGURE 9.33 Accompanies Exercise 1.

2. Find the electrostatic potential $\phi(x, y)$ in the upper half-plane Im $z > 0$ that satisfies the boundary values as shown in Figure 9.34.

$$\phi(x, 0) = 100 \quad \text{for } x > 1$$

$$\phi(x, 0) = 0 \quad \text{for } -1 < x < 1$$

$$\phi(x, 0) = -100 \quad \text{for } x < -1$$

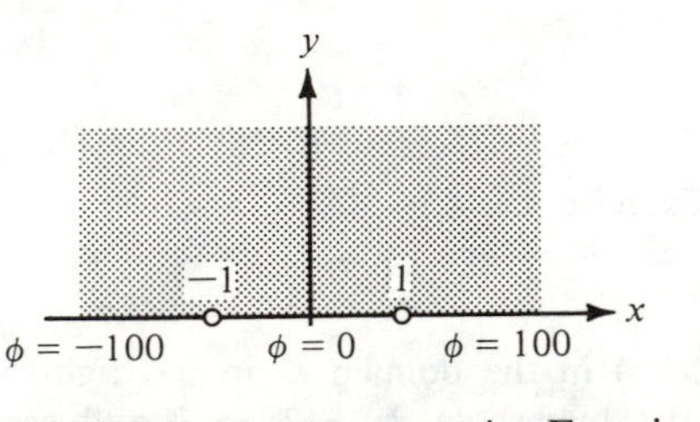

FIGURE 9.34 Accompanies Exercise 2.

3. Find the electrostatic potential $\phi(x, y)$ in the crescent-shaped region that lies inside the disk $|z - 2| < 2$ and outside the circle $|z - 1| = 1$ that satisfies the boundary values as shown in Figure 9.35. *Hint*: Use $w = 1/z$.

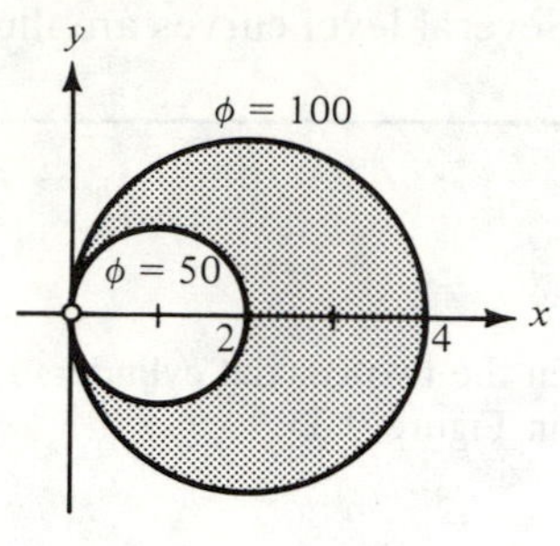

FIGURE 9.35 Accompanies Exercise 3.

$$\phi(x, y) = 100 \qquad \text{for } |z-2| = 2,\ z \neq 0$$

$$\phi(x, y) = 50 \qquad \text{for } |z-1| = 1,\ z \neq 0$$

4. Find the electrostatic potential $\phi(x, y)$ in the semi-infinite strip $-\pi/2 < x < \pi/2$, $y > 0$ that has the boundary values as shown in Figure 9.36.

$$\phi\left(\frac{\pi}{2}, y\right) = 0 \qquad \text{for } y > 0$$

$$\phi(x, 0) = 50 \qquad \text{for } \frac{-\pi}{2} < x < \frac{\pi}{2}$$

$$\phi\left(\frac{-\pi}{2}, y\right) = 100 \qquad \text{for } y > 0$$

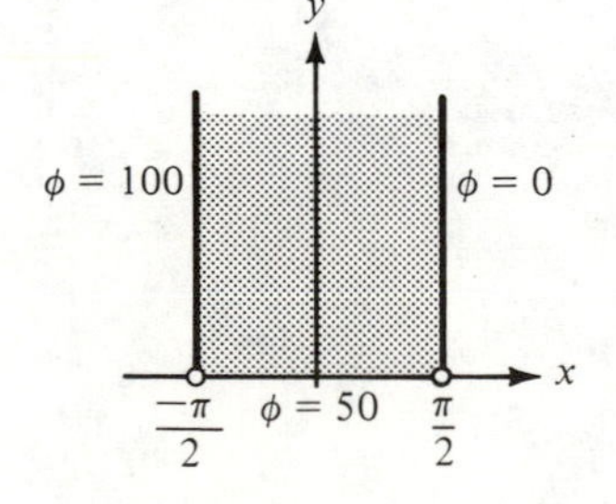

FIGURE 9.36 Accompanies Exercise 4.

5. Find the electrostatic potential $\phi(x, y)$ in the domain D in the right half-plane $\text{Re}\, z > 0$ that lies to the left of the hyperbola $2x^2 - 2y^2 = 1$ and satisfies the boundary values as shown in Figure 9.37. *Hint*: Use $w = \text{Arcsin}\, z$.

$$\phi(0, y) = 50 \qquad \text{for all } y$$

$$\phi(x, y) = 100 \qquad \text{when } 2x^2 - 2y^2 = 1$$

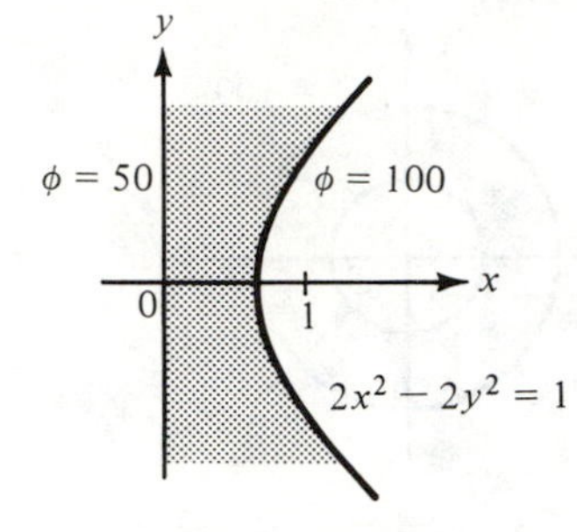

FIGURE 9.37 Accompanies Exercise 5.

6. Find the electrostatic potential $\phi(x, y)$ in the infinite strip $0 < x < \pi/2$ that satisfies the boundary values as shown in Figure 9.38. *Hint*: Use $w = \sin z$.

$$\phi(0, y) = 100 \qquad \text{for } y > 0$$

$$\phi\left(\frac{\pi}{2}, y\right) = 0 \qquad \text{for all } y$$

$$\phi(0, y) = -100 \qquad \text{for } y < 0$$

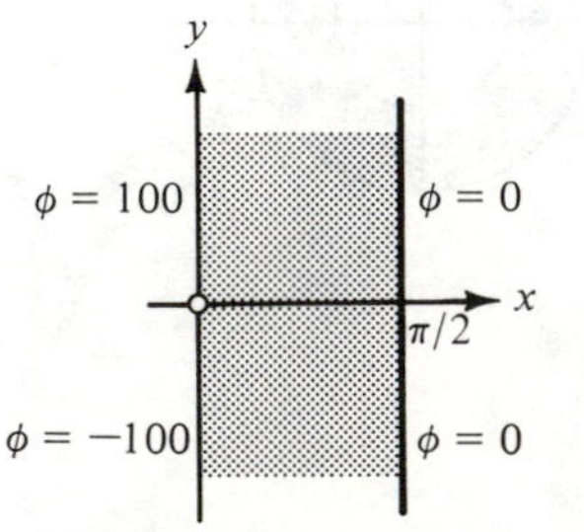

FIGURE 9.38 Accompanies Exercise 6.

7. **(a)** Show that the conformal mapping $w = S(z) = (2z - 6)/(z + 3)$ maps the domain D that is the portion of the right half-plane $\text{Re } z > 0$ that lies exterior to the circle $|z - 5| = 4$ onto the annulus $1 < |w| < 2$.

   (b) Find the electrostatic potential $\phi(x, y)$ in the domain D that satisfies the boundary values as shown in Figure 9.39.

$$\phi(0, y) = 100 \quad \text{for all } y, \qquad \phi(x, y) = 200 \quad \text{when } |z - 5| = 4$$

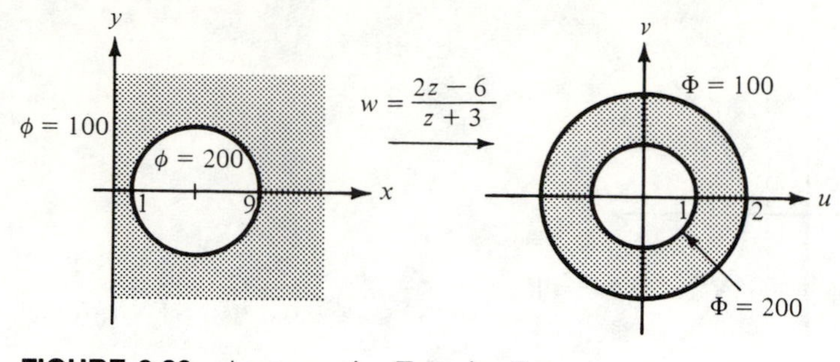

FIGURE 9.39 Accompanies Exercise 7.

8. **(a)** Show that the conformal mapping $w = S(z) = (z - 10)/(2z - 5)$ maps the domain D that is the portion of the disk $|z| < 5$ that lies outside the circle $|z - 2| = 2$ onto the annulus $1 < |w| < 2$.

(b) Find the electrostatic potential $\phi(x, y)$ in the domain D that satisfies the boundary values as shown in Figure 9.40.

$$\phi(x, y) = 100 \quad \text{when } |z| = 5, \qquad \phi(x, y) = 200 \quad \text{when } |z - 2| = 2$$

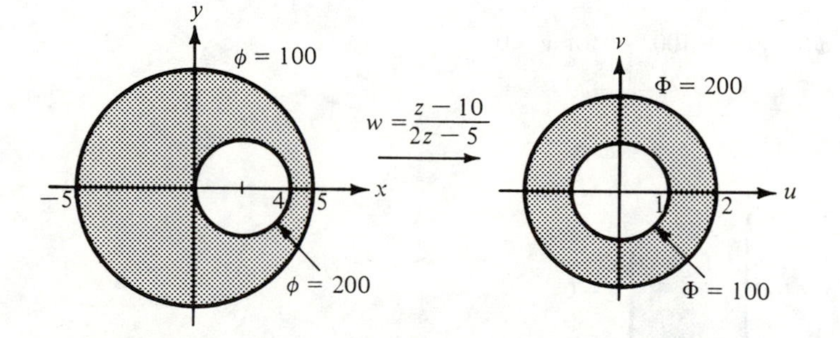

FIGURE 9.40 Accompanies Exercise 8.

9.7 Two-Dimensional Fluid Flow

Suppose that a fluid flows over the complex plane and that the velocity at the point $z = x + iy$ is given by the velocity vector

(1) $\quad \mathbf{V}(x, y) = p(x, y) + iq(x, y).$

We also require that the velocity does not depend on time and that the components $p(x, y)$ and $q(x, y)$ have continuous partial derivatives. The divergence of the vector field in (1) is given by

(2) $\quad \operatorname{div} \mathbf{V}(x, y) = p_x(x, y) + q_y(x, y)$

and is a measure of the extent to which the velocity field diverges near the point. We will consider only fluid flows for which the divergence is zero. This is more precisely characterized by requiring that the net flow through any simply closed contour be identically zero.

If we consider the flow out of the small rectangle in Figure 9.41, then the rate of flow out of this rectangle is the line integral of the exterior normal component of $\mathbf{V}(x, y)$ taken over the sides of the rectangle. The exterior normal component is given by $-q$ on the bottom edge, p on the right edge, q on the top edge, and $-p$ on the left edge. Integrating and setting the resulting net flow equal to zero yield

$$(3)\qquad \int_y^{y+\Delta y} [p(x+\Delta x, t) - p(x, t)]\, dt + \int_x^{x+\Delta x} [q(t, y+\Delta y) - q(t, y)]\, dt = 0.$$

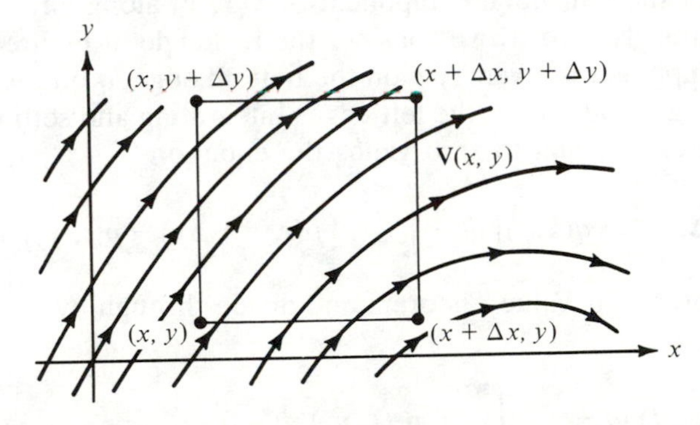

FIGURE 9.41 A two-dimensional vector field.

Since p and q are continuously differentiable, the Mean Value Theorem can be used to show that

$$(4)\qquad \begin{aligned} &p(x+\Delta x, t) - p(x, t) = p_x(x_1, t)\,\Delta x \quad \text{and} \\ &q(t, y+\Delta y) - q(t, y) = q_y(t, y_2)\,\Delta y \end{aligned}$$

where $x < x_1 < x + \Delta x$ and $y < y_2 < y + \Delta y$. Substitution of the expressions in (4) into (3) and subsequently dividing through by $\Delta x\, \Delta y$ results in

$$(5)\qquad \frac{1}{\Delta y}\int_y^{y+\Delta y} p_x(x_1, t)\, dt + \frac{1}{\Delta x}\int_x^{x+\Delta x} q_y(t, y_2)\, dt = 0.$$

The Mean Value Theorem for integrals can be used with (5) to show that

$$(6)\qquad p_x(x_1, y_1) + q_y(x_2, y_2) = 0$$

where $y < y_1 < y + \Delta y$ and $x < x_2 < x + \Delta x$. Letting $\Delta x \to 0$ and $\Delta y \to 0$ in (6) results in

$$(7) \qquad p_x(x, y) + q_y(x, y) = 0,$$

which is called the *equation of continuity*.

The curl of the vector field in (1) is

$$(8) \qquad \text{curl } \mathbf{V}(x, y) = q_x(x, y) - p_y(x, y)$$

and is an indication of how the field swirls in the vicinity of a point. Imagine that a "fluid element" at the point (x, y) is suddenly frozen and then moves freely in the fluid. It can be shown that the fluid element will rotate with an angular velocity given by

$$(9) \qquad \tfrac{1}{2} p_x(x, y) - \tfrac{1}{2} q_y(x, y) = \tfrac{1}{2} \text{ curl } \mathbf{V}(x, y).$$

We will consider only fluid flows for which the curl is zero. Such fluid flows are called *irrotational*. This is more precisely characterized by requiring that the line integral of the tangential component of $\mathbf{V}(x, y)$ along any simply closed contour be identically zero. If we consider the rectangle in Figure 9.41, then the tangential component is given by p on the bottom edge, q on the right edge, $-p$ on the top edge, and $-q$ on the left edge. Integrating and setting the resulting *circulation* integral equal to zero yields the equation

$$(10) \qquad \int_y^{y+\Delta y} [q(x + \Delta x, t) - q(x, t)]\, dt - \int_x^{x+\Delta x} [p(t, y + \Delta y) - p(t, y)]\, dt = 0.$$

As before, we apply the Mean Value Theorem and divide through by $\Delta x\, \Delta y$ and obtain the equation

$$(11) \qquad \frac{1}{\Delta y} \int_y^{y+\Delta y} q_x(x_1, t)\, dt - \frac{1}{\Delta x} \int_x^{x+\Delta x} p_y(t, y_2)\, dt = 0.$$

The mean value for integrals can be used with (11) to deduce the equation $q_x(x_1, y_1) - p_y(x_2, y_2) = 0$. Letting $\Delta x \to 0$ and $\Delta y \to 0$ yields

$$(12) \qquad q_x(x, y) - p_y(x, y) = 0.$$

Equations (7) and (12) show that the complex function $f(z) = p(x, y) - iq(x, y)$ satisfies the Cauchy-Riemann equations and is an analytic function. Let $F(z)$ denote the antiderivative of $f(z)$. Then

$$(13) \qquad F(z) = \phi(x, y) + i\psi(x, y)$$

is called the *complex potential* of the flow and has the property

$$(14) \qquad \overline{F'(z)} = \phi_x(x, y) - i\psi_x(x, y) = p(x, y) + iq(x, y) = \mathbf{V}(x, y).$$

Since $\phi_x = p$ and $\phi_y = q$, we also have

$$(15) \qquad \text{grad } \phi(x, y) = p(x, y) + iq(x, y) = \mathbf{V}(x, y),$$

so $\phi(x, y)$ is the *velocity potential* for the flow, and the curves

$$\phi(x, y) = \text{constant} \tag{16}$$

are called *equipotentials*. The function $\psi(x, y)$ is called the *stream function*, and the curves

$$\psi(x, y) = \text{constant} \tag{17}$$

are called *streamlines* and describe the paths of the fluid particles. To see this fact, we can implicitly differentiate $\psi(x, y) = C$ and find that the slope of the tangent is given by

$$\frac{dy}{dx} = \frac{-\psi_x(x, y)}{\psi_y(x, y)}. \tag{18}$$

Using the fact that $\psi_y = \phi_x$ and equation (18), we find that the tangent vector to the curve is

$$\mathbf{T} = \phi_x(x, y) - i\psi_x(x, y) = p(x, y) + iq(x, y) = \mathbf{V}(x, y). \tag{19}$$

The salient idea of the above discussion is the conclusion that if

$$F(z) = \phi(x, y) + i\psi(x, y) \tag{20}$$

is an analytic function, then the family of curves

$$\{\psi(x, y) = \text{constant}\} \tag{21}$$

represents the streamlines of a fluid flow.

The boundary condition for an ideal fluid flow is that $\mathbf{V}$ should be parallel to the boundary curve containing the fluid (the fluid flows parallel to the walls of a containing vessel). This means that if equation (20) is the complex potential for the flow, then the boundary curve must be given by $\psi(x, y) = K$ for some constant K; that is, the boundary curve must be a streamline.

Theorem 9.5 (Invariance of Flow) *Let*

$$F_1(w) = \Phi(u, v) + i\Psi(u, v) \tag{22}$$

denote the complex potential for a fluid flow in a domain G in the w plane where the velocity is

$$\mathbf{V}_1(u, v) = \overline{F_1'(w)}. \tag{23}$$

If the function

$$w = S(z) = u(x, y) + iv(x, y) \tag{24}$$

is a one-to-one conformal mapping from a domain D in the z plane onto G, then the composite function

$$F_2(z) = F_1(S(z)) = \Phi(u(x, y), v(x, y)) + i\psi(u(x, y), v(x, y)) \tag{25}$$

is the complex potential for a fluid flow in D where the velocity is

(26) $\mathbf{V}_2(x, y) = \overline{F_2'(z)}$.

The situation is shown in Figure 9.42.

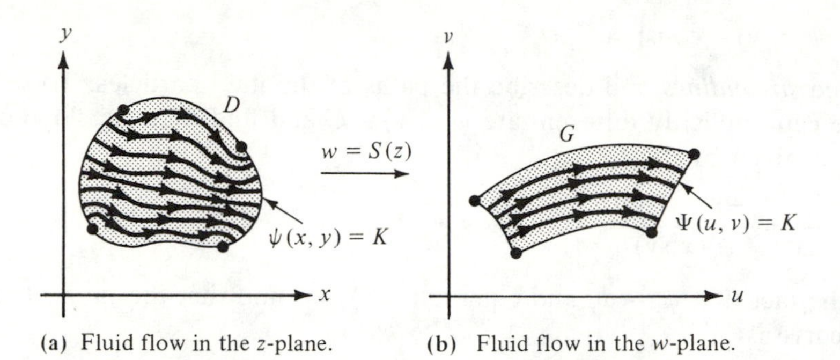

(a) Fluid flow in the z-plane. (b) Fluid flow in the w-plane.

FIGURE 9.42 The image of a fluid flow under conformal mapping.

Proof From equation (13) we see that $F_1(w)$ is an analytic function. Since the composition in (25) is an analytic function, $F_2(z)$ is the complex potential for an ideal fluid flow in D.

We note that the functions

(27) $\phi(x, y) = \Phi(u(x, y), v(x, y))$ and $\psi(x, y) = \Psi(u(x, y), v(x, y))$

are the new velocity potential and stream function, respectively, for the flow in D. A streamline or natural boundary curve

(28) $\psi(x, y) = K$

in the z plane is mapped onto a streamline or natural boundary curve

(29) $\Psi(u, v) = K$

in the w plane by the transformation $w = S(z)$. One method for finding a flow inside a domain D in the z plane is to conformally map D onto a domain G in the w plane in which the flow is known.

For an ideal fluid with uniform density ρ the fluid pressure $P(x, y)$ and speed $|\mathbf{V}(x, y)|$ are related by the following special case of *Bernoulli's equation*:

(30) $$\frac{P(x, y)}{\rho} + \frac{1}{2}|\mathbf{V}(x, y)| = \text{constant}.$$

It is of importance to notice that the pressure is greatest when the speed is least.

EXAMPLE 9.20 The complex potential $F(z) = (a + ib)z$ has the velocity potential and stream function given by

$$(31) \qquad \phi(x, y) = ax - by \qquad \text{and} \qquad \psi(x, y) = bx + ay,$$

respectively, and gives rise to the fluid flow defined in the entire complex plane that has a uniform parallel velocity given by

$$(32) \qquad \mathbf{V}(x, y) = \overline{F'(z)} = a - ib.$$

The streamlines are parallel lines given by the equation $bx + ay = \text{constant}$ and are inclined at an angle $\alpha = -\arctan(b/a)$ as indicated in Figure 9.43.

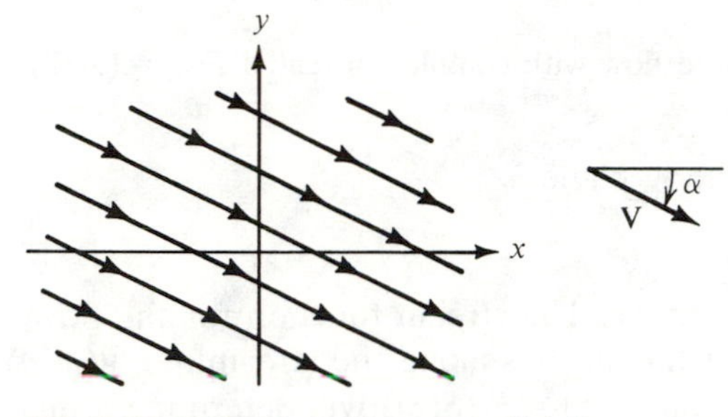

FIGURE 9.43 A uniform parallel flow.

EXAMPLE 9.21 Consider the complex potential $F(z) = (A/2)z^2$ where A is a positive real number. The velocity potential and stream function are given by

$$(33) \qquad \phi(x, y) = \frac{A}{2}(x^2 - y^2) \qquad \text{and} \qquad \psi(x, y) = Axy,$$

respectively. The streamlines $\psi(x, y) = \text{constant}$ form a family of hyperbolas with asymptotes along the coordinate axes. The velocity vector $\mathbf{V} = A\bar{z}$ indicates that in the upper half-plane $\operatorname{Im} z > 0$ the fluid flows down along the streamlines and spreads out along the x axis. This depicts the flow against a wall and is illustrated in Figure 9.44.

EXAMPLE 9.22 Find the complex potential for an ideal fluid flowing from left to right across the complex plane and around the unit circle $|z| = 1$.

Solution We will use the fact that the conformal mapping

$$(34) \qquad w = S(z) = z + \frac{1}{z}$$

maps the domain $D = \{z: |z| < 1\}$ one-to-one and onto the w plane slit along the segment $-2 \leqq u \leqq 2$, $v = 0$. The complex potential for a uniform horizontal

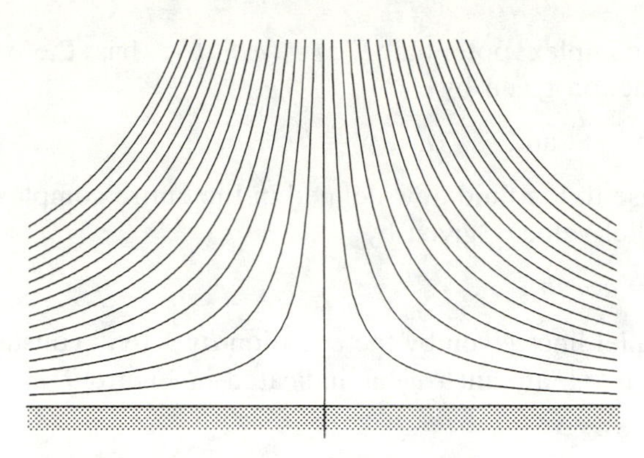

FIGURE 9.44 The fluid flow with complex potential $F(z) = (A/2)z^2$.

flow parallel to this slit in the w plane is

$$(35) \qquad F_1(w) = Aw$$

where A is a positive real number. The stream function for the flow in the w plane is $\psi(u, v) = Av$ so that the slit lies along the streamline $\Psi(u, v) = 0$.

The composite function $F_2(z) = F_1(S(z))$ will determine a fluid flow in the domain D where the complex potential is

$$(36) \qquad F_2(z) = A\left(z + \frac{1}{z}\right) \qquad \text{where } A > 0.$$

Polar coordinates can be used to express $F_2(z)$ by the equation

$$(37) \qquad F_2(z) = A\left(r + \frac{1}{r}\right)\cos\theta + iA\left(r - \frac{1}{r}\right)\sin\theta.$$

The streamline $\psi(r, \theta) = A(r - 1/r)\sin\theta = 0$ consists of the rays

$$(38) \qquad r > 1, \quad \theta = 0 \qquad \text{and} \qquad r > 1, \theta = \pi$$

along the x axis and the curve $r - 1/r = 0$, which is easily seen to be the unit circle $r = 1$. This shows that the unit circle can be considered as a boundary curve for the fluid flow.

Since the approximation $F_2(z) = A(z + 1/z) \approx Az$ is valid for large values of z, we see that the flow is approximated by a uniform horizontal flow with speed $|\mathbf{V}| = A$ at points that are distant from the origin. The streamlines $\psi(x, y) = \text{constant}$ and their images $\Psi(u, v) = \text{constant}$ under the conformal mapping $w = S(z) = z + 1/z$ are illustrated in Figure 9.45.

EXAMPLE 9.23 Find the complex potential for an ideal fluid flowing from left to right across the complex plane and around the segment from $-i$ to i.

Solution We will use the conformal mapping

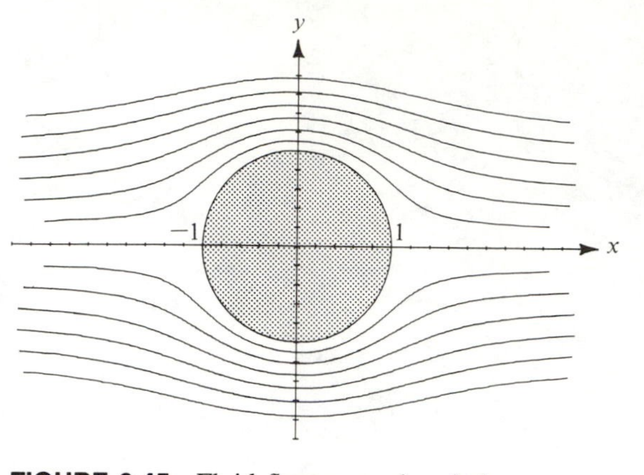

FIGURE 9.45 Fluid flow around a circle.

(39) $\quad w = S(z) = (z^2 + 1)^{1/2} = (z + i)^{1/2}(z - i)^{1/2}$

where the branch of the square root of $Z = z \pm i$ in each factor is $Z^{1/2} = R^{1/2}e^{i\theta/2}$ where $R = |Z|$ and $\theta = \arg Z$ where $-\pi/2 < \arg Z \leqq 3\pi/2$. The function $w = S(z)$ is a one-to-one conformal mapping of the domain D consisting of the z plane slit along the segment $x = 0$, $-1 \leqq y \leqq 1$ onto the domain G consisting of the w plane slit along the segment $-1 \leqq u \leqq 1$, $v = 0$.

The complex potential for a uniform horizontal flow parallel to the slit in the w plane is given by $F_1(w) = Aw$ where A is a positive real number and where the slit lies along the streamline $\Psi(u, v) = Au = 0$. The composite function

(40) $\quad F_2(z) = F_1(S(z)) = A(z^2 + 1)^{1/2}$

is the complex potential for a fluid flow in the domain D. The streamlines $\psi(x, y) = cA$ for the flow in D are obtained by finding the preimage of the streamline $\Psi(u, v) = cA$ in G given by the parametric equations

(41) $\quad v = c, \quad u = t \qquad \text{for } -\infty < t < \infty.$

The corresponding streamline in D is found by solving the equation

(42) $\quad t + ic = (z^2 + 1)^{1/2}$

for x and y in terms of t. Squaring both sides of (42) yields

(43) $\quad t^2 - c^2 - 1 + i2ct = x^2 - y^2 + i2xy.$

Equating the real and imaginary parts leads to the system of equations

(44) $\quad x^2 - y^2 = t^2 - c^2 - 1 \qquad \text{and} \qquad xy = ct.$

Eliminating the parameter t in (44) results in $c^2 = (x^2 + c^2)(y^2 - c^2)$, and we can solve for y in terms of x to obtain

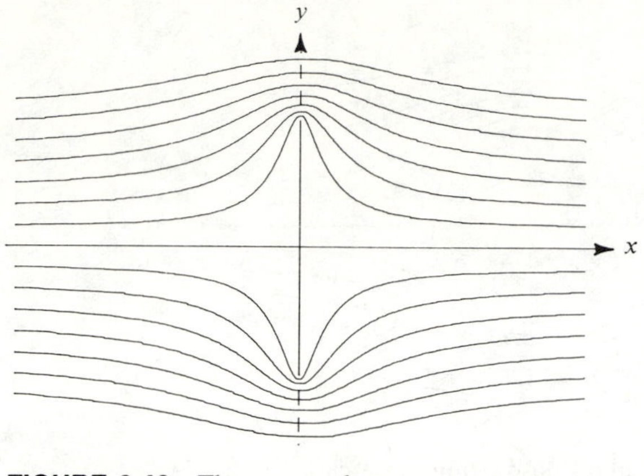

FIGURE 9.46 Flow around a segment.

$$y = c\sqrt{\frac{1 + c^2 + x^2}{c^2 + x^2}} \tag{45}$$

for streamlines in D. For large values of x this streamline approaches the asymptote $y = c$ and approximates a horizontal flow, as shown in Figure 9.46.

EXERCISES FOR SECTION 9.7

1. Consider the ideal fluid flow where the complex potential is $F(z) = A(z + 1/z)$ where A is a positive real number.

(a) Show that the velocity vector at the point $(1, \theta)$, $z = re^{i\theta}$ on the unit circle is given by $\mathbf{V}(1, \theta) = A(1 - \cos 2\theta - i \sin 2\theta)$.

(b) Show that the velocity vector $\mathbf{V}(1, \theta)$ is tangent to the unit circle $|z| = 1$ at all points except -1 and $+1$. *Hint*: Show that $\mathbf{V} \cdot \mathbf{P} = 0$ where $\mathbf{P} = \cos\theta + i \sin\theta$.

(c) Show that the speed at the point $(1, \theta)$ on the unit circle is given by $|\mathbf{V}| = 2A|\sin\theta|$ and that the speed attains the maximum of $2A$ at the points $\pm i$ and is zero at the points ± 1. Where is the pressure the greatest?

2. Show that the complex potential $F(z) = ze^{-i\alpha} + e^{i\alpha}/z$ determines the ideal fluid flow around the unit circle $|z| = 1$ where the velocity at points distant from the origin is given approximately by $\mathbf{V} \approx e^{i\alpha}$; that is, the direction of the flow for large values of z is inclined at an angle α with the x axis as shown in Figure 9.47.

3. Consider the ideal fluid flow in the channel bounded by the hyperbolas $xy = 1$ and $xy = 4$ in the first quadrant where the complex potential is given by $F(z) = (A/2)z^2$ and where A is a positive real number.

(a) Find the speed at each point, and find the point on the boundary where the speed attains a minimum value.

(b) Where is the pressure greatest?

4. Show that the stream function is given by $\psi(r, \theta) = Ar^3 \sin 3\theta$ for an ideal fluid flow around the angular region $0 < \theta < \pi/3$ indicated in Figure 9.48. Sketch several streamlines of the flow. *Hint*: Use the conformal mapping $w = z^3$.

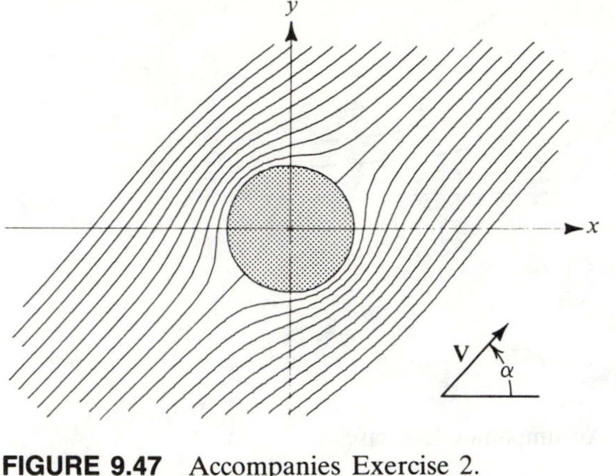

FIGURE 9.47 Accompanies Exercise 2.

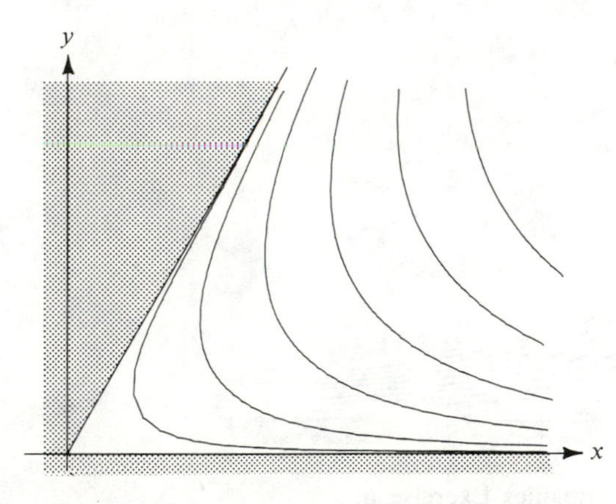

FIGURE 9.48 Accompanies Exercise 4.

5. Consider the ideal fluid flow where the complex potential is

$$F(z) = Az^{3/2} = Ar^{3/2}\left(\cos\frac{3\theta}{2} + i\sin\frac{3\theta}{2}\right) \qquad \text{where } 0 \leqq \theta < 2\pi.$$

   (a) Find the stream function $\psi(r, \theta)$.
   (b) Sketch several streamlines of the flow in the angular region $0 < \theta < 4\pi/3$ as indicated in Figure 9.49.

6. **(a)** Let A be a positive real number. Show that the complex potential $F(z) = A(z^2 + 1/z^2)$ determines an ideal fluid flow around the domain $r > 1$, $0 < \theta < \pi/2$ indicated in Figure 9.50, which shows the flow around a circle in the first quadrant. *Hint*: Use the conformal mapping $w = z^2$.
   (b) Show that the speed at the point $(1, \theta)$, $z = re^{i\theta}$ on the quarter circle $r = 1$, $0 < \theta < \pi/2$ is given by $\mathbf{V} = 4A|\sin 2\theta|$.
   (c) Determine the stream function for the flow and sketch several streamlines.

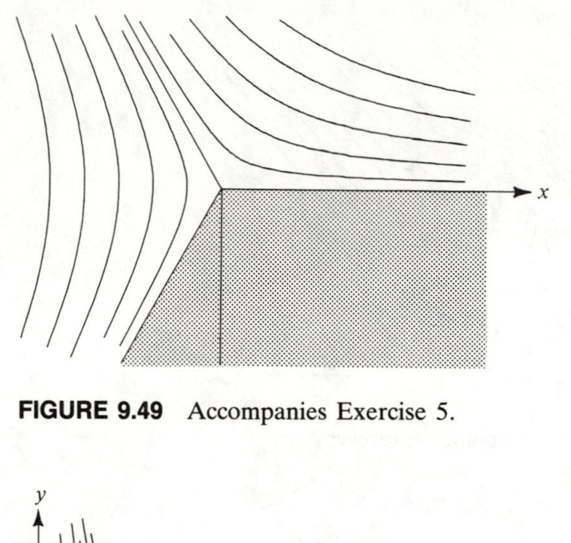

FIGURE 9.49 Accompanies Exercise 5.

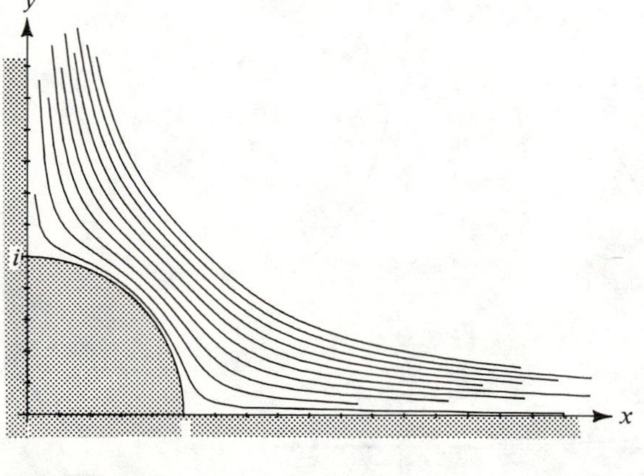

FIGURE 9.50 Accompanies Exercise 6.

7. Show that $F(z) = \sin z$ is the complex potential for the ideal fluid flow inside the semi-infinite strip $-\pi/2 < x < \pi/2$, $y > 0$ as indicated in Figure 9.51. Find the stream function.

8. Let $w = S(z) = \frac{1}{2}[z + (z^2 - 4)^{1/2}]$ denote the branch of the inverse of $z = w + 1/w$ that defines a one-to-one conformal mapping of the z plane slit along the segment $-2 \leqq x \leqq 2$, $y = 0$ onto the domain $|w| > 1$. Use the complex potential $F_2(w) = we^{-i\alpha} + (e^{i\alpha}/w)$ in the w plane to show that the complex potential $F_1(z) = z \cos \alpha - i(z^2 - 4)^{1/2} \sin \alpha$ determines the ideal fluid flow around the segment $-2 \leqq x \leqq 2$, $y = 0$ where the velocity at points distant from the origin is given approximately by $\mathbf{V} \approx e^{i\alpha}$ as shown in Figure 9.52.

9. **(a)** Show that the complex potential $F(z) = -i \operatorname{Arcsin} z$ determines the ideal fluid flow through the aperture from -1 to $+1$ as indicated in Figure 9.53.
   (b) Show that the streamline $\psi(x, y) = c$ for the flow is a portion of the hyperbola $(x^2/\sin^2 c) - (y^2/\cos^2 c) = 1$.

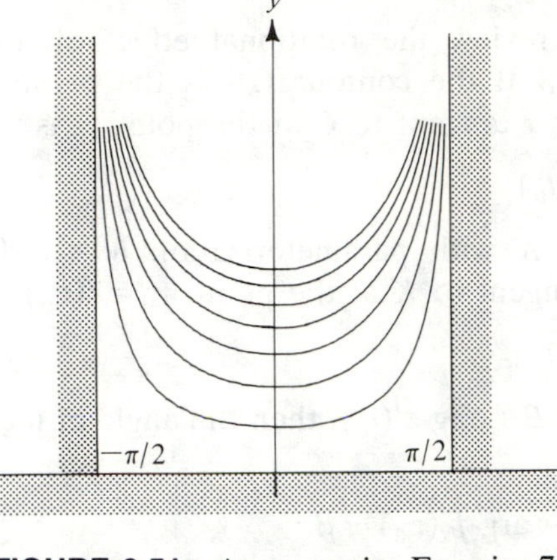

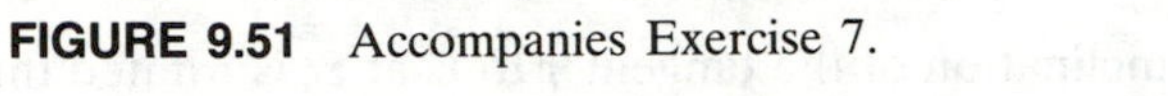
FIGURE 9.51 Accompanies Exercise 7.

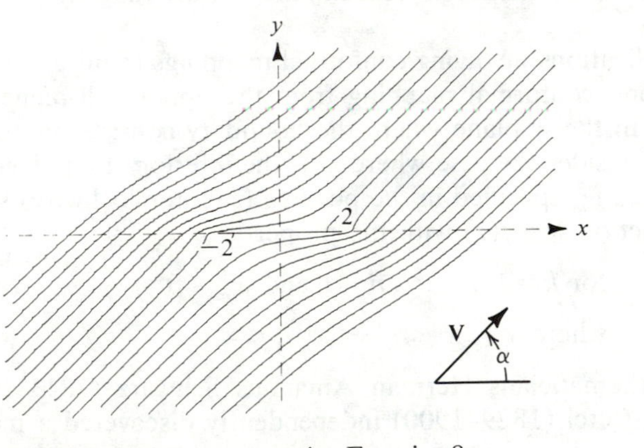

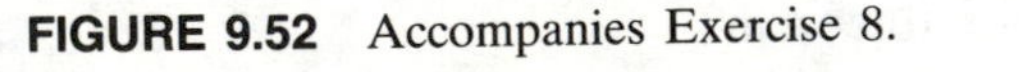
FIGURE 9.52 Accompanies Exercise 8.

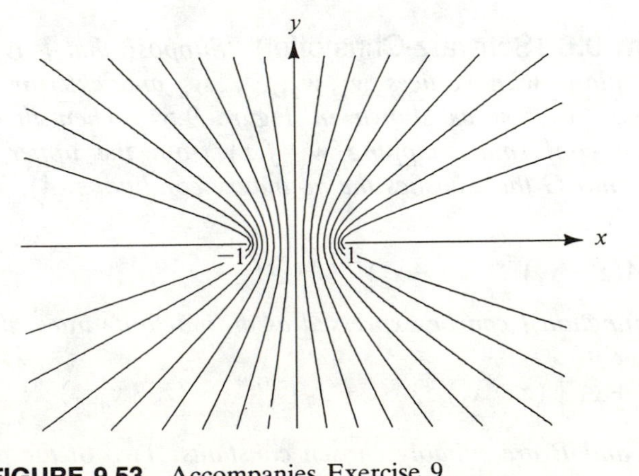

FIGURE 9.53 Accompanies Exercise 9.

9.8 The Schwarz-Christoffel Transformation

To proceed further, we must review the rotational effect of a conformal mapping $w = f(z)$ at a point z_0. If the contour C has the parameterization $z(t) = x(t) + iy(t)$, then a vector $\boldsymbol{\tau}$ tangent to C at the point z_0 is

$$(1) \qquad \boldsymbol{\tau} = z'(t_0) = x'(t_0) + iy'(t_0)$$

The image of C is a contour K with parameterization $w = u(x(t), y(t)) + iv(x(t), y(t))$, and a vector $\mathbf{T}$ tangent to K at the point $w_0 = f(z_0)$ is

$$(2) \qquad \mathbf{T} = w'(z_0) = f'(z_0)z'(t_0).$$

If the angle of inclination of $\boldsymbol{\tau}$ is $\beta = \arg z'(t_0)$, then the angle of inclination of $\mathbf{T}$ is

$$(3) \qquad \arg \mathbf{T} = \arg f'(z_0)z'(t_0) = \arg f'(z_0) + \beta.$$

Hence the angle of inclination of the tangent $\boldsymbol{\tau}$ to C at z_0 is rotated through the angle $\arg f'(z_0)$ to obtain the angle of inclination of the tangent $\mathbf{T}$ to K at the point w_0.

Many applications involving conformal mappings require the construction of a one-to-one conformal mapping from the upper half-plane $\operatorname{Im} z > 0$ onto a domain G in the w plane where the boundary consists of straight line segments. Let us consider the case where G is the interior of a polygon P with vertices $w_1, w_2, \ldots, w_n$ specified in the positive (counterclockwise) sense. We want to find a function $w = f(z)$ with the property

$$(4) \qquad w_k = f(x_k) \quad \text{for } k = 1, 2, \ldots, n-1 \text{ and } w_n = f(\infty)$$
$$\text{where } x_1 < x_2 < \cdots < x_{n-1} < \infty.$$

Two German mathematicians Herman Amandus Schwarz (1843–1921) and Elwin Bruno Christoffel (1829–1900) independently discovered a method for finding f, and that is our next theorem.

Theorem 9.6 (Schwarz-Christoffel) *Suppose that P is a polygon in the w plane with vertices $w_1, w_2, \ldots, w_n$ and exterior angles α_k, where $-\pi < \alpha_k < \pi$ as shown in Figure 9.54. Then there exists a one-to-one conformal mapping $w = f(z)$ from the upper half-plane $\operatorname{Im} z > 0$ onto G that satisfies the boundary conditions (4). The derivative $f'(z)$ is*

$$(5) \qquad f'(z) = A(z - x_1)^{-\alpha_1/\pi}(z - x_2)^{-\alpha_2/\pi} \cdots (z - x_{n-1})^{-\alpha_{n-1}/\pi},$$

and the function f can be expressed as an indefinite integral

$$(6) \qquad f(z) = B + A \int (z - x_1)^{-\alpha_1/\pi}(z - x_2)^{-\alpha_2/\pi} \cdots (z - x_{n-1})^{-\alpha_{n-1}/\pi}\, dz.$$

where A and B are suitably chosen constants. Two of the points $\{x_k\}$

may be chosen arbitrarily, and the constants A and B determine the size and position of P.

Proof The proof relies on finding how much the tangent

(7) $\tau_j = 1 + 0i$

(which always points to the right) at the point $(x, 0)$ must be rotated by the mapping $w = f(z)$ so that the line segment $x_{j-1} < x < x_j$ is mapped onto the edge of P that lies between the points $w_{j-1} = f(x_{j-1})$ and $w_j = f(x_j)$. Since the amount of rotation is determined by $\arg f'(x)$, formula (5) specifies $f'(z)$ in terms of the values x_j and the amount of rotation α_j that is required at the vertex $f(x_j)$.

If we let $x_0 = -\infty$ and $x_n = \infty$, then, for values of x in the interval $x_{j-1} < x < x_j$, the amount of rotation is

(8) $$\arg f'(x) = \arg A - \frac{1}{\pi}\left[\alpha_1 \arg(x - x_1) + \alpha_2 \arg(x - x_2) + \cdots + \alpha_{n-1} \arg(x - x_{n-1})\right].$$

Since we have $\text{Arg}(x - x_k) = 0$ for $1 \leqq k < j$ and $\text{Arg}(x - x_k) = \pi$ for $j \leqq k \leqq n - 1$, we can write (8) as

(9) $$\arg f'(x) = \arg A - \alpha_j - \alpha_{j+1} - \cdots - \alpha_{n-1}.$$

The angle of inclination of the tangent vector $\mathbf{T}_j$ to the polygon P at the point $w = f(x)$ for $x_{j-1} < x < x_j$ is

(10) $$\gamma_j = \text{Arg } A - \alpha_j - \alpha_{j+1} - \cdots - \alpha_{n-1}.$$

The angle of inclination of the tangent vector $\mathbf{T}_{j+1}$ to the polygon P at the point $w = f(x)$ for $x_j < x < x_{j+1}$ is

(11) $$\gamma_{j+1} = \text{Arg } A - \alpha_{j+1} - \alpha_{j+2} - \cdots - \alpha_{n-1}.$$

The angle of inclination of the vector tangent to the polygon P jumps abruptly by the amount α_j as the point $w = f(x)$ moves along the side $\widehat{w_{j-1}w_j}$ through the vertex w_j to the side $\widehat{w_jw_{j+1}}$. Therefore the exterior angle to the polygon P at the vertex w_j is given by the angle α_j and satisfies the inequality $-\pi < \alpha_j < \pi$ for $j = 1, 2, \ldots, n - 1$. Since the sum of the exterior angles of a polygon is 2π, we have $\alpha_n = 2\pi - \alpha_1 - \alpha_2 - \cdots - \alpha_{n-1}$ so that only $n - 1$ angles need to be specified. This case with $n = 5$ is indicated in Figure 9.54.

If the case $\alpha_1 + \alpha_2 + \cdots + \alpha_{n-1} \leqq \pi$ occurs, then $\alpha_n > \pi$, and the vertices $w_1, w_2, \ldots, w_n$ cannot form a closed polygon. For this case, formulas (5) and (6) will determine a mapping from the upper half-plane $\text{Im } z > 0$ onto an infinite region in the w plane where the vertex w_n is at infinity. The case $n = 5$ is illustrated in Figure 9.55.

Formula (6) gives a representation for f in terms of an indefinite

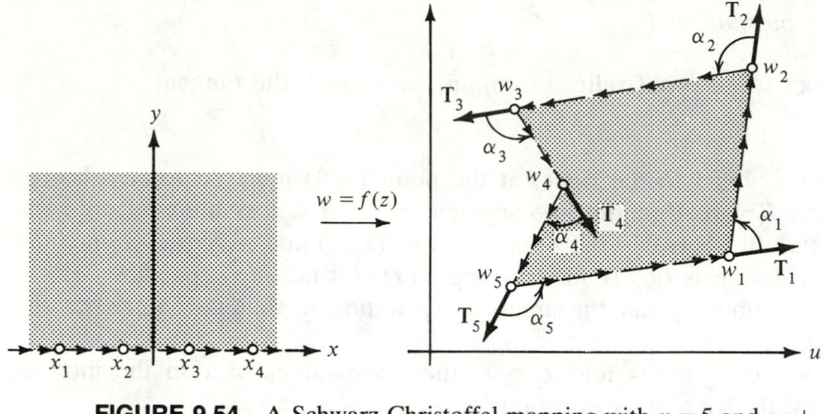

FIGURE 9.54 A Schwarz-Christoffel mapping with $n = 5$ and $\alpha_1 + \alpha_2 + \cdots + \alpha_4 > \pi$.

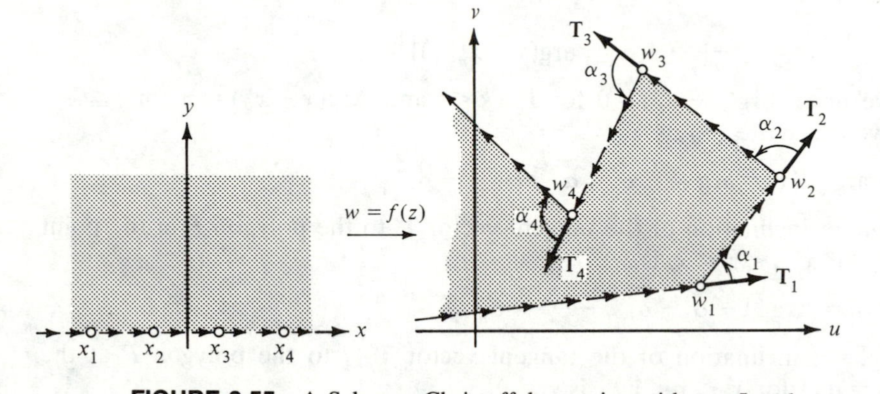

FIGURE 9.55 A Schwarz-Christoffel mapping with $n = 5$ and $\alpha_1 + \alpha_2 + \cdots + \alpha_4 \leqq \pi$.

integral. It is important to note that these integrals do not represent elementary functions unless the image is an infinite region. Also, the integral will involve a multivalued function, and a specific branch must be selected to fit the boundary values specified in the problem. Table 9.2 is useful for our purposes.

EXAMPLE 9.24 Use the Schwarz-Christoffel Formula to verify that the function $w = f(z) = \text{Arcsin } z$ maps the upper half-plane $\text{Im } z > 0$ onto the semi-infinite strip $-\pi/2 < u < \pi/2$, $v > 0$ shown in Figure 9.56.

Solution If we choose $x_1 = -1$, $x_2 = 1$, $w_1 = -\pi/2$, $w_2 = \pi/2$, then $\alpha_1 = \pi/2$ and $\alpha_2 = \pi/2$, and equation (5) for $f'(z)$ becomes

$$(12) \qquad f'(z) = A(z+1)^{-(\pi/2)/\pi}(z-1)^{-(\pi/2)/\pi} = \frac{A}{(z^2-1)^{1/2}}.$$

TABLE 9.2 Indefinite Integrals

$$\int \frac{dz}{(z^2-1)^{1/2}} = i \arcsin z = \log(z + (z^2-1)^{1/2})$$

$$\int \frac{dz}{z^2+1} = \arctan z = \frac{i}{2}\log\left(\frac{i+z}{i-z}\right)$$

$$\int \frac{dz}{z(z^2-1)^{1/2}} = -\arcsin\frac{1}{z} = i\log\left(\frac{1}{z} + \left(\frac{1}{z^2}-1\right)^{1/2}\right)$$

$$\int \frac{dz}{z(z+1)^{1/2}} = -2\,\text{arctanh}((z+1)^{1/2}) = \log\left(\frac{1-(z+1)^{1/2}}{1+(z+1)^{1/2}}\right)$$

$$\int (1-z^2)^{1/2}\,dz = \frac{1}{2}[z(1-z^2)^{1/2} + \arcsin z] = \frac{i}{2}[z(z^2-1)^{1/2} + \log(z+(z^2-1)^{1/2})]$$

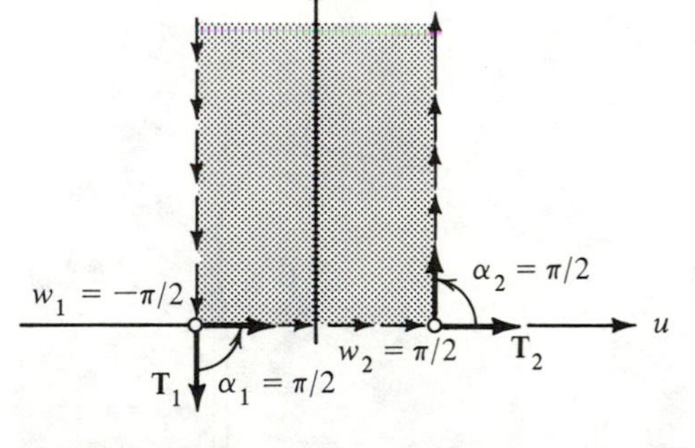

FIGURE 9.56

Using Table 9.2 we see that the solution to equation (12) is

(13) $\quad f(z) = Ai \operatorname{Arcsin} z + B.$

Using the image values $f(-1) = -\pi/2$ and $f(1) = \pi/2$, we obtain the system

(14) $\quad \dfrac{-\pi}{2} = A\dfrac{-i\pi}{2} + B \quad$ and $\quad \dfrac{\pi}{2} = A\dfrac{i\pi}{2} + B,$

which can be solved to obtain $B = 0$ and $A = -i$. Hence the required function is

(15) $\quad f(z) = \operatorname{Arcsin} z.$

EXAMPLE 9.25 Verify that $w = f(z) = (z^2-1)^{1/2}$ maps the upper half-plane $\operatorname{Im} z > 0$ onto the upper half-plane $\operatorname{Im} w > 0$ slit along the segment from 0 to i.

Solution If we choose $x_1 = -1$, $x_2 = 0$, $x_3 = 1$, $w_1 = -d$, $w_2 = i$, and $w_3 = d$, then we see that the formula

$$(16) \qquad g'(z) = A(z+1)^{-\alpha_1/\pi}(z)^{-\alpha_2/\pi}(z-1)^{-\alpha_3/\pi}$$

will determine a mapping $w = g(z)$ from the upper half-plane $\text{Im } z > 0$ onto the portion of the upper half-plane $\text{Im } w > 0$ that lies outside the triangle with vertices $\pm d$, i as indicated in Figure 9.57(a). If we let $d \to 0$, then $w_1 \to 0$, $w_3 \to 0$, $\alpha_1 \to \pi/2$, $\alpha_2 \to -\pi$, and $\alpha_3 \to \pi/2$. The limiting formula for the derivative in (16) becomes

$$(17) \qquad f'(z) = A(z+1)^{-1/2}(z)(z-1)^{-1/2},$$

which will determine a mapping $w = f(z)$ from the upper half-plane $\text{Im } z > 0$ onto the upper half-plane $\text{Im } w > 0$ slit from 0 to i as indicated in Figure 9.57(b). An easy computation reveals that $f(z)$ is

$$(18) \qquad f(z) = A \int \frac{z\,dz}{(z^2-1)^{1/2}} = A(z^2-1)^{1/2} + B,$$

and the boundary values $f(\pm 1) = 0$ and $f(0) = i$ lead to the solution

$$(19) \qquad f(z) = (z^2-1)^{1/2}.$$

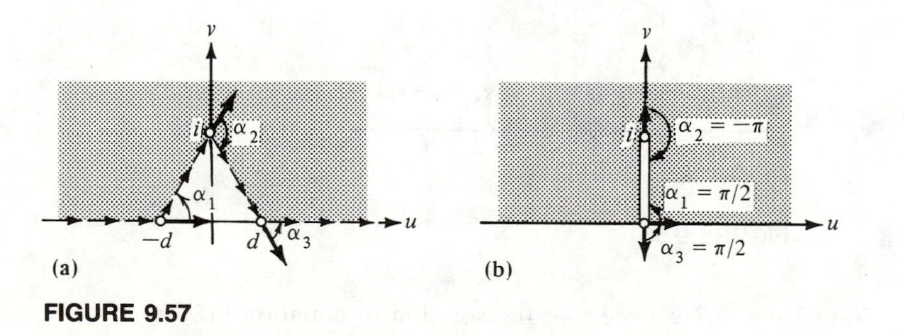

FIGURE 9.57

EXAMPLE 9.26 Show that the function

$$(20) \qquad w = f(z) = \frac{i}{\pi}\text{ Arcsin } z + \frac{1}{\pi}\text{ Arcsin }\frac{1}{z} + \frac{1+i}{2}$$

maps the upper half-plane $\text{Im } z > 0$ onto the right angle channel in the first quadrant, which is bounded by the coordinate axes and the rays $x \geqq 1$, $y = 1$ and $y \geqq 1$, $x = 1$ in Figure 9.58(b).

Solution If we choose $x_1 = -1$, $x_2 = 0$, $x_3 = 1$, $w_1 = 0$, $w_2 = d$, and $w_3 = 1 + i$, then the formula

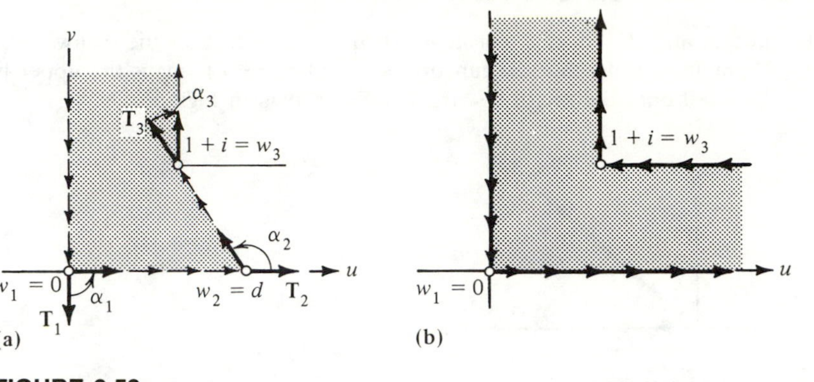

FIGURE 9.58

$$(21) \qquad g'(z) = A(z+1)^{-\alpha_1/\pi}(z)^{-\alpha_2/\pi}(z-1)^{-\alpha_3/\pi}$$

will determine a mapping of the upper half-plane onto the domain indicated in Figure 9.58(a).

If we let $d \to \infty$, then $\alpha_2 \to \pi$ and $\alpha_3 \to -\pi/2$. Then the limiting formula for the derivative in (21) becomes

$$(22) \qquad f'(z) = A(z+1)^{-(\pi/2)/\pi}(z)^{-(\pi)/\pi}(z-1)^{-(-\pi/2)/\pi}$$

$$= A\,\frac{1}{z}\,\frac{(z-1)^{1/2}}{(z+1)^{1/2}} = A\,\frac{z-1}{z(z^2-1)^{1/2}},$$

which will determine a mapping $w = f(z)$ from the upper half-plane onto the channel as indicated in Figure 9.58(b). Using Table 9.2, we obtain

$$(23) \qquad f(z) = A\left[\int \frac{dz}{(z^2-1)^{1/2}} - \int \frac{dz}{z(z^2-1)^{1/2}}\right]$$

$$= A\left[i \arcsin z + \arcsin \frac{1}{z}\right] + B.$$

If the principal branch of the inverse sine function is used, then the boundary values $f(-1) = 0$ and $f(1) = 1 + i$ lead to the system

$$A\left[i\left(\frac{-\pi}{2}\right) - \frac{\pi}{2}\right] + B = 0, \qquad A\left[i\left(\frac{\pi}{2}\right) + \frac{\pi}{2}\right] + B = 1 + i,$$

which can be solved to obtain $A = 1/\pi$ and $B = (1+i)/2$. Hence the required solution is

$$(24) \qquad w = f(z) = \frac{i}{\pi}\,\text{Arcsin}\, z + \frac{1}{\pi}\,\text{Arcsin}\,\frac{1}{z} + \frac{1+i}{2}.$$

EXERCISES FOR SECTION 9.8

1. Let a and K be real constants with $0 < K < 2$. Use the Schwarz-Christoffel Formula to show that the function $w = f(z) = (z - a)^K$ maps the upper half plane $\text{Im } z > 0$ onto the sector $0 < \arg w < K\pi$, shown in Figure 9.59.

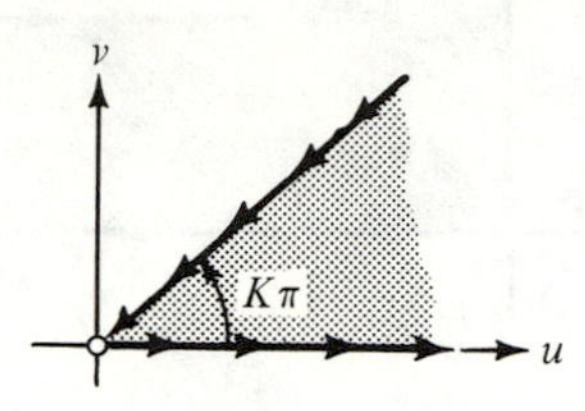

FIGURE 9.59 Accompanies Exercise 1.

2. Let a be a real constant. Use the Schwarz-Christoffel Formula to show that the function $w = f(z) = \text{Log}(z - a)$ maps the upper half-plane $\text{Im } z > 0$ onto the infinite strip $0 < v < \pi$ in Figure 9.60. *Hint*: Set $x_1 = a - 1$, $x_2 = a$, $w_1 = i\pi$, $w_2 = -d$, and let $d \to \infty$.

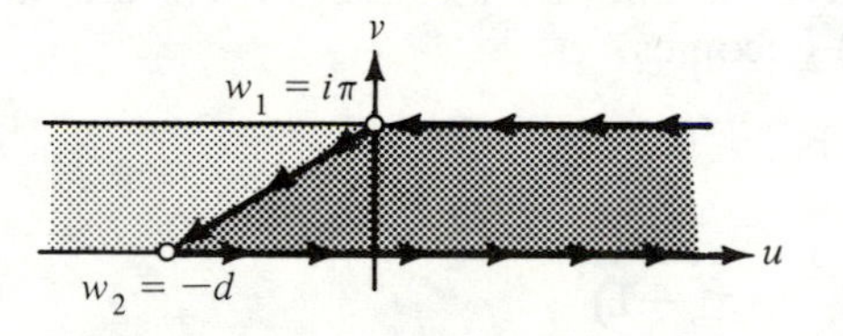

FIGURE 9.60 Accompanies Exercise 2.

3. Use the Schwarz-Christoffel Formula to show that the function

$$w = f(z) = \frac{1}{\pi}[(z^2 - 1)^{1/2} + \text{Log}(z + (z^2 - 1)^{1/2})] - i$$

maps the upper half-plane onto the domain indicated in Figure 9.61. *Hint*: Set $x_1 = -1$, $x_2 = 1$, $w_1 = 0$, and $w_2 = -i$.

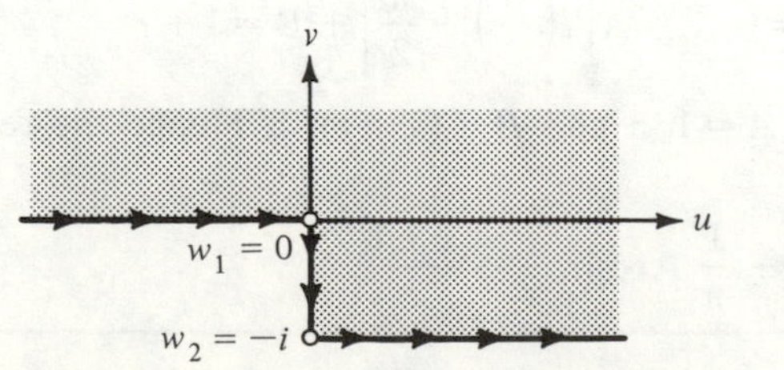

FIGURE 9.61 Accompanies Exercise 3.

4. Use the Schwarz-Christoffel Formula to show that the function

$$w = f(z) = \frac{2}{\pi}\left[(z^2 - 1)^{1/2} + \text{Arcsin}\,\frac{1}{z}\right]$$

maps the upper half-plane onto the domain indicated in Figure 9.62. *Hint*: Set $x_1 = w_1 = -1$, $x_2 = 0$, $x_3 = w_3 = 1$, $w_2 = -id$, and let $d \to \infty$.

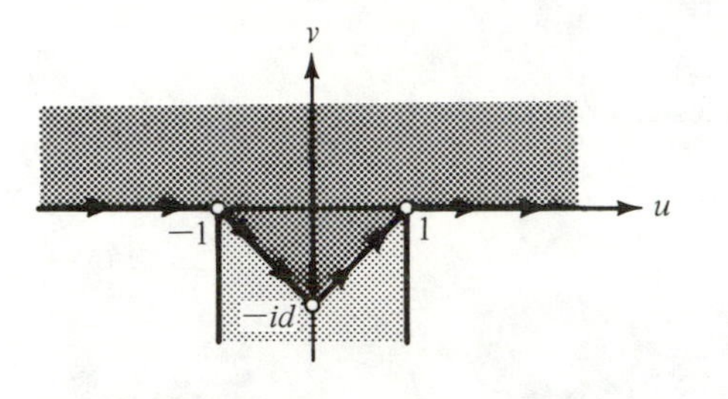

FIGURE 9.62 Accompanies Exercise 4.

5. Use the Schwarz-Christoffel Formula to show that the function

$$w = f(z) = \tfrac{1}{2}\log(z^2 - 1) = \text{Log}[(z^2 - 1)^{1/2}]$$

maps the upper half-plane $\text{Im}\, z > 0$ onto the infinite strip $0 < v < \pi$ slit along the ray $u \leqq 0$, $v = \pi/2$, shown in Figure 9.63. *Hint*: Set $x_1 = -1$, $x_2 = 0$, $x_3 = 1$, $w_1 = i\pi - d$, $w_2 = i\pi/2$, $w_3 = -d$, and let $d \to \infty$.

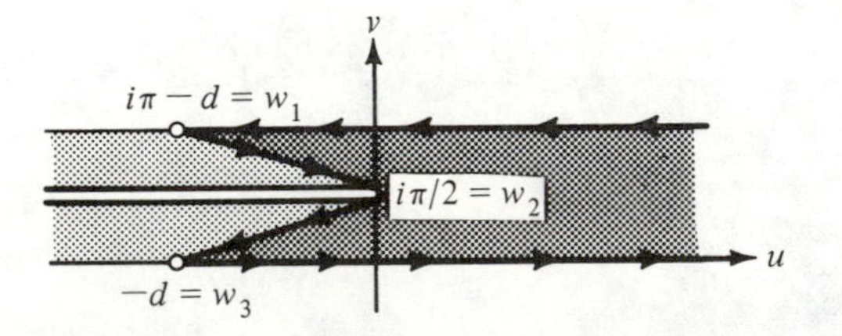

FIGURE 9.63 Accompanies Exercise 5.

6. Use the Schwarz-Christoffel Formula to show that the function

$$w = f(z) = \frac{-2}{\pi}\,[z(1 - z^2)^{1/2} + \text{Arcsin}\, z]$$

maps the upper half-plane onto the domain indicated in Figure 9.64. *Hint*: Set $x_1 = -1$, $x_2 = 1$, $w_1 = 1$, $w_2 = -1$.

7. Use the Schwarz-Christoffel Formula to show that the function $w = f(z) = z + \text{Log}\, z$ maps the upper half-plane $\text{Im}\, z > 0$ onto the upper half-plane $\text{Im}\, w > 0$ slit along the ray $u \leqq -1$, $v = \pi$, shown in Figure 9.65. *Hint*: Set $x_1 = -1$, $x_2 = 0$, $w_1 = -1 + i\pi$, $w_2 = -d$, and let $d \to \infty$.

8. Use the Schwarz-Christoffel Formula to show that the function

$$w = f(z) = 2(z + 1)^{1/2} + \text{Log}\left(\frac{1 - (z + 1)^{1/2}}{1 + (z + 1)^{1/2}}\right) + i\pi$$

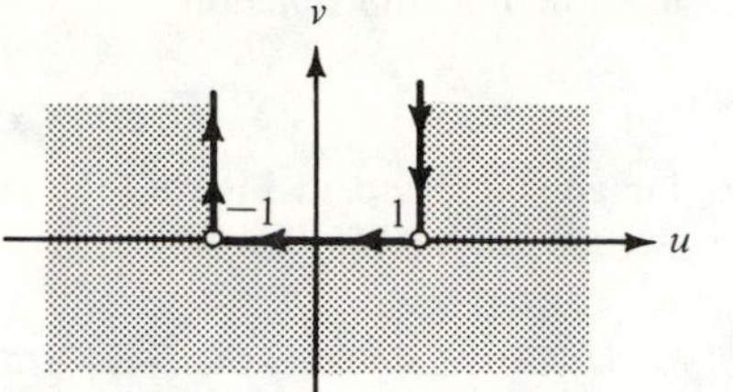

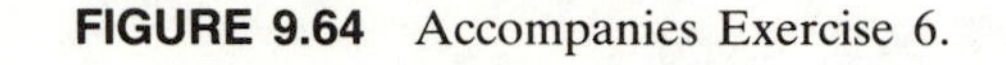
FIGURE 9.64 Accompanies Exercise 6.

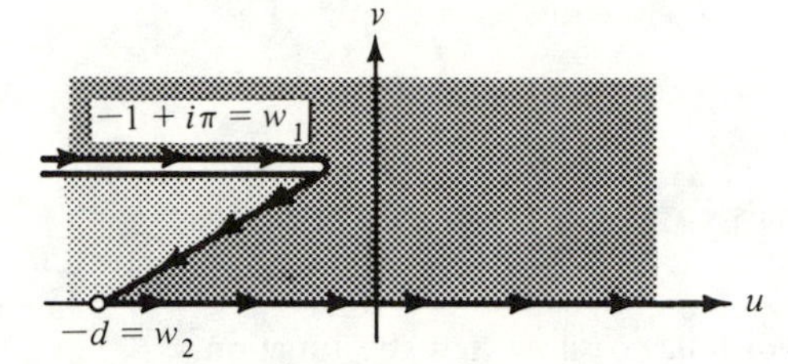

FIGURE 9.65 Accompanies Exercise 7.

maps the upper half-plane onto the domain indicated in Figure 9.66. *Hint*: Set $x_1 = -1$, $x_2 = 0$, $w_1 = i\pi$, $w_2 = -d$, and let $d \to \infty$.

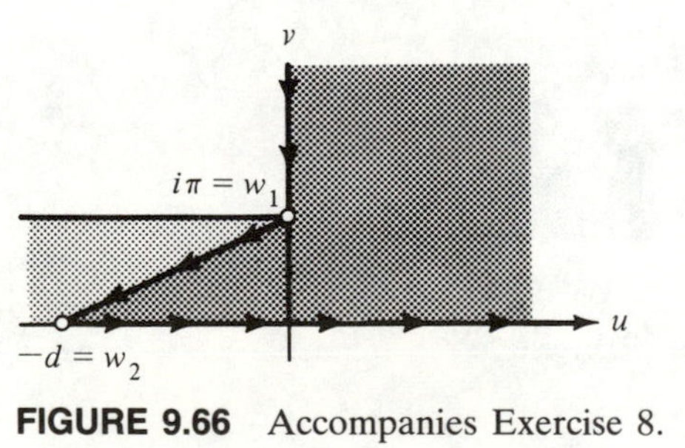

FIGURE 9.66 Accompanies Exercise 8.

9. Show that the function $w = f(z) = (z-1)^{\alpha}(1 + \alpha z/(1-\alpha))^{1-\alpha}$ maps the upper half-plane $\operatorname{Im} z > 0$ onto the upper half-plane $\operatorname{Im} w > 0$ slit along the segment

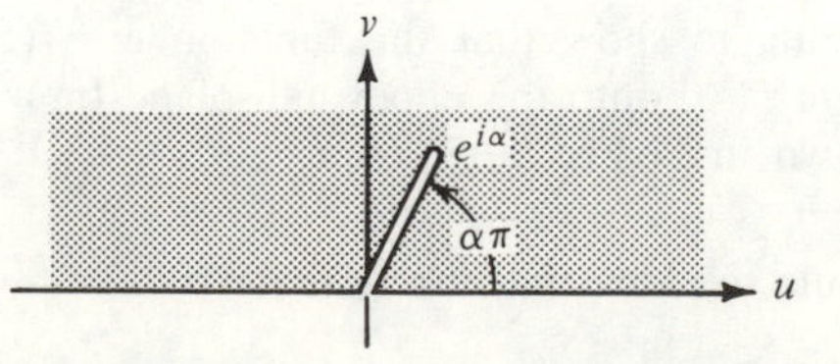

FIGURE 9.67 Accompanies Exercise 9.

from 0 to $e^{i\alpha\pi}$, as shown in Figure 9.67. *Hint*: Show that $f'(z) = A(z + (1-\alpha)/\alpha)^{-\alpha}(z)(z-1)^{\alpha-1}$.

10. Use the Schwarz-Christoffel Formula to show that the function

$$w = f(z) = 4(z+1)^{1/4} + \log\left(\frac{(z+1)^{1/4} - 1}{(z+1)^{1/4} + 1}\right) + i\log\left(\frac{i - (z+1)^{1/4}}{i + (z+1)^{1/4}}\right)$$

maps the upper half-plane onto the domain indicated in Figure 9.68. *Hint*: Set $z_1 = -1$, $z_2 = 0$, $w_1 = i\pi$, $w_2 = -d$, and let $d \to \infty$. Use the change of variable $z + 1 = s^4$ in the resulting integral.

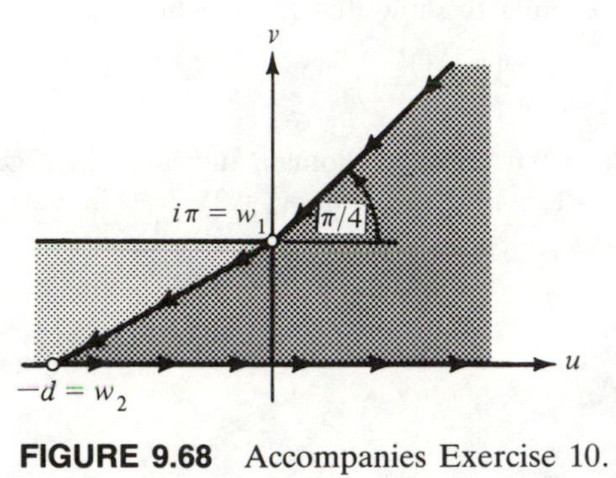

FIGURE 9.68 Accompanies Exercise 10.

11. Use the Schwarz-Christoffel Formula to show that the function

$$w = f(z) = \frac{-i}{2}\, z^{1/2}(z-3)$$

maps the upper half-plane onto the domain indicated in Figure 9.69. *Hint*: Set $x_1 = 0$, $x_2 = 1$, $w_1 = -d$, $w_2 = i$, and let $d \to 0$.

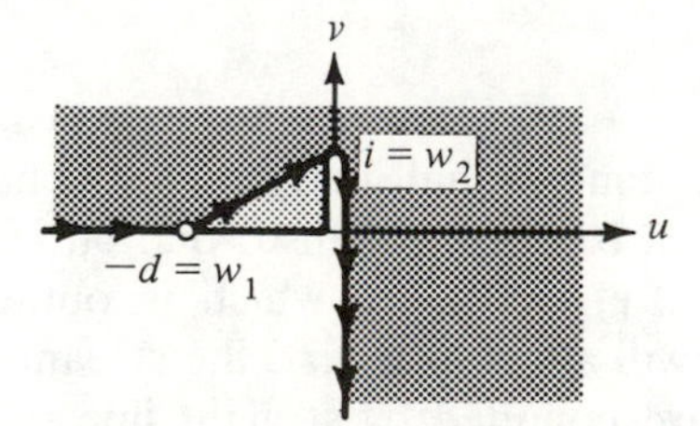

FIGURE 9.69 Accompanies Exercise 11.

12. Show that the function

$$w = f(z) = \int \frac{dz}{(1-z^2)^{3/4}}$$

maps the upper half-plane $\operatorname{Im} z > 0$ onto a right triangle with angles $\pi/2$, $\pi/4$, $\pi/4$.

13. Show that the function

$$w = f(z) = \int \frac{dz}{(1 - z^2)^{2/3}}$$

maps the upper half-plane onto an equilateral triangle.

14. Show that the function

$$w = f(z) = \int \frac{dz}{(z - z^3)^{1/2}}$$

maps the upper half-plane onto a square.

15. Use the Schwarz-Christoffel Formula to show that the function

$$w = f(z) = 2(z + 1)^{1/2} - \mathrm{Log}\left(\frac{1 - (z + 1)^{1/2}}{1 + (z + 1)^{1/2}}\right)$$

maps the upper half-plane $\mathrm{Im}\, z > 0$ onto the domain indicated in Figure 9.70. *Hint*: Set $x_1 = -1$, $x_2 = 0$, $x_3 = 1$, $w_1 = 0$, $w_2 = d$, $w_3 = 2\sqrt{2} - 2\ln(\sqrt{2} - 1) + i\pi$, and let $d \to \infty$.

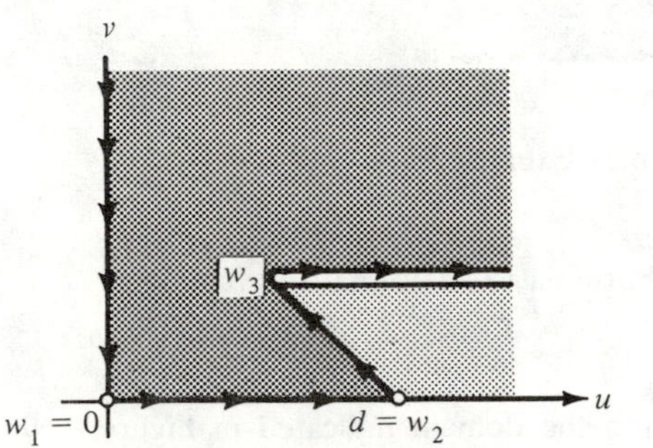

FIGURE 9.70 Accompanies Exercise 15.

9.9 Image of a Fluid Flow

We have already examined several two-dimensional fluid flows and have discovered that the image of a flow under a conformal transformation is a flow. The conformal mapping $w = f(z) = u(x, y) + iv(x, y)$, which is obtained by using the Schwarz-Christoffel Formula, will allow us to find the streamlines for flows in domains in the w plane that are bounded by straight line segments.

The first technique is finding the image of a fluid flowing horizontally from left to right across the upper half-plane $\mathrm{Im}\, z > 0$. The image of the streamline $-\infty < t < \infty$, $y = c$ will be a streamline given by the parametric equations

$$(1) \qquad u = u(t, c), \quad v = v(t, c) \qquad \text{for } -\infty < t < \infty$$

and will be oriented in the counterclockwise (positive) sense. The streamline $u = u(t, 0)$, $v = v(t, 0)$ is considered to be a boundary wall for a containing vessel for the fluid flow.

EXAMPLE 9.27 Consider the conformal mapping

$$(2) \qquad w = f(z) = \frac{1}{\pi}\left[(z^2-1)^{1/2} + \operatorname{Log}(z + (z^2-1)^{1/2})\right],$$

which is obtained by using the Schwarz-Christoffel Formula, to map the upper half-plane $\operatorname{Im} z > 0$ onto the domain in the w plane that lies above the boundary curve consisting of the rays $u \leqq 0$, $v = 1$ and $u \geqq 0$, $v = 0$ and the segment $u = 0$, $-1 \leqq v \leqq 0$.

The image of horizontal streamlines in the z plane are curves in the w plane given by the parametric equation

$$(3) \qquad w = f(t + ic) = \frac{1}{\pi}(t^2 - c^2 - 1 + i2ct)^{1/2} + \frac{1}{\pi}\operatorname{Log}[t + ic + (t^2 - c^2 - 1 + i2ct)^{1/2}]$$

for $-\infty < t < \infty$. The new flow is that of a step in the bed of a deep stream and is illustrated in Figure 9.71(a). The function $w = f(z)$ is also defined for values of z in the lower half-plane, and the images of horizontal streamlines that lie above or below the x axis are mapped onto streamlines that flow past a long rectangular obstacle. This is illustrated in Figure 9.71(b).

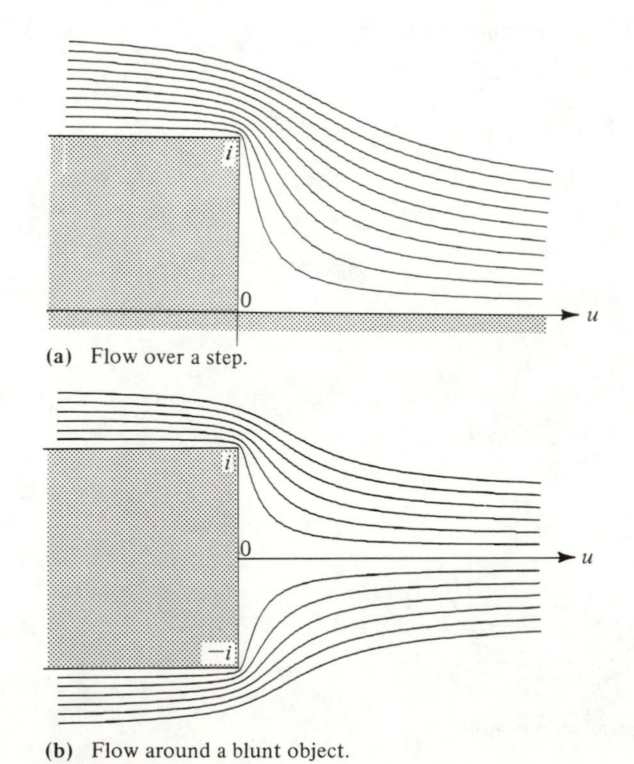

(a) Flow over a step.

(b) Flow around a blunt object.

FIGURE 9.71

EXERCISES FOR SECTION 9.9

For Exercises 1–4, use the Schwarz-Christoffel Formula to find a conformal mapping $w = f(z)$ that will map the flow in the upper half-plane $\text{Im } z > 0$ onto the flow indicated in each of the following figures.

1.

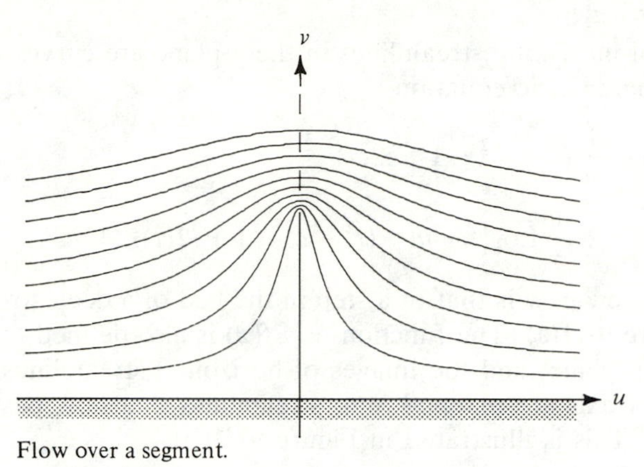

Flow over a segment.

FIGURE 9.72 Accompanies Exercise 1.

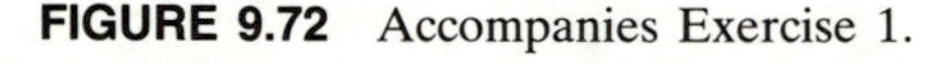

2.

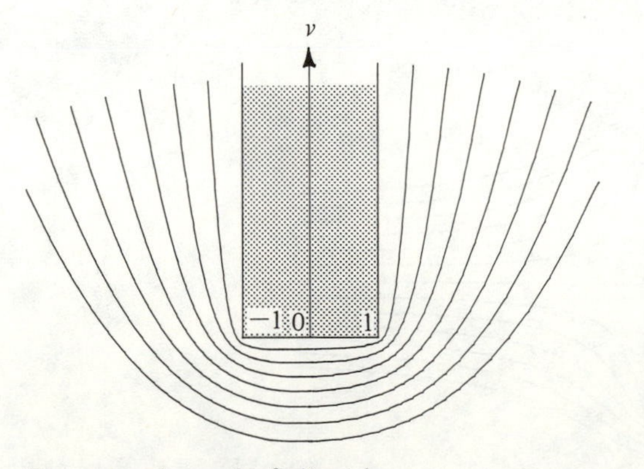

Flow around a semi-infinite strip.

FIGURE 9.73 Accompanies Exercise 2.

3.

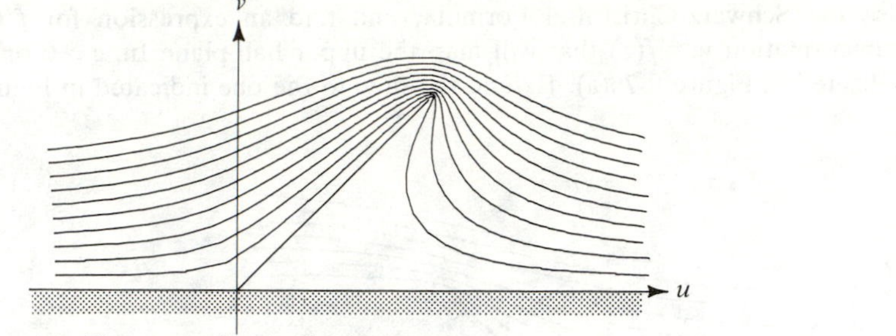

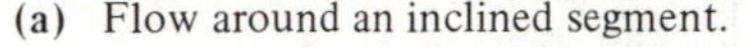
(a) Flow around an inclined segment.

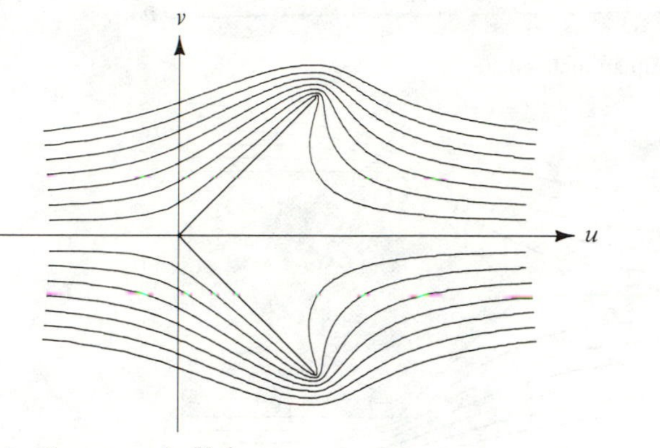

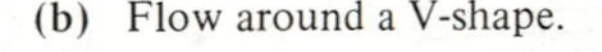
(b) Flow around a V-shape.

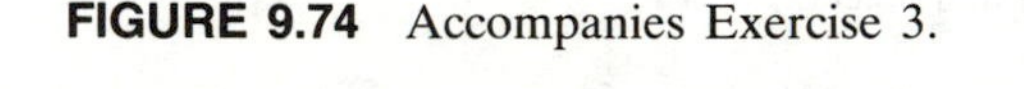
FIGURE 9.74 Accompanies Exercise 3.

4.

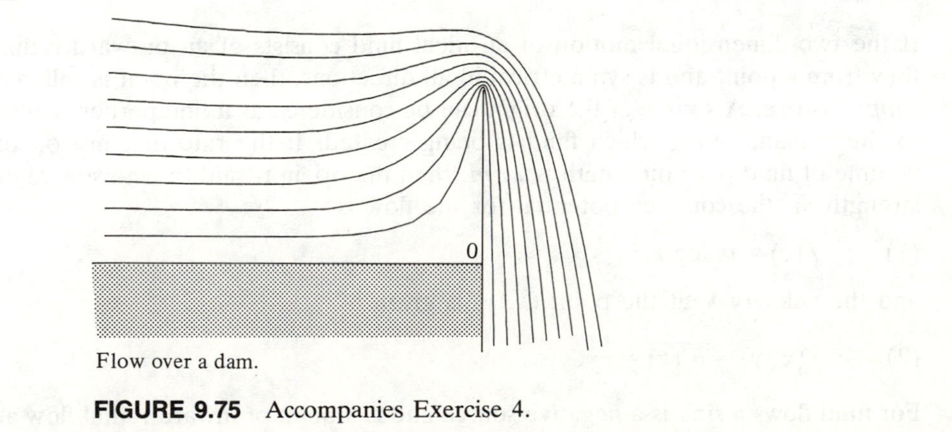

Flow over a dam.

FIGURE 9.75 Accompanies Exercise 4.

5. Use the Schwarz-Christoffel Formula, and find an expression for $f'(z)$ for the transformation $w = f(z)$ that will map the upper half-plane $\text{Im } z > 0$ onto the flow indicated in Figure 9.76(a). Extend the flow to the one indicated in Figure 9.76(b).

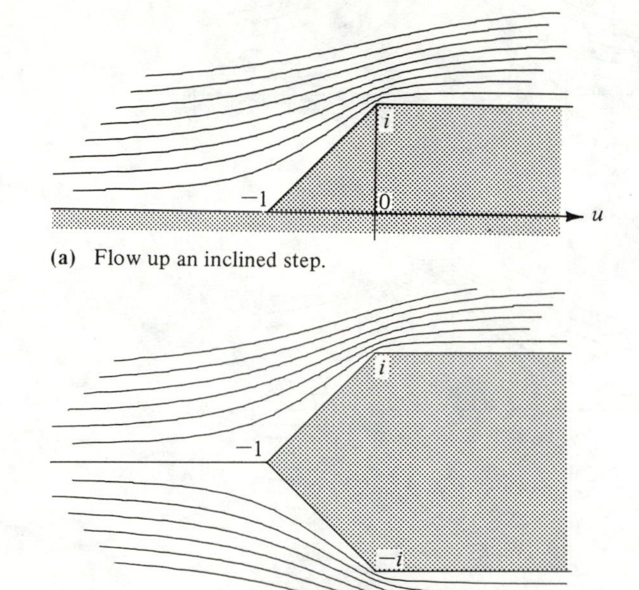

(a) Flow up an inclined step.

(b) Flow around a pointed object.

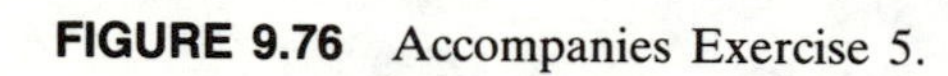

FIGURE 9.76 Accompanies Exercise 5.

9.10 Sources and Sinks

If the two-dimensional motion of an ideal fluid consists of an outward radial flow from a point and is symmetrical in all directions, then the point is called a *simple source*. A source at the origin can be considered as a line perpendicular to the z plane along which fluid is being created. If the rate of emission of volume of fluid per unit length is $2\pi m$, then the origin is said to be a source of strength m, the complex potential for the flow is

$$(1) \qquad F(z) = m \log z,$$

and the velocity $\mathbf{V}$ at the point (x, y) is given by

$$(2) \qquad \mathbf{V}(x, y) = \overline{F'(z)} = \frac{m}{\bar{z}}.$$

For fluid flows a sink is a negative source and is a point of inward radial flow at which the fluid is considered to be absorbed or annihilated. Sources and sinks for flows are illustrated in Figure 9.77.

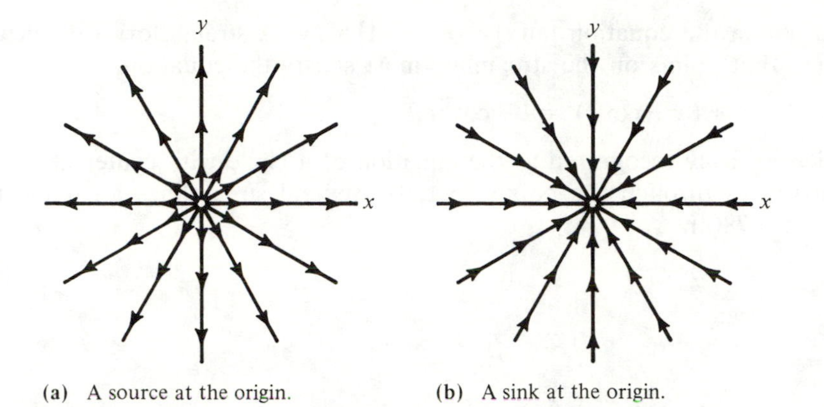

(a) A source at the origin. (b) A sink at the origin.

FIGURE 9.77

In the case of electrostatics a source will correspond to a uniformly positively charged line perpendicular to the z plane. If the line carries a charge $2\pi\varepsilon_0 q$ per unit length, then the electric field $\mathbf{E}$ has strength $q/(x^2 + y^2)^{1/2}$ at the point $z = x + iy$ so that $\mathbf{E}$ is given by the equation

$$(3) \qquad E(x, y) = \frac{qz}{|z|^2} = \frac{q}{\bar{z}}.$$

It is easy to verify that the complex potential

$$(4) \qquad F(z) = -q \log z$$

will satisfy the equation $\mathbf{E}(x, y) = -\overline{F'(z)}$ relating $\mathbf{E}$ and $F(z)$. A sink for electrostatics is a negatively charged line perpendicular to the z plane. The electric field for electrostatic problems corresponds to the velocity field for fluid flow problems, except that their corresponding complex potentials differ by a sign change.

EXAMPLE 9.28 (Source and Sink of Equal Strength) Let a source and sink of unit strength be located at the points $+1$ and -1, respectively. The complex potential for a fluid flowing from the source at $+1$ to the sink at -1 is

$$(5) \qquad F(z) = \log(z - 1) - \log(z + 1) = \log\left(\frac{z-1}{z+1}\right).$$

The velocity potential and stream function are

$$(6) \qquad \phi(x, y) = \ln\left|\frac{z-1}{z+1}\right| \qquad \text{and} \qquad \psi(x, y) = \arg\left(\frac{z-1}{z+1}\right),$$

respectively. Solving for the streamline $\psi(x, y) = c$, we start with

$$(7) \qquad c = \arg\left(\frac{z-1}{z+1}\right) = \arg\left(\frac{x^2 + y^2 - 1 + i2y}{(x+1)^2 + y^2}\right) = \arctan\left(\frac{2y}{x^2 + y^2 - 1}\right)$$

and obtain the equation $\tan c[x^2 + y^2 - 1] = 2y$. A straightforward calculation shows that points on the streamline must satisfy the equation

$$(8) \qquad x^2 + (y - \cot c)^2 = 1 + \cot^2 c,$$

which is easily recognized as the equation of a circle with center at $(0, \cot c)$ that passes through the points $(\pm 1, 0)$. Several streamlines are indicated in Figure 9.78(a).

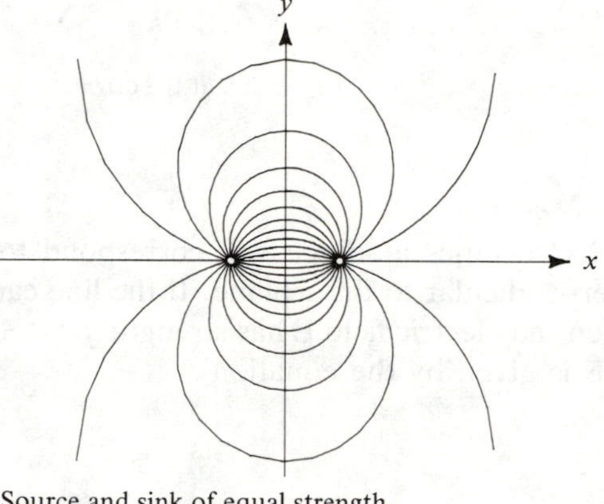

(a) Source and sink of equal strength.

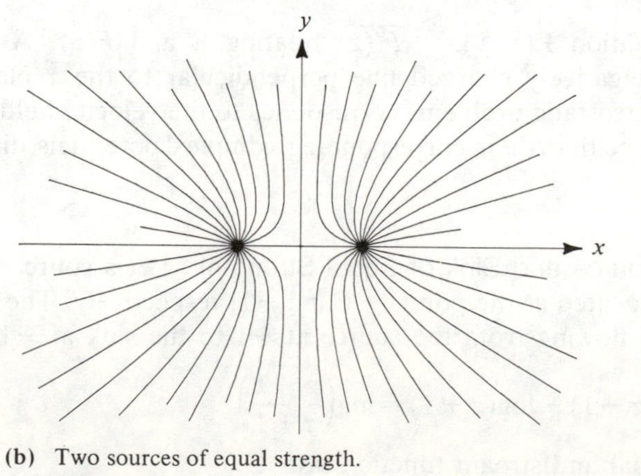

(b) Two sources of equal strength.

FIGURE 9.78

EXAMPLE 9.29 (Two Sources of Equal Strength) Let two sources of unit strength be located at the points ± 1. The resulting complex potential for a fluid flow is

$$(9) \qquad F(z) = \log(z - 1) + \log(z + 1) = \log(z^2 - 1).$$

The velocity potential and stream function are

$$(10) \qquad \phi(x, y) = \ln|z^2 - 1| \qquad \text{and} \qquad \psi(x, y) = \arg(z^2 - 1),$$

respectively. Solving for the streamline $\psi(x, y) = c$, we start with

$$(11) \qquad c = \arg(z^2 - 1) = \arg(x^2 - y^2 - 1 + i2xy) = \arctan\left(\frac{2xy}{x^2 - y^2 - 1}\right)$$

and obtain the equation $x^2 + 2xy \cot c - y^2 = 1$. If we express this in the form $(x - y \tan(c/2))(x + y \cot(c/2)) = 1$ or

$$(12) \qquad \left(x \cos\frac{c}{2} - y \sin\frac{c}{2}\right)\left(x \sin\frac{c}{2} + y \cos\frac{c}{2}\right) = \sin\frac{c}{2} \cos\frac{c}{2} = \frac{\sin c}{2}$$

and use the rotation of axes

$$(13) \qquad x^* = x \cos\frac{-c}{2} + y \sin\frac{-c}{2} \qquad \text{and} \qquad y^* = -x \sin\frac{-c}{2} + y \cos\frac{-c}{2},$$

then the streamlines must satisfy the equation $x^*y^* = (\sin c)/2$ and are easily recognized to be rectangular hyperbolas with centers at the origin that pass through the points ± 1. Several streamlines are indicated in Figure 9.78(b).

Let an ideal fluid flow in a domain in the z plane be effected by a source located at the point z_0. Then the flow at points z, which lie in a small neighborhood of the point z_0, is approximated by that of a source with complex potential

$$(14) \qquad \log(z - z_0) + \text{constant}.$$

If $w = S(z)$ is a conformal mapping and $w_0 = S(z_0)$, then $S(z)$ has a nonzero derivative at z_0, and

$$(15) \qquad w - w_0 = (z - z_0)[S'(z_0) + \eta(z)]$$

where $\eta(z) \to 0$ as $z \to z_0$. Taking logarithms yields

$$(16) \qquad \log(w - w_0) = \log(z - z_0) + \text{Log}[S'(z_0) + \eta(z)].$$

Since $S'(z_0) \neq 0$, the term $[\text{Log}\, S'(z_0) + \eta(z)]$ approaches the constant value $\text{Log}[S'(z_0)]$ as $z \to z_0$. Since $\log(z - z_0)$ is the complex potential for a source located at the point z_0, we see that the image of a source under a conformal mapping is a source.

The technique of conformal mapping can be used to determine the fluid flow in a domain D in the z plane that is produced by sources and sinks. If a conformal mapping $w = S(z)$ can be constructed so that the image of sources, sinks, and boundary curves for the flow in D are mapped onto sources, sinks, and boundary curves in a domain G where the complex potential is known to be $F_1(w)$, then the complex potential in D is given by $F_2(z) = F_1(S(z))$.

EXAMPLE 9.30 Suppose that the lines $x = \pm \pi/2$ are considered as walls of a containing vessel for a fluid flow produced by a single source of unit strength located at the origin. The conformal mapping $w = S(z) = \sin z$ maps the infinite strip bounded by the lines $x = \pm \pi/2$ onto the w plane slit along the boundary rays $u \leqq -1$, $v = 0$ and $u \geqq 1$, $v = 0$, and the image of the source at $z_0 = 0$ is a source located at $w_0 = 0$. It is easy to see that the complex potential

(17) $$F_1(w) = \log w$$

will determine a fluid flow in the w plane past the boundary curves $u \leqq -1$, $v = 0$ and $u \geqq 1$, $v = 0$, which lie along streamlines of the flow. Therefore the complex potential for the fluid flow in the infinite strip in the z plane is

(18) $$F_2(z) = \log(\sin z).$$

Several streamlines for the flow are illustrated in Figure 9.79.

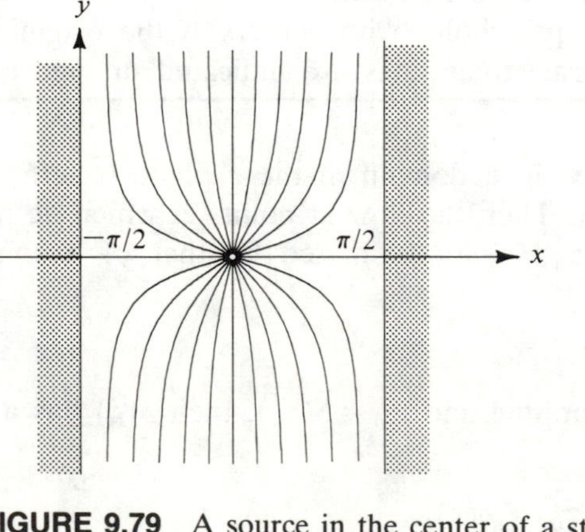

FIGURE 9.79 A source in the center of a strip.

EXAMPLE 9.31 Suppose that the lines $x = \pm \pi/2$ are considered as walls of a containing vessel for the fluid flow produced by a source of unit strength located at the point $z_1 = \pi/2$ and a sink of unit strength located at the point $z_2 = -\pi/2$. The conformal mapping $w = S(z) = \sin z$ maps the infinite strip bounded by the lines $x = \pm \pi/2$ onto the w plane slit along the boundary rays K_1: $u \leqq -1$, $v = 0$ and K_2: $u \geqq 1$, $v = 0$. The image of the source at z_1 is a source at $w_1 = 1$, and the image of the sink at z_2 is a sink at $w_2 = -1$. It is easy to verify that the complex potential

$$(19)\qquad F_1(w) = \log\left(\frac{w-1}{w+1}\right)$$

will determine a fluid flow in the w plane past the boundary curves K_1 and K_2, which lie along streamlines of the flow. Therefore the complex potential for the fluid flow in the infinite strip in the z plane is

$$(20)\qquad F_2(z) = \log\left(\frac{\sin z - 1}{\sin z + 1}\right).$$

Several streamlines for the flow are illustrated in Figure 9.80.

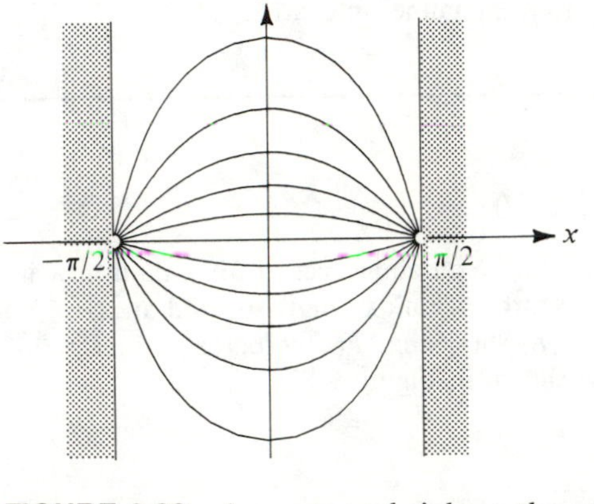

FIGURE 9.80 A source and sink on the edges of a strip.

The technique of transformation of a source can be used to determine the effluence from a channel extending from infinity. In this case a conformal mapping $w = S(z)$ from the upper half-plane $\operatorname{Im} z > 0$ is constructed so that the single source located at $z_0 = 0$ is mapped to the point w_0 at infinity that lies along the channel. The streamlines emanating from $z_0 = 0$ in the upper half-plane are mapped onto streamlines issuing from the channel.

EXAMPLE 9.32 Consider the conformal mapping

$$(21)\qquad w = S(z) = \frac{2}{\pi}\left[(z^2-1)^{1/2} + \operatorname{Arcsin}\frac{1}{z}\right],$$

which maps the upper half-plane $\operatorname{Im} z > 0$ onto the domain consisting of the upper half-plane $\operatorname{Im} w > 0$ joined to the channel $-1 \leqq u \leqq 1$, $v = 0$. The point $z_0 = 0$ is mapped onto the point $w_0 = -i\infty$ along the channel. Images of the rays $r > 0$, $\theta = \alpha$ are streamlines issuing from the channel as indicated in Figure 9.81.

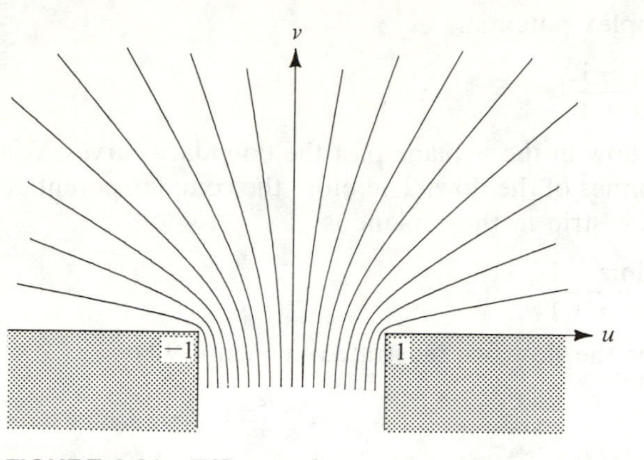

FIGURE 9.81 Effluence from a channel into a half-plane.

EXERCISES FOR SECTION 9.10

1. Let the coordinate axes be walls of a containing vessel for a fluid flow in the first quadrant that is produced by a source of unit strength located at $z_1 = 1$ and a sink of unit strength located at $z_2 = i$. Show that $F(z) = \log((z^2 - 1)/(z^2 + 1))$ is the complex potential for the flow shown in Figure 9.82.

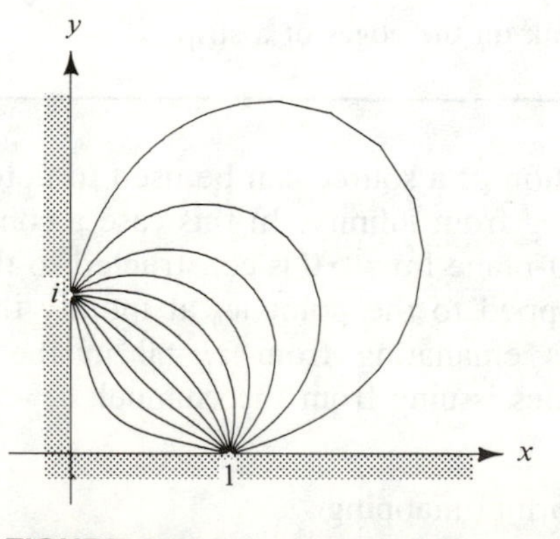

FIGURE 9.82 Accompanies Exercise 1.

2. Let the coordinate axes be walls of a containing vessel for a fluid flow in the first quadrant that is produced by two sources of equal strength located at the points $z_1 = 1$ and $z_2 = i$. Find the complex potential $F(z)$ for the flow in Figure 9.83.
3. Let the lines $x = 0$ and $x = \pi/2$ form the walls of a containing vessel for a fluid flow in the infinite strip $0 < x < \pi/2$ that is produced by a single source located at the point $z_0 = 0$. Find the complex potential for the flow in Figure 9.84.

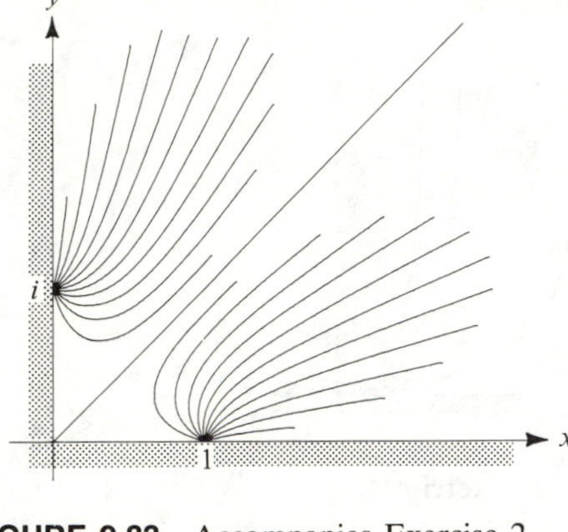

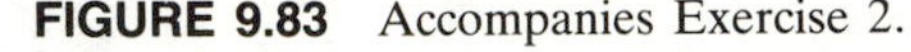

FIGURE 9.83 Accompanies Exercise 2.

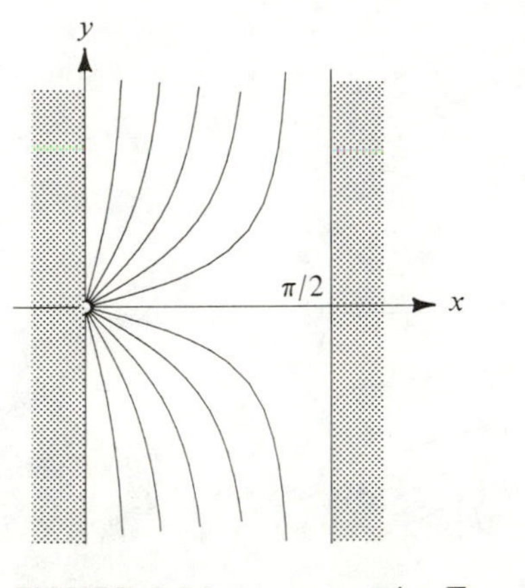

FIGURE 9.84 Accompanies Exercise 3.

4. Let the rays $x = 0$, $y > 0$ and $x = \pi$, $y > 0$ and the segment $y = 0$, $0 < x < \pi$ form the walls of a containing vessel for a fluid flow in the semi-infinite strip $0 < x < \pi$, $y > 0$ that is produced by two sources of equal strength located at the points $z_1 = 0$ and $z_2 = \pi$. Find the complex potential for the flow shown in Figure 9.85. *Hint*: Use the fact that $\sin(\pi/2 + z) = \sin(\pi/2 - z)$.

5. Let the y axis be considered a wall of a containing vessel for a fluid flow in the right half-plane $\text{Re } z > 0$ that is produced by a single source located at the point $z_0 = 1$. Find the complex potential for the flow shown in Figure 9.86.

6. The complex potential $F(z) = 1/z$ determines an electrostatic field that is referred to as a dipole.
   (a) Show that

$$F(z) = \lim_{a \to 0} \frac{\log(z) - \log(z - a)}{a}$$

and conclude that a dipole is the limiting case of a source and sink.

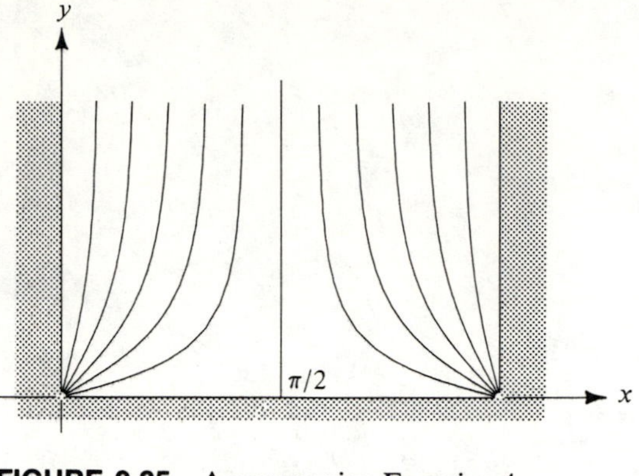

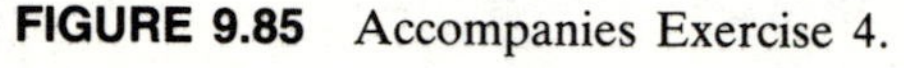

FIGURE 9.85 Accompanies Exercise 4.

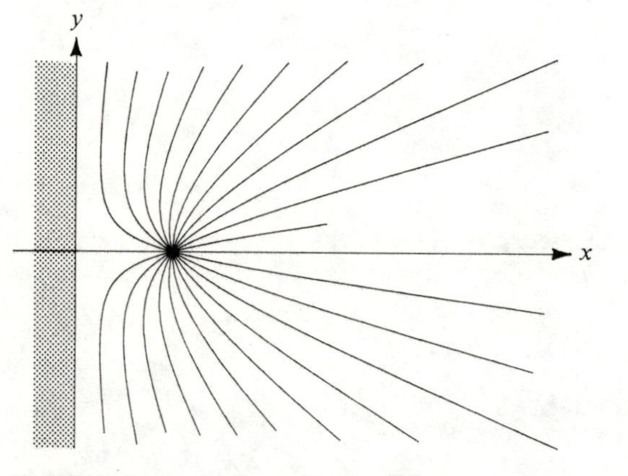

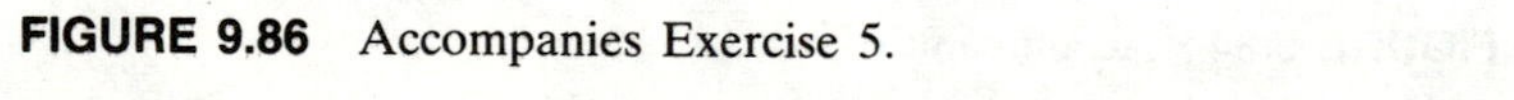

FIGURE 9.86 Accompanies Exercise 5.

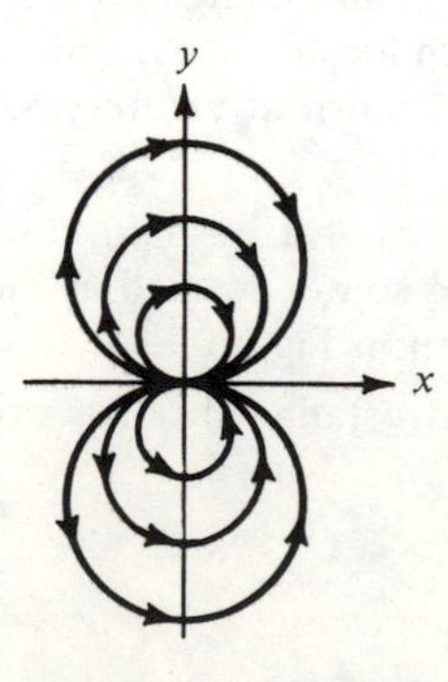

FIGURE 9.87 Accompanies Exercise 6.

(b) Show that the lines of flux of a dipole are circles that pass through the origin as shown in Figure 9.87.

7. Use a Schwarz-Christoffel transformation to find a conformal mapping $w = S(z)$ that will map the flow in the upper half-plane onto the flow from a channel into a quadrant as indicated in Figure 9.88.

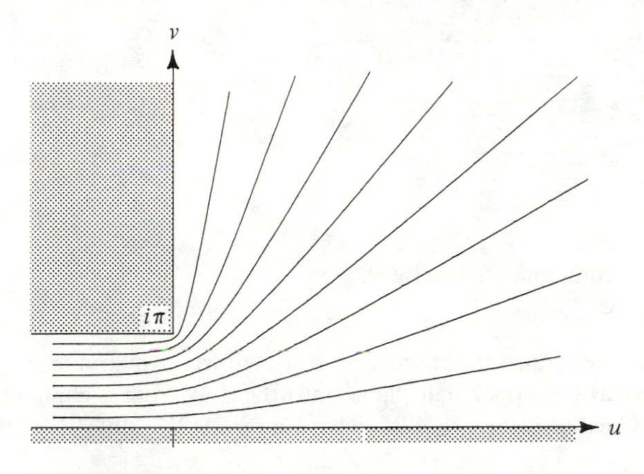

FIGURE 9.88 Accompanies Exercise 7.

8. Use a Schwarz-Christoffel transformation to find a conformal mapping $w = S(z)$ that will map the flow in the upper half-plane onto the flow from a channel into a sector as indicated in Figure 9.89.

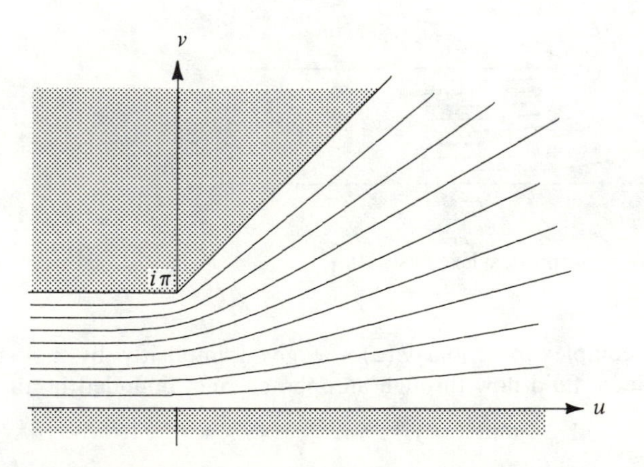

FIGURE 9.89 Accompanies Exercise 8.

9. Use a Schwarz-Christoffel transformation to find a conformal mapping $w = S(z)$ that will map the flow in the upper half-plane onto the flow in a right-angled channel indicated in Figure 9.90.

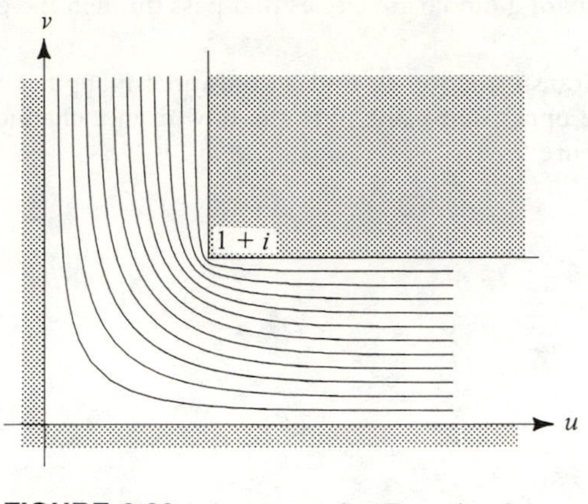

FIGURE 9.90 Accompanies Exercise 9.

10. Use a Schwarz-Christoffel transformation to find a conformal mapping $w = S(z)$ that will map the flow in the upper half-plane onto the flow from a channel back into a quadrant as indicated in Figure 9.91, where $w_0 = 2\sqrt{2} - 2\ln(\sqrt{2} - 1) + i\pi$.

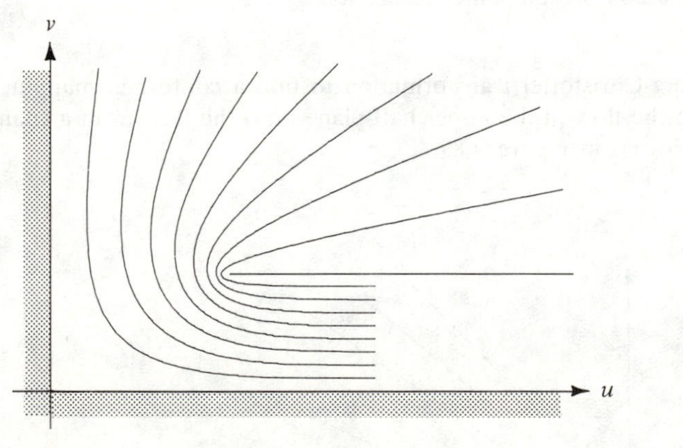

FIGURE 9.91 Accompanies Exercise 10.

11. **(a)** Show that the complex potential $F(z) = w$ given implicitly by $z = w + e^w$ determines the ideal fluid flow through an open channel bounded by the rays

$$y = \pi, \quad -\infty < x < -1 \qquad \text{and} \qquad y = -\pi, \quad -\infty < x < -1$$

into the plane.

(b) Show that the streamline $\psi(x, y) = c$ of the flow is given by the parametric equations

$$x = t + e^t \cos c, \qquad y = c + e^t \sin c \qquad \text{for } -\infty < t < \infty$$

as shown in Figure 9.92.

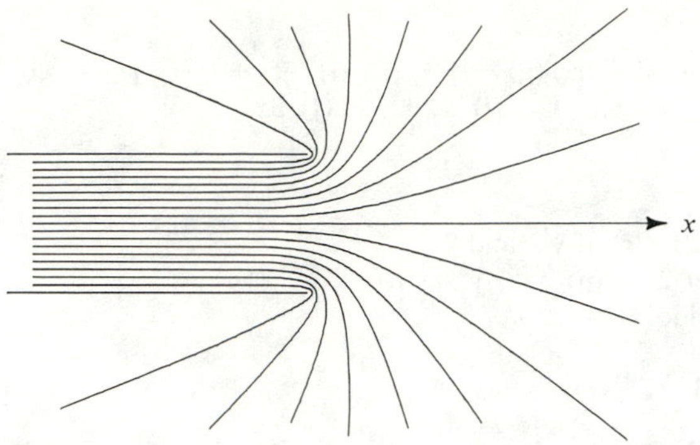

x

FIGURE 9.92 Accompanies Exercise 11.

Answers to Selected Problems

Section 1.2, Page 4

1. **(a)** $8-6i$ **(c)** $10i$ **(e)** $2+2i$ **(g)** $\frac{-12}{5}+\frac{4}{5}i$ **(i)** $\frac{-27}{5}+\frac{11}{5}i$ **(j)** $4i$

2. **(a)** 1 **(c)** $\frac{11}{5}$ **(e)** 2 **(g)** $x_1^2-y_1^2$ **(i)** $x_1^2+y_1^2$ **(j)** $3x_1^2y_1-y_1^3$

Section 1.3, Page 8

1. **(a)** $6+4i$ and $-2+2i$ **(c)** $i2\sqrt{3}$ and 2

2. **(a)** $\sqrt{10}$ **(c)** $\sqrt{5}$ **(e)** 2^{25} **(g)** x^2+y^2

3. **(a)** Inside **(b)** Outside

Section 1.4, Page 14

1. **(a)** $-\pi/4$ **(c)** $2\pi/3$ **(e)** $-\pi/3$ **(g)** $-\pi/6$

2. **(a)** $4(\cos\pi+i\sin\pi)=4e^{i\pi}$ **(c)** $7\left(\cos\frac{-\pi}{2}+i\sin\frac{-\pi}{2}\right)=7e^{-i\pi/2}$

(e) $\frac{1}{2}\left(\cos\frac{\pi}{2}+i\sin\frac{\pi}{2}\right)=\frac{1}{2}e^{i\pi/2}$ **(h)** $5(\cos\theta+i\sin\theta)=5e^{i\theta}$ where $\theta=\arctan\frac{4}{3}$

3. **(a)** i **(c)** $4+i4\sqrt{3}$ **(e)** $\sqrt{2}-i\sqrt{2}$ **(g)** $-e^2$

6. $\text{Arg}(iz)=\text{Arg}\,z+(\pi/2)$, $\text{Arg}(-z)=\text{Arg}\,z-\pi$, $\text{Arg}(-iz)=\text{Arg}\,z-(\pi/2)$ when $z=\sqrt{3}+i$.

11. An example is $z_1=-4i$ and $z_2=-\sqrt{3}+i$.

12. All z except $z=0$ and the negative real numbers.

Section 1.5, Page 20

2. **(a)** $-16-i16\sqrt{3}$ **(c)** -64

7. **(a)** $(5-12i)^{1/2}=\pm(3-2i)$ **(b)** $(-3+4i)^{1/2}=\pm(1+2i)$

8. $\sqrt{2}\cos\left(\frac{\pi}{4}+\frac{2\pi k}{3}\right)+i\sqrt{2}\sin\left(\frac{\pi}{4}+\frac{2\pi k}{3}\right)$ for $k=0,1,2$

9. $2\pm 2i,\ -2\pm 2i$

11. $2\cos\left(\frac{\pi}{8}+\frac{k\pi}{2}\right)+i2\sin\left(\frac{\pi}{8}+\frac{k\pi}{2}\right)$ for $k=0,1,2,3$

14. $1-2i$ and $-2+i$ **16.** 1 and $-1\pm 2i$ **19.** $\pm i$ and $2\pm i$

21. $2\sqrt{3}+2i,\ -4i,\ -2\sqrt{3}+2i$

Section 1.6, Page 25

2. **(a)** $z(t)=t+it$ for $0\leqq t\leqq 1$
(b) $z(t)=t+i$ for $0\leqq t\leqq 1$
(d) $z(t)=2-t+it$ for $0\leqq t\leqq 1$

3. **(a)** $z(t)=t+it^2$ for $0\leqq t\leqq 2$
(c) $z(t)=1-t+i(1-t)^2$ for $0\leqq t\leqq 1$

4. **(a)** $z(t)=\cos t+i\sin t$ for $-\pi/2\leqq t\leqq\pi/2$
(b) $z(t)=-\cos t+i\sin t$ for $-\pi/2\leqq t\leqq\pi/2$

5. **(a)** $z(t)=\cos t+i\sin t$ for $0\leqq t\leqq\pi/2$
(b) $z(t)=\cos t-i\sin t$ for $0\leqq t\leqq 3\pi/2$

7. The sets (a), (d), (e), (f), (g) are open. **8.** The sets (a)–(f) are connected.
9. The sets (a), (d), (e), (f) are domains. **10.** The sets (a)–(f) are regions.
11. The set (c) is a closed region. **12.** The sets (c), (e), (g) are bounded.

Section 2.1, Page 29

1. **(a)** $2-12i$ **(b)** $1-33i$ **2.** **(a)** $74-12i$ **(b)** $24-4i$

3. **(a)** $6+\frac{i}{2}$ **(b)** $\frac{1}{5}-\frac{2i}{5}$ **5.** **(a)** $1028-984i$

6. $x^2+2x+3y-y^2+i(-3x-2xy+2y)$
9. $r^5\cos 5\theta+r^3\cos 3\theta+i(r^5\sin 5\theta-r^3\sin 3\theta)$

10. **(a)** 1 **(b)** e **(c)** $\frac{\sqrt{2}}{2}+i\frac{\sqrt{2}}{2}$ **(d)** $\frac{e}{\sqrt{2}}+i\frac{e}{\sqrt{2}}$ **(e)** $-\frac{1}{2}+i\frac{1}{2}\sqrt{3}$ **(f)** $-e^2$

11. **(a)** 0 **(b)** $\frac{1}{2}\ln 2+\frac{i\pi}{4}$ **(c)** $\frac{1}{2}\ln 3$ **(d)** $\ln 2+\frac{i\pi}{6}$ **(e)** $\ln 2+\frac{i\pi}{3}$

(f) $\ln 5+i\arctan\frac{4}{3}$

13. **(a)** 0 **(b)** $\ln\sqrt{2}+\frac{i\pi}{4}$ **(c)** $\ln 2+i\pi$ **(d)** $\ln 2+\frac{i5\pi}{6}$

Section 2.2, Page 36

1. **(a)** the half-plane $v>1-u$ **2.** the line $u=-4+4t$, $v=6-3t$
3. **(a)** the disk $|w-1-5i|<5$
4. The circle $u=-3+3\cos t-4\sin t$, $v=8+4\cos t+3\sin t$

5. The triangle with vertices $-5-2i$, -6, $3+2i$. **6.** $w=f(z)=\frac{3+2i}{13}z+\frac{7+9i}{13}$

7. $w=f(z)=-5z+3-2i$ **8.** $w=f(z)=\frac{i}{5}z+\frac{7+4i}{5}$

Section 2.3, Page 41

3. The region in the upper half-plane $\text{Im}\, w>0$ that lies between the parabolas $u=4-(v^2/16)$ and $u=(v^2/4)-1$.
4. The region in the first quadrant that lies under the parabola $u=4-(v^2/16)$.
7. **(a)** The points that lie to the extreme right or left of the branches of the hyperbola $x^2-y^2=4$.
(b) The points in quadrant I above the hyperbola $xy=3$, and the points in quadrant III below $xy=3$.
10. **(a)** $\rho>1$, $\pi/6<\phi<\pi/4$ **(b)** $1<\rho<3$, $0<\phi<\pi/3$
(c) $\rho<2$, $-\pi/2<\phi<\pi/4$
11. The region in the w plane that lies to the right of the parabola $u=4-(v^2/16)$.
13. The horizontal strip $1<v<8$.
15. **(a)** $1<\rho<8$, $-3\pi/4<\phi<\pi$ **(b)** $\rho>27$, $2\pi<\phi<9\pi/4$
16. **(a)** $\rho<8$, $3\pi/4<\phi<\pi$ **(c)** $\rho<64$, $3\pi/2<\phi<2\pi$
17. **(a)** $\rho>0$, $-\pi/2<\phi<\pi/3$ **(c)** $\rho>0$, $-\pi/4<\phi<\pi/6$

Section 2.4, Page 47

1. $-3+5i$ **2.** $(5+3i)/2$ **3.** $-4i$ **4.** $1-4i$ **5.** $1-\frac{3}{2}i$
10. **(a)** i **(b)** $(-3+4i)/5$ **(c)** 1 **(d)** The limit does *not* exist.
12. Yes. The limit is zero.
14. No. Arg z is discontinuous along the negative real axis.
15. **(a)** For all z. **(b)** All z except $\pm i$. **(c)** All z except -1, -2.
19. No. The limit does not exist.

Section 2.5, Page 54

1. The circle $|w+\frac{5}{2}i|=\frac{5}{2}$. **3.** The circle $|w+\frac{1}{6}|=\frac{1}{6}$.
5. The circle $|w-1+i|=\sqrt{2}$. **7.** The circle $|w-\frac{6}{5}|=\frac{4}{5}$.

Section 3.1, Page 60

1. **(a)** $f'(z)=15z^2-8z+7$ **(c)** $h'(z)=3/(z+2)^2$ for $z\neq -2$.
3. Parts (a), (b), (e), (f) are entire, and (c) is entire provided that $g(z)\neq 0$ for all z.
7. **(a)** $-4i$ **(c)** 3 **(e)** -16

Section 3.2, Page 66

1. **(c)** $u_x=v_y=-2(y+1)$ and $u_y=-v_x=-2x$. Then $f'(z)=u_x+iv_x=-2(y+1)+i2x$.
2. $f'(z)=f''(z)=e^x\cos y+ie^x\sin y$ **3.** $a=1$ and $b=2$
4. $f(z)=i/z$ and $f'(z)=-i/z^2$
5. $u_x=v_y=2e^{2xy}[y\cos(y^2-x^2)+x\sin(y^2-x^2)]$,
$u_y=-v_x=2e^{2xy}[x\cos(y^2-x^2)-y\sin(y^2-x^2)]$
6. **(c)** $u_x=-e^y\sin x$, $v_y=e^y\sin x$, $u_y=e^y\cos x$, $-v_x=-e^y\cos x$. The Cauchy-Riemann equations hold if and only if both $\sin x=0$ and $\cos x=0$, which is impossible.
8. $u_x=v_y=2x$, $u_y=2y$, and $v_x=2y$. The Cauchy-Riemann equations hold if and only if $y=0$.
10.

$$u_r=\frac{2\ln r}{r}=\frac{1}{r}(2\ln r)=\frac{1}{r}v_\theta,\ v_r=\frac{2\theta}{r}=\frac{-1}{r}(-2\theta)=\frac{-1}{r}u_\theta.$$

$$f'(z)=e^{-i\theta}[u_r+iv_r]=\frac{2}{r}e^{-i\theta}[\ln r+i\theta].$$

Section 3.3, Page 74

3. f is differentiable only at points on the coordinate axes. f is nowhere analytic.
4. f is differentiable only at points on the circle $|z|=2$. f is nowhere analytic.
5. f is differentiable in quadrants I and III. f is analytic in quadrants I and III.
8. $c=-a$ **9.** No. v is *not* harmonic.
10. **(a)** $v(x,y)=x^3-3xy^2$ **(c)** $u(x,y)=-e^y\cos x$
12. $U_x(x,y)=u_x(x,-y)$, $U_{xx}(x,y)=u_{xx}(x,-y)$, $U_y(x,y)=-u_y(x,-y)$, $U_{yy}(x,y)=u_{yy}(x,-y)$. Hence, $U_{xx}+U_{yy}=u_{xx}+u_{yy}=0$.

Section 4.1, Page 82

1. **(a)** The sector $\rho > 0$, $\pi/4 < \phi < \pi/2$. **(b)** The sector $\rho > 0$, $5\pi/4 < \phi < 3\pi/2$.
(c) The sector $\rho > 0$, $-\pi/4 < \phi < \pi/4$. **(d)** The sector $\rho > 0$, $3\pi/4 < \phi < 5\pi/4$.
2. **(a)** $\frac{1}{2}$ **(b)** $(1-i)/(2\sqrt{2})$ **(c)** No. f_1 is not continuous at -1.
3. **(a)** $-\frac{1}{2}$ **(b)** $(-1+i)/(2\sqrt{2})$ **4.** The sector $\rho > 0$, $0 < \phi \leqq \pi$.
6. For example, $f(z) = r^{1/2}\cos(\theta/2) + ir^{1/2}\sin(\theta/2)$ where $r > 0$, $0 < \theta \leqq 2\pi$.
7. **(b)** The sector $\rho > 0$, $-\pi/3 < \phi \leqq \pi/3$.
(c), (e) Everywhere except at the origin and at points that lie on the negative x axis.
(f) $(1 - i\sqrt{3})/6$

Section 4.2, Page 87

2. **(a)** $\frac{1}{2} - i\frac{1}{2}\sqrt{3}$ **(c)** $e^{-4}\cos 5 + ie^{-4}\sin 5$ **(e)** $-e\sqrt{2}(1+i)/2$
3. **(a)** $\ln 4 + i(1+2n)\pi$ where n is an integer.
(c) $\ln 2 + i\left(\frac{-1}{6} + 2n\right)\pi$ where n is an integer
4. $\exp(z^2) = e^{x^2-y^2}[\cos 2xy + i \sin 2xy]$,

$$\exp\left(\frac{1}{z}\right) = e^{x/(x^2+y^2)}\left[\cos\left(\frac{y}{x^2+y^2}\right) - i\sin\left(\frac{y}{x^2+y^2}\right)\right]$$

9. **(b)** $(4z^3 + 3z^6)\exp(z^3)$ **(d)** $(-\exp(1/z))/z^2$

Section 4.3, Page 94

1. **(a)** $\cos(1+i) = \cos 1 \cosh 1 - i \sin 1 \sinh 1$ **(c)** $\sin(2i) = i \sinh 2$
(e) $\tan\left(\frac{\pi + 2i}{4}\right) = \frac{1 + i\sinh 1}{\cosh 1}$
5. **(a)** $(-\cos(1/z))/z^2$ **(c)** $2z \sec(z^2)\tan(z^2)$
10. **(a)** $z = (\frac{1}{2} + 2n)\pi \pm 4i$ where n is an integer.
(c) $z = 2\pi n + i$ and $z = (2n+1)\pi - i$ where n is an integer.
19. **(a)** $\sinh(1 + i\pi) = -\sinh 1$ **(c)** $\cosh\left(\frac{4 - i\pi}{4}\right) = \frac{\cosh 1 - i\sinh 1}{\sqrt{2}}$
22. **(a)** $z = (\pi/6 + 2\pi n)i$ and $z = (5\pi/6 + 2\pi n)i$ where n is an integer.
23. **(a)** $\sinh z + z\cosh z$ **(c)** $\tanh z + z\,\mathrm{sech}^2 z$

Section 4.4, Page 100

1. **(a)** $2 + i\pi/2$ **(c)** $\ln 2 + 3\pi i/4$
2. **(a)** $\ln 3 + i(1+2n)\pi$ where n is an integer.
(b) $\ln 4 + i(\frac{1}{2} + 2n)\pi$ where n is an integer.
3. **(a)** $(e\sqrt{2}/2)(1-i)$ **(c)** $1 + i(-1/2 + 2n)\pi$ where n is an integer.
8. **(a)** $(2z-1)/(z^2 - z + 2)$ **(b)** $1 + \log z$
12. **(a)** $\ln(x^2 + y^2) = 2\,\mathrm{Re}(\log z)$. Hence it is harmonic. **14.** **(b)** No.
15. **(a)** No. Not along the negative x axis.
16. **(a)** $f(z) = \ln|z+2| + i\arg(z+2)$ where $0 < \arg(z+2) < 2\pi$.
(c) $h(z) = \ln|z+2| + i\arg(z+2)$ where $-\pi/2 < \arg(z+2) < 3\pi/2$.

Section 4.5, Page 104

1. **(a)** $\cos(\ln 4) + i\sin(\ln 4)$ **(b)** $e^{-\pi^2/4}[\cos(\pi \ln \sqrt{2}) + i\sin(\pi \ln \sqrt{2})]$
   (c) $\cos 1 + i \sin 1$
2. **(a)** $e^{-(1/2+2n)\pi}$ where n is an integer.
   (b) $\cos(\sqrt{2}(1+2n)\pi) + i\sin(\sqrt{2}(1+2n)\pi)$ where n is an integer.
   (c) $\cos(1+4n) + i\sin(1+4n)$ where n is an integer.
4. $(-1)^{3/4} = \dfrac{1 \pm i}{\sqrt{2}}, \dfrac{-1 \pm i}{\sqrt{2}}$; $(i)^{2/3} = -1, \dfrac{1}{2} \pm i\dfrac{1}{2}\sqrt{3}$.
6. $\alpha r^{\alpha-1}\cos(\alpha-1)\theta + i\alpha r^{\alpha-1}\sin(\alpha-1)\theta$ where $-\pi < \theta < \pi$.
13. No. $1^{a+ib} = e^{a2\pi n}\cos b2\pi n + ie^{a2\pi n}\sin b2\pi n$ where n is an integer.

Section 4.6, Page 107

1. **(a)** $(\frac{1}{2} + 2n)\pi \pm i \ln 2$ where n is an integer.
   (b) $2\pi n \pm i \ln 3$ where n is an integer.
   (c) $(\frac{1}{2} + 2n)\pi \pm i\ln(3 + 2\sqrt{2})$ where n is an integer.
   (e) $-(\frac{1}{2} + n)\pi + i \ln\sqrt{3}$ where n is an integer.
2. **(a)** $i(\frac{1}{2} + 2n)\pi$ where n is an integer. **(b)** $\pm \ln 2 + i2\pi n$ where n is an integer.
   (c) $\ln(\sqrt{2}+1) + i(\frac{1}{2} + 2n)\pi$ and $\ln(\sqrt{2}-1) + i(-\frac{1}{2} + 2n)\pi$ where n is an integer.
   (e) $i(\frac{1}{4} + n)\pi$ where n is an integer.

Section 5.1, Page 111

1. $2 - 3i$ **2.** $-\frac{23}{4} - 6i$ **3.** 1 **4.** $2 - \arctan 2 - i\ln\sqrt{5}$
5. $\sqrt{2}\pi/8 + \sqrt{2}/2 - 1 + i(\sqrt{2}/2 - \sqrt{2}\pi/8)$

Section 5.2, Page 121

2. C_1: $z_1(t) = 2\cos t + i2\sin t$ for $0 \leqq t \leqq \pi/2$, C_2: $z_2(t) = -t + i(2-t)$ for $0 \leqq t \leqq 2$.
3. C_1: $z_1(t) = (-2+t) + it$ for $0 \leqq t \leqq 2$,
   C_2: $z_2(t) = t + 2i$ for $0 \leqq t \leqq 2$,
   C_3: $z_3(t) = 2 + i(2-t)$ for $0 \leqq t \leqq 2$.
4. **(a)** $3/2$ **(b)** $\pi/2$ **5.** **(a)** $-32i$ **(b)** $-8\pi i$ **6.** 0 **7.** $32\pi i$ **8.** $i - 2$
9. 1 **10.** $-1 + 2i/3$ **13.** $-4 - \pi i$ **14.** $-2\pi i$ **15.** 0 **16.** $-2e$

Section 5.3, Page 134

4. **(a)** 0 **(b)** $2\pi i$ **5.** **(a)** $4\pi i$ **(b)** $2\pi i$ **6.** $4\pi i$ **7.** 0
8. **(a)** $\pi i/4$ **(b)** $-\pi i/4$ **(c)** 0 **9.** **(a)** 0 **(b)** $-2\pi i$ **11.** $-4i/3$ **12.** 0

Section 5.4, Page 139

1. $\frac{4}{3} + 3i$ **2.** $-1 + i((\pi+2)/2)$ **3.** $i - e^2$ **4.** $-7/6 + i/2$
6. $2 - i2\sinh 1$ **7.** $(\pi/2e) - e^2 - i(e^2\pi + 2/e)$ **9.** $-1 - \sinh 1 + \cosh 1$
10. $i(1/2 - (\sinh 2)/4)$ **11.** $\ln\sqrt{2} - \pi/4 + i(\ln\sqrt{2} + \pi/4 - 1)$
13. $\ln\sqrt{10} - \ln 2 + i\arctan 3 = \ln\sqrt{5/2} + i\arctan 3$ or $\ln\sqrt{5/2} + i(\pi/4 + \arctan 1/2)$

Section 5.5, Page 145

1. $4\pi i$ **2.** πi **3.** $-\pi i/2$ **4.** $2\pi i/3$ **6.** $-\pi i/3$ **7.** $2\pi i$
9. $2\pi i/(n-1)!$ **10.** **(a)** $-\pi i/8$ **(b)** $e^4\, i\pi/64$ **11.** $(\pi - i\pi)/8$
12. **(a)** π **(b)** $-\pi$ **13.** **(a)** $i\pi \sinh 1$ **(b)** $i\pi \sinh 1$ **14.** $\pi/2$

Section 5.6, Page 152

1. $(z+1+i)(z+1-i)(z-1+i)(z-1-i)$ **2.** $(z-1+2i)(z+2-i)$
3. $(z+i)(z-i)(z-2+i)(z-2-i)$ **4.** $(z-i)(z-1-i)(z-2-i)$

7. **(a)** 18 **(b)** 5 **(c)** 8 **(d)** 4 **8.** $\sqrt{1+\sinh^2(2)}$ **9.** $|f^{(3)}(1)| \leqq \dfrac{3!(10)}{3^3} = \dfrac{20}{9}$

10. $|f^{(3)}(0)| \leqq \dfrac{3!(10)6\pi}{2^4(2\pi)} = \dfrac{45}{4}$

Section 5.7, Page 158

5. $U(\theta) = \dfrac{\pi}{2} - \dfrac{4}{\pi} \sum_{n=1}^{\infty} \dfrac{r^{(2n-1)} \cos(2n-1)\theta}{(2n-1)^2}$

7. $U(\theta) = \dfrac{\pi^2}{3} + 4 \sum_{n=1}^{\infty} \dfrac{r^n(-1)^n \cos n\theta}{n^2}$

9. $U(\theta) = \dfrac{4}{\pi} \sum_{n=1}^{\infty} \dfrac{r^{2n-1} \cos(2n-1)\theta}{(2n-1)^2}$

Section 6.1, Page 171

1. **(a)** 0 **(b)** 1 **(c)** i **(d)** i **8.** No. **12.** Yes. **16.** Yes.

Section 6.2, Page 178

6. **(a)** $0.999959 - i0.015625$ **(b)** $0.999959 + i0.015625$
7. **(a)** $0.100333 + i0.099667$ **(b)** $0.099667 + i0.100333$
10. $\arctan z = z - \frac{1}{3}z^3 + \frac{1}{5}z^5 - \frac{1}{7}z^7 + \cdots$ for $|z| < 1$

16. $z - \dfrac{z^3}{(3!)3} + \dfrac{z^5}{(5!)5} - \dfrac{z^7}{(7!)7} + \cdots$

Section 6.3, Page 184

1. **(a)** $\sum_{n=0}^{\infty} z^{n-3}$ for $|z| < 1$ **(b)** $-\sum_{n=1}^{\infty} \dfrac{1}{z^{n+3}}$ for $|z| > 1$

2. $\sum_{n=0}^{\infty} \dfrac{(-1)^n 2^{2n+1} z^{2n-3}}{(2n+1)!}$ for $|z| > 0$ **6.** $\sum_{n=0}^{\infty} \dfrac{(-1)^n}{(2n+1)!z^{2n+1}}$ for $|z| > 0$

7. $\sum_{n=1}^{\infty} \dfrac{2z^{4n-7}}{(4n-2)!}$ for $|z| > 0$

9. $\dfrac{1}{16z} + \sum_{n=0}^{\infty} \dfrac{(n+2)z^n}{4^{n+3}}$ for $|z| < 4$, $\sum_{n=1}^{\infty} \dfrac{n(4)^{n-1}}{z^{n+2}}$ for $|z| > 4$

Section 6.4, Page 197

1. $R=\sqrt{2}$ **2.** $R=5$ **9.** $R=2$ **10.** **(b)** $\overline{f(\overline{z})}=\sum_{n=0}^{\infty}\overline{a}_n z^n$
13. No. Since $\sqrt{13}=|z_2|<|z_1|=\sqrt{17}$, it would contradict Theorem 6.11.
14. **(a)** $f'''(0)=48$ **(b)** $f'''(0)=-4+4i$ **(c)** $f'''(0)=-3i/4$

Section 6.5, Page 204

1. **(a)** Zeros of order 4 at $\pm i$. **(c)** Simple zeros at $-1\pm i$.
(e) Simple zeros at $\pm i$ and $\pm 3i$.
2. **(a)** Simple zeros at $(\sqrt{3}\pm i)/2$, $(-\sqrt{3}\pm i)/2$ and $\pm i$.
(c) Zeros of order 2 at $(1\pm i\sqrt{3})/2$ and -1.
(e) Simple zeros at $(1\pm i)/\sqrt{2}$ and $(-1\pm i)/\sqrt{2}$, and a zero of order 4 at the origin.
3. **(a)** Poles of order 3 at $\pm i$, and a pole of order 4 at 1.
(c) Simple poles at $(\sqrt{3}\pm i)/2$, $(-\sqrt{3}\pm i)/2$ and $\pm i$.
(e) Simple poles at $\pm\sqrt{3}i$ and $\pm i/\sqrt{3}$.
4. **(a)** Simple poles at $z=n\pi$ for $n=\pm 1, \pm 2, \ldots$.
(c) Simple poles at $z=n\pi$ for $n=\pm 1, \pm 2, \ldots$, and a pole of order 3 at the origin.
(e) Simple poles at $z=2n\pi i$ for $n=0, \pm 1, \pm 2, \ldots$.
5. **(a)** Removable singularity at the origin. **(c)** Essential singularity at the origin.
6. **(a)** Removable singularity at the origin, and a simple pole at -1.
(c) Removable singularity at the origin.
7. $(-1-i)/16$ **8.** $-1/4$ **9.** 3
20. A nonisolated singularity at the origin.
21. Simple poles at $z=1/n\pi$ for $n=\pm 1, \pm 2, \ldots$, and a nonisolated singularity at the origin.

Section 6.6, Page 207

2. No
3. Yes
4. No

Section 7.2, Page 215

1. **(a)** 1 **(b)** 8 **(c)** 1 **(d)** 5 **2.** **(a)** 1 **(b)** $-1/2$ **(c)** 0 **(d)** 1
3. **(a)** e **(b)** $1/5!$ **(c)** 0 **4.** **(a)** $\frac{1}{6}$ **(b)** 4 **(c)** $\frac{1}{3}$
5. $(\pi+i\pi)/8$ **6.** $(\pi+i\pi)/2$
7. $(1-\cos 1)2\pi i$ **8.** i **9.** $i2\pi\sinh 1$ **10.** **(a)** 0 **(b)** $-4\pi i/25$
11. **(a)** $\pi/3$ **(b)** $(\pi/6)(3-i\sqrt{3})$ **12.** **(a)** $-\pi/(8\sqrt{3})$ **(b)** $\pi\sqrt{3}/8$
13. **(a)** $\pi i/2$ **(b)** $-\pi i/6$ **14.** $\pi i/3$ **15.** $2\pi i/3$

18. **(a)** $\dfrac{1}{z+1}-\dfrac{1}{z+2}$ **(b)** $\dfrac{2}{z+1}+\dfrac{1}{z-2}$ **(c)** $\dfrac{1}{z^2}-\dfrac{2}{z}+\dfrac{3}{z+4}$

(d) $\dfrac{2z}{z^2+4}-\dfrac{2z}{z^2+9}$ **(e)** $\dfrac{2}{z-1}+\dfrac{1}{(z-1)^2}-\dfrac{2}{(z-1)^3}$

Section 7.3, Page 219

2. $2\pi/3$ **4.** $\pi/3$ **6.** $2\pi/9$ **8.** $10\pi/27$ **10.** $8\pi/45$ **12.** $10\pi/27$
14. $4\pi/27$

Section 7.4, Page 223

2. $\pi/4$ **4.** $\pi/18$ **6.** $\pi/4$ **8.** $\pi/64$ **10.** $\pi/15$ **12.** $2\pi/3$

Section 7.5, Page 227

2. 0 and $\dfrac{\pi}{e^3}$ **4.** $\dfrac{3\pi}{16e^2}$ **6.** $\dfrac{\pi}{3}\left(\dfrac{1}{e}-\dfrac{1}{2e^2}\right)$ **8.** $\dfrac{\pi\cos 2}{e}$ **10.** $\dfrac{\pi\cos 1}{e}$

12. $\dfrac{\pi\cos 2}{e^2}$

Section 7.6, Page 231

2. 0 **4.** $-\pi/\sqrt{3}$ **6.** $\pi/\sqrt{3}$ **8.** $\pi(1-\sin 1)$ **10.** $\frac{1}{2}$
12. $\pi(\sin 2-\sin 1)$

Section 7.7, Page 235

2. π **4.** $\pi/\sqrt{2}$ **6.** $-\pi/4$ **8.** $(\ln a)/a$

Section 8.1, Page 248

1. **(b)** All z except $z=\dfrac{\pi}{2}+2n\pi$ **(d)** All z except $z=0$ **(f)** All z except $z=1$.

2. $\alpha=\pi$, $|-1|=1$; $\alpha=\pi/2$, $\left|\dfrac{i}{2}\right|=\dfrac{1}{2}$; $\alpha=0$, $|1|=1$

3. $\alpha=0$, $|1|=1$; $\alpha=-\pi/4$, $|(1-i)/2|=\sqrt{2}/2$; $\alpha=-\pi/2$, $|-i|=1$; $\alpha=\pi$, $|-1|=1$
5. $\alpha=-\pi/2$, $|-i\sinh 1|=\sinh 1$; $\alpha=0$, $|1|=1$; $\alpha=\pi/2$, $|i\sinh 1|=\sinh 1$

Section 8.2, Page 255

1. $S^{-1}(w)=\dfrac{-2w+2}{(1+i)w-1+i}$ **2.** $S^{-1}(w)=\dfrac{iw-i}{w+1}$ **3.** The disk $|w|<1$

5. The disk $|w|>1$ **6.** $w=\dfrac{z+i}{3z-i}$ **7.** $w=\dfrac{-iz+i}{z+1}$ **9.** $w=\dfrac{i-iz}{1+z}$

11. The disk $|w|<1$ **12.** $S_1(S_2(z))=\dfrac{-z-6}{2z+3}$

13. The portion of the disk $|w|<1$ that lies in the upper half-plane $\text{Im } w>0$.

Section 8.3, Page 259

1. The portion of the disk $|w|<1$ that lies in the first quadrant $u>0$, $v>0$.
2. $\{\rho e^{i\phi}\colon 1<\rho<2,\ 0<\phi<\pi/2\}$ **3.** The horizontal strip $0<\text{Im } w<1$.
4. $\{u+iv\colon 0<u<1,\ -\pi<v\leqq\pi\}$ **9.** The horizontal strip $0<\text{Im } w<\pi$.
10. The horizontal strip $\pi/2<\text{Im } w<\pi$.

12. The horizontal strip $|v| < \pi$ slit along the ray $u \leqq 0$, $v = 0$.
13. $Z = z^2 + 1$, $w = Z^{1/2}$ where the principal branch of the square root $Z^{1/2}$ is used.
15. The unit disk $|w| < 1$.

Section 8.4, Page 265

1. The portion of the disk $|w| < 1$ that lies in the second quadrant $u < 0$, $v > 0$.
3. The right branch of the hyperbola $u^2 - v^2 = \frac{1}{2}$.
5. The region in the first quadrant $u > 0$, $v > 0$ that lies inside the ellipse $(u^2/(\cosh^2 1)) + (v^2/(\sinh^2 1)) = 1$ and to the left of the hyperbola $u^2 - v^2 = \frac{1}{2}$.
7. **(a)** $\pi/3$ **(b)** $-5\pi/6$
8. **(a)** $0.754249145 + i1.734324521$ **(c)** $0.307603649 - i1.864161544$
10. The right half-plane $\operatorname{Re} w > 0$ slit along the ray $v = 0$, $u \geqq 1$.
12. The vertical strip $0 < u < \pi/2$.
14. The semi-infinite strip $-\pi/2 < u < \pi/2$, $v > 0$.
16. The horizontal strip $0 < v < \pi$.

Section 9.2, Page 276

1. $15 - 9y$ 3. $5 + (3/\ln 2) \ln |z|$

5. $4 - \frac{4}{\pi} \operatorname{Arg}(z+3) + \frac{6}{\pi} \operatorname{Arg}(z+1) - \frac{3}{\pi} \operatorname{Arg}(z-2)$

$$= 4 - \frac{4}{\pi} \operatorname{Arctan} \frac{y}{x+3} + \frac{6}{\pi} \operatorname{Arctan} \frac{y}{x+1} - \frac{3}{\pi} \operatorname{Arctan} \frac{y}{x-2}$$

6. $\frac{-1}{\pi} \operatorname{Arg}(z^2+1) + \frac{1}{\pi} \operatorname{Arg}(z^2-1) = \frac{-1}{\pi} \operatorname{Arctan} \frac{2xy}{x^2-y^2+1} + \frac{1}{\pi} \operatorname{Arctan} \frac{2xy}{x^2-y^2-1}$

8. $8 - \frac{4}{\pi} \operatorname{Arctan} \frac{1-x^2-y^2}{2y}$

10. $1 - \frac{2}{\pi} \operatorname{Arg}\left(i \frac{1+1/z}{1-1/z}\right) = 1 - \frac{2}{\pi} \operatorname{Arctan} \frac{x^2+y^2-1}{2y}$

12. $\frac{-1}{\pi} \operatorname{Arg}\left(i \frac{1-z}{1+z} + 1\right) + \frac{1}{\pi} \operatorname{Arg}\left(i \frac{1-z}{1+z} - 1\right)$

$$= \frac{-1}{\pi} \operatorname{Arctan} \frac{1-x^2-y^2}{2y+(1+x)^2+y^2} + \frac{1}{\pi} \operatorname{Arctan} \frac{1-x^2-y^2}{2y-(1+x)^2-y^2}$$

Section 9.3, Page 280

1. $\frac{y}{2\pi} \ln \frac{(x-1)^2+y^2}{(x+1)^2+y^2} + \frac{x-1}{\pi} \operatorname{Arctan} \frac{y}{x-1} - \frac{x+1}{\pi} \operatorname{Arctan} \frac{y}{x+1} + 1$

2. $\frac{y}{2\pi} \ln \frac{(x-1)^2+y^2}{x^2+y^2} + \frac{x}{\pi} \operatorname{Arctan} \frac{y}{x-1} - \frac{x}{\pi} \operatorname{Arctan} \frac{y}{x}$

Section 9.5, Page 289

2. $25 + 50(x+y)$ 4. $60 + \frac{40}{\pi} \operatorname{Arg}\left(i \frac{1-z}{1+z}\right) - \frac{40}{\pi} \operatorname{Arg}\left(i \frac{1-z}{1+z} - 1\right)$

6. $100 - \frac{200}{\pi}\text{ Arctan }\frac{x^2+y^2-1}{2y}$ **8.** $100 - \frac{150}{\alpha}\text{ Arg } z$

10. $\frac{100}{\pi}\text{ Re}[\text{Arcsin}(e^z)]$ **12.** $50 + \frac{200}{\pi}\text{ Re}[\text{Arcsin}(iz)]$

Section 9.6, Page 299

1. $100 + \frac{100}{\ln 2}\ln|z|$ **2.** $100 - \frac{100}{\pi}\text{ Arg}(z-1) - \frac{100}{\pi}\text{ Arg}(z+1)$

3. $150 - \frac{200x}{x^2+y^2}$ **4.** $\frac{50}{\pi}\text{ Arg}(\sin z - 1) + \frac{50}{\pi}\text{ Arg}(\sin z + 1)$

5. $50 + \frac{200}{\pi}\text{ Re}[\text{Arcsin } z]$ **6.** $\frac{200}{\pi}\text{ Arg}(\sin z)$

Section 9.7, Page 310

3. **(a)** Speed $= A|\bar{z}|$. The minimum speed is $A|1-i| = A\sqrt{2}$.
(b) The maximum pressure in the channel occurs at the point $1+i$.

5. $\psi(r,\theta) = Ar^{3/2}\sin(3\theta/2)$

Section 9.9, Page 326

1. $w = (z^2-1)^{1/2}$ **2.** $w = \frac{-2}{\pi}[z(1-z^2)^{1/2} + \text{Arcsin } z]$

3. $w = (z-1)^{\alpha}\left(1 + \frac{\alpha z}{1-\alpha}\right)^{1-\alpha}$ **4.** $w = \frac{-i}{2} z^{1/2}(z-3)$

5. $w = -1 + \int_{-1}^{z} \frac{(\xi-1)^{1/4}}{\xi^{1/4}}\,d\xi$

Section 9.10, Page 334

2. $F(z) = \log(z^4-1)$ **3.** $F(z) = \log(\sin z)$ **4.** $F(z) = \log(\sin z)$

5. $F(z) = \log(z^2-1)$ **7.** $w = 2(z+1)^{1/2} + \text{Log}\left(\frac{1-(z+1)^{1/2}}{1+(z+1)^{1/2}}\right) + i\pi$

8. $w = 4(z+1)^{1/4} + \text{Log}\left(\frac{(z+1)^{1/4}-1}{(z+1)^{1/4}+1}\right) - i\,\text{Log}\left(\frac{i+(z+1)^{1/4}}{i-(z+1)^{1/4}}\right)$

9. $w = \frac{1}{\pi}\text{ Arcsin } z + \frac{i}{\pi}\text{ Arcsin }\frac{1}{z} + \frac{1+i}{2}$

10. $w = 2(z+1)^{1/2} - \text{Log}\left(\frac{1-(z+1)^{1/2}}{1+(z+1)^{1/2}}\right)$

Bibliography of Articles

Brandt, Siegmund, and Schneider, Hermann. "Computer Simulation of Experiments in University Physics Teaching." *Computers in Education* (1975): 443–48.

Brandt, Siegmund, and Schneider, Hermann. "Computer-Drawn Field Lines and Potential Surfaces for a Wide Range of Field Configurations." *American Journal of Physics* 44 (Dec, 1976): 1160–71.

Bruch, John C. "The Use of Interactive Computer Graphics in the Conformal Mapping Area." *Computers and Graphics* 1 (1975): 361–374.

Bruch, John C., and Wood, Roger C. "Teaching Complex Variables with an Interactive Computer System." *IEEE Transaction on Education* E-15, No. 1 (Feb, 1972): 73–80.

Bryngdahl, Olof. "Geometrical Transformations in Optics." *Journal of the Optical Society of America* 64, No. 8 (Aug, 1974): 1092–99.

Celik, I.; Cekirge, H. M.; and Esen, I. I. "Pollution Control in Bays with Currents." *Letters in Applied Engineering Science* 5, No. 3 (1977): 241–252.

Choudary, A. R.; Kellett, B. H., and Verzegnassi, C. "A Conformal-Mapping Evaluation of the Pion Form Factor in the Spacelike Region." *Lettere al Nuovo Cimento* Ser. 2, Vol. 15, No. 16 (Apr, 1976): 579–582.

Foster, K., and Anderson, R. "Transmission-Line Properties by Conformal Mapping." *Proceedings IEE* 121, No. 5 (May, 1974): 337–339.

Goldstein, Marvin, and Siegel, Robert. "Conformal Mapping for Heat Conduction in a Region with an Unknown Boundary." *International Journal of Heat Mass Transfer* 13 (Apr, 1970): 1632–36.

Ives, David C., and Liutermoza, John F. "Analysis of Transonic Cascade Flow Using Conformal Mapping and Relaxation Techniques." AIAA 9th Fluid and Plasma Dynamics Conference, San Diego (July, 1976): 1–10.

Laura, Patricio A.; Shahady, Paul A.; and Passarelli, Ralph. "Application of Complex-Variable Theory to the Determination of the Fundamental Frequency of Vibrating Plates." *Journal of the Acoustical Society of America* 42, No. 4 (Oct, 1967): 806–9.

Miyamoto, S.; Warrick, A. W.; and Bohn, H. L. "Land Disposal of Waste Gases: I. Flow Analysis of Gas Injection Systems." *Journal of Environmental Quality* 3, No. 1 (1974): 49–60.

Nilson, R. H., and Tsuei, Y. G. "Inverted Cauchy Problem for the Laplace Equation in Engineering Design." *Journal of Engineering Mathematics* 8, No. 4 (Oct, 1974): 329–37.

Piele, Donald T.; Firebaugh, Morris W.; and Manulik, Robert. "Applications of Conformal Mapping to Potential Theory Through Computer Graphics." *American Mathematical Monthly* 84, No. 9 (Nov, 1977): 677–92.

Potter, Robert I.; Schmulian, Robert J.; and Hartman, Keith. "Fringe Field and Readback Voltage Computations for Finite Pole-Tip Length Recording Heads." *IEEE Transactions on Magnets* Mag-7, No. 3 (Sept, 1971): 689–95.

Van Der Veer, P. "Exact Solutions for Two-Dimensional Groundwater Flow Problems Involving a Semi-Pervious Boundary." *Journal of Hydrology* 37 (1978): 159–68.

Bibliography of Books

Ash, Robert B. *Complex Variables*. New York: Academic Press, 1971.

Callander, R. A., and Raudkivi, A. J. *Analysis of Groundwater Flow*. New York: Halsted Press, 1976.

Carrier, George F.; Krook, Max; and Pearson, Carl E. *Functions of a Complex Variable*. New York: McGraw-Hill Book Company, Inc., 1966.

Churchill, Ruel V.; Brown, James W.; and Verhey, Roger F. *Complex Variables and Applications*. 3rd ed. New York: McGraw-Hill Book Company, Inc., 1974.

Colwell, Peter, and Mathews, Jerold C. *Introduction to Complex Variables*. Columbus: Charles E. Merrill Publishing, 1973.

Derrick, William R. *Introductory Complex Analysis and Applications*. New York: Academic Press, 1972.

Dettman, John W. *Applied Complex Variables*. New York: The Macmillan Company, 1966.

Flanigan, Francis J. *Complex Variables: Harmonic and Analytic Functions*. Boston: Allyn and Bacon, Inc., 1972.

Fuchs, B. A., and Shabat, B. V. *Functions of a Complex Variable*. Vol. 1. New York: Pergamon Press, Inc., 1964.

Greenleaf, Frederick P. *Introduction to Complex Variables*. Philadelphia: W. B. Saunders Company, 1972.

Grove, E. A., and Ladas, G. *Introduction to Complex Variables*. Boston: Houghton Mifflin Company, 1974.

Hecht, Eugene, and Zajac, Alfred. *Optics*. Reading, Mass: Addison-Wesley Publishing Company, 1974.

Irving, J., and Mullineux, N. *Mathematics in Physics and Engineering*. New York: Academic Press, 1959.

Kip, Arthur F. *Fundamentals of Electricity*. New York: McGraw-Hill Book Company, Inc., 1969.

Kyrala, A. *Applied Functions of a Complex Variable*. New York: John Wiley & Sons, Inc., 1972.

Levinson, Norman, and Redheffer, Raymond M. *Complex Variables*. San Francisco: Holden-Day, Inc., 1970.

Mackey, George W. *Lectures on the Theory of Functions of a Complex Variable*. New York: Van Nostrand Reinhold Company, 1967.

McCullough, Thomas, and Phillips, Keith. *Foundations of Analysis in the Complex Plane*. New York: Holt, Rinehart and Winston, Inc., 1973.

Marsden, Jerrold E. *Basic Complex Analysis*. San Francisco: W. H. Freeman and Company, 1973.

Milne-Thompson, L. M. *Theoretical Hydrodynamics*. 5th ed. New York: The Macmillan Company, 1968.

Motteler, Zane C. *Functions of a Complex Variable*. New York: Intext Educational Publishers, 1975.

Nehari, Zeev. *Introduction to Complex Analysis*. Revised ed. Boston: Allyn and Bacon, Inc., 1968.

Paliouras, John D. *Complex Variables for Scientists and Engineers*. New York: The Macmillan Publishing Company, Inc., 1975.

Pennisi, Louis L. with the collaboration of Louis I. Gordon and Sim Lasher. *Elements of Complex Variables*. 2nd ed. New York: Holt, Rinehart and Winston, 1976.

Phillips, E. G. *Some Topics in Complex Analysis*. New York: Pergamon Press, Inc., 1966.

Priestley, H. A. *Introduction to Complex Analysis*. Oxford Clarendon Press, 1985.

Raven, Francis H. *Mathematics of Engineering Systems*. New York: McGraw-Hill Book Company, Inc., 1966.

Saff, E. B., and Snider, A. D. *Fundamentals of Complex Analysis for Mathematics, Science and Engineering*. Englewood Cliffs, N.J.: Prentice-Hall, Inc., 1976.

Silverman, Richard A. *Complex Analysis with Applications*. Englewood Cliffs, N.J.: Prentice-Hall, Inc., 1974.

Spiegel, M. R. *Problems in Complex Variables*. New York: McGraw-Hill Book Company, Inc., 1964.

Wayland, Harold. *Complex Variables Applied in Science and Engineering*. New York: Van Nostrand Reinhold Company, 1970.

Index

Absolute convergence, 166
Absolute value, 6, 10
d'Alembert's ratio test, 171, 189
Analytic function, 67, 68, 153
  derivative formula, 61, 64
  identity theorem for, 195, 205
  maximum principle for, 147, 149
  singularities of, 67, 199, 208
  zero of, 200, 205
Angle of rotation, 244, 315
Antiderivative, 135, 137
Arc
  simple, 21, 111
  smooth, 112
Arc length, 114
arcsin z, 105, 263, 288
arctan(y/x), 10, 75, 101, 270
arctan z, 105
Argument
  of a complex number, 10
  of conjugate, 11
  of quotient, 14
  principle value of, 11
Argument principle, 236
Arg z, 11, 96, 270

Bernoulli's equation, 306
Bessel function, 185, 194
Bilinear transformation, 249, 273
  conformal, 255
  fixed point of, 225
  implicit formula for, 251, 254
  inverse of, 249
Binomial series, 179
Boundary
  insulated, 287, 291
  point, 23
  value problem, 154, 270, 272
Bounded function, 150
Bounded set, 25
Branch
  of function, 77, 96
  integral around, 232
  of logarithm, 96, 98
  of square root, 78
Branch cut, 78
Branch point, 79

Cartesian plane, 5
Cauchy-Goursat Theorem, 127, 131, 158
Cauchy-Hadamard formula, 189
Cauchy principal value, 220, 224, 227, 232
Cauchy product of series, 196
Cauchy-Riemann equations, 61, 64
  in polar form, 64
Cauchy's inequalities, 149
Cauchy's integral formulas, 140, 143, 174
Cauchy's Residue Theorem, 209
Cauchy's root test, 189
Chain rule, 58, 119, 243
Charged line, 29, 295, 328
Circle, 22, 53, 117, 130, 250
  of convergence, 167, 173, 189
  equation of, 22, 117, 130
Closed contour, 21, 111
Closed curve, 21, 111
Closed region, 25
Closed set, 24
Comparison test for series, 170
Complex conjugate, 7
Complex derivative, 56
Complex differential, 114
Complex exponents, 101
Complex function, 27
Complex integral, 108, 115
Complex number, 1
  absolute value of, 6
  argument of, 10
  conjugate of, 7
  imaginary part of, 1, 6
  modulus of, 6
  polar form of, 10
  powers of, 15, 101
  real part of, 1, 6
  roots of, 18
Complex plane, 5
Complex potential, 281, 304, 328
Complex variable, 27
Conformal mapping, 243, 272, 287, 305
  angle of rotation, 244, 315
  applications of, 282, 283, 295, 302
  bilinear, 249, 251, 254
  inverse, 247, 249
  properties of, 244
  scale factor of, 247
  Schwarz-Christoffel formula, 314
Conjugate
  of complex number, 7
  of harmonic function, 69, 72, 153, 282
Connected set, 24, 123
Continuity, equation of, 304
Continuous function, 45, 46, 58
Contour, 111, 113
  closed, 21, 111
  deformation of, 128

Contour (*Cont.*)
indented, 227
integral along, 115
length of, 114
simple closed, 21, 111
Convergence
absolute, 164
circle of, 167, 173, 189
pointwise, 166
of power series, 191, 192
radius of, 167, 173, 189
of sequences, 162
of series, 164, 173, 180, 188, 192
Cosine function, 88, 178
cos z, 88, 89, 178, 216
Critical point, 245
Curve, 21, 111, 114, 119
closed, 21, 111
equipotential, 73, 295, 305
exterior of, 123
interior of, 123
length of, 114
level, 76, 281
opposite, 113, 124
orthogonal families of, 73, 281
simple, 21, 111
smooth, 112

d'Alembert's ratio test, 171, 189
Definite integral, 137
Deformation of contour, 128
DeMoivre's Formula, 15
Derivative
definition of, 56
of exponential function, 83
of hyperbolic function, 92
implicit, 79
of inverse hyperbolic functions, 106
of inverse trigonometric functions, 105, 288
of logarithm function, 98
normal, 288
rules for, 58, 176, 192
of series, 176, 192
of square root function, 79
of trigonometric functions, 89
Differential, 114
Differentiation, 56, 57, 61, 176
definition of, 56
implicit, 79, 83, 99
rules for, 57, 176, 192
of series, 176, 192
Diffusion, 282
Dipole, 76, 336
Dirichlet problem, 154, 268
N-value, 270, 272
for the unit disk $|z| < 1$, 154
for the upper half-plane, 270
Distance between points, 7
Divergence
of sequences, 162
of series, 164
Domain, 24
of definition of function, 27
multiply connected, 124
simply connected, 124

$e^{i\theta}$, 13
Elasticity, 282
Electrostatic potential, 282, 295, 328
Entire function, 67, 150
Equation
Bernoulli's, 306
Cauchy-Riemann, 61, 64
of circle, 22
of continuity, 304
of curve, 21
Laplace's, 68, 268, 289
of line, 21, 116
parametric, 21, 111, 114, 119, 324
Equipotential curves, 73, 295, 305
Essential singularity, 199
Euler's formula, 13
Exponential function, 83, 175, 255
derivative of, 83
fundamental period strip, 86, 256
inverse of, 86, 96
mapping by, 85, 255
period of, 85
exp z, 85, 175, 255
Extended complex plane, 50
Exterior, of curve, 123
Exterior point, 23

Fixed point of bilinear transformation, 255
Fluid flow, 72, 282, 302, 305, 324
in a channel, 337
complex potential, 304, 328
about a cylinder, 309
about a plate, 310
through a slit, 313
over a step, 325
at a wall, 336
Fluid, velocity of, 282
Fourier integral, 224
Fourier series, 156, 186
Fresnel integral, 180, 235
Function
analytic, 67, 68, 153
Bessel, 185, 194
bilinear, 249, 251, 254

Function (*Cont.*)
  bounded, 150
  branch of, 77, 96
  complex, 27
  continuous, 45, 46, 58
  cosine, 88, 89
  differentiable, 56, 61, 64
  domain of definition, 27
  entire, 67, 150
  exponential, 83
  harmonic, 68, 72, 145, 154, 266
  hyperbolic, 92
  image of, 27, 77
  implicit, 78, 251
  integral of rational, 220
  inverse of, 33, 104, 247
  limit of, 42, 43, 47
  linear, 33, 249
  logarithmic, 96
  meromorphic, 236
  multivalued, 77
  one-to-one, 31, 249
  period of, 85, 90
  principal value of, 37, 40, 78, 96
  range of, 27, 77
  rational, 220, 236
  sine, 88, 90
  stream, 305
  tangent, 89
  trigonometric, 88
  zeros of, 151, 200
Fundamental period strip, 86, 256
Fundamental theorem
  of algebra, 151
  of integration, 135, 137

Gauss's Mean Value Theorem, 147
Geometric series, 167
Goursat
  Cauchy's Theorem, 127, 131, 158
  proof by, 158
Green's Theorem, 125

Hadamard, Cauchy, formula, 189
Harmonic conjugate, 69, 72, 153, 282
Harmonic function, 68, 145, 154, 266
  applications of, 72, 153, 266
  conjugate of, 69, 153, 282
  maximum principle for, 152
Heat flow, 282, 284
L'Hôpital's Rule, 59, 179, 212, 225
Hyperbolic functions, 92
  derivatives of, 92
  identities for, 92, 93
  inverses of, 106

Ideal fluid, 72, 282, 302, 324
Identity theorem
  for analytic function, 195, 205
  for series, 195, 198
Image
  of flow, 305, 324
  of function, 27, 77
  of source, 331
Imaginary axis, 5
Imaginary part of complex number, 1
Imaginary unit, 1
Implicit differentiation, 79, 83, 99
Implicit form of bilinear transformation, 251, 254
Implicit function, 78, 251
Im z, 1, 6
Indefinite integrals
  table of, 317
  theorem of, 135
Indented contour integral, 227
Inequality
  Cauchy's, 149
  ML, 120
  triangle, 7, 9, 119
Infinity, 50, 206, 249, 254
Initial point, 111
Insulated boundary, 287, 291
Integral
  around branch points, 232
  Cauchy principal value of, 220, 224, 227, 232
  Cauchy's, Formula, 140, 143, 174
  complex, 108
  contour, 115
  definite, 137
  Fourier, 224
  Fresnel, 180, 235
  improper, 219, 220, 224
  indefinite, 135
  Leibniz's rule for, 142
  line, 115
  Poisson, 154, 277, 279
  of rational function, 220
  representation for $f(z)$, 140
  table of indefinite, 317
  theorem of indefinite, 135
  trigonometric, 216, 224
Interior of curve, 123
Interior point, 23
Invariance
  of flow, 305, 324
  of Laplace's equation, 268
Inverse of
  bilinear transformation, 249, 251, 254
  function, 31, 104, 247

Inverse of (*Cont.*)
  hyperbolic function, 106
  sine function, 105, 288
  trigonometric functions, 105, 288
Inversion mapping, 49
Irrotational vector field, 67, 72, 304
Isolated point, 205
Isolated singularity, 199
Isolated zeros, 205
Isothermals, 282, 284

Jacobian determinant, 247
Jordan's Lemma, 225

Lagrange's identity, 20
Laplace's equation, 68, 268, 289
  invariance of, 268
  in polar form, 75
Laurent series, 180, 208
Legendre polynomial, 146
Leibniz's Rule, 122, 196
  for integrals, 142
Length of contour, 114
Level curves, 73, 281, 282, 295, 305
L'Hôpital's Rule, 59, 179, 212, 225
Limit
  of complex function, 43, 44, 47
  at infinity, 50, 55
  of sequence, 162
  superior, 188
Line
  of charge, 29, 295, 328
  equation of, 21, 116
  of flux, 282, 295
  heat flow, 282, 284
  integral of, 115
Linear approximation, 243
Linear fractional transformation, 249, 251, 254
Linear function, 33, 249
Linear transformation, 33, 249
Liouville's Theorem, 150
Logarithmic function
  branch of, 96, 98
  derivative of, 98
  mapping by, 99, 257
  principal branch of, 96
  Riemann surface, 83
Log z, 96, 177, 185, 257
LRC circuit, 93

Maclaurin series, 173, 176
Magnetism, 282
Magnification, 33, 247
Mapping
  bilinear, 249, 251, 254
  conformal, 243, 272, 287, 305
  by exp z, 85, 255
  linear, 33, 249
  by Log z, 99, 257
  Möbius, 249, 251, 254
  one-to-one, 31, 249
  by $1/z$, 49
  by sin z, 261
  by trigonometric functions, 260
  by z^n, 36, 40, 246, 258
  by $z^{1/n}$, 36, 39, 77, 83, 259
Mathematical models, 280, 295, 302, 328
  electrostatics, 295, 328
  ideal fluid flow, 172, 302, 324, 328
  steady state temperatures, 282
Maximum Principle
  for analytic functions, 147, 149
  for harmonic functions, 152
ML inequality, 120
Möbius transformation, 249, 251, 254
Modulus, 6, 10, 71, 147, 152
Morera's Theorem, 146
Multiply connected domain, 124
Multivalued function, 77

Negative orientation, 111, 124
Neighborhood, 22, 43, 67
Normal derivative, 288
*n*th root, 18
  principal value of, 40, 83
N-value Dirichlet problem, 270, 272
Nyquist stability criterion, 239

One-to-one function, 31, 249
$1/z$, 3, 49
Open neighborhood, 22, 43, 67
Open set, 24
Opposite curve, 113, 124
Order of pole, 199
Order of zero, 200
Orientation, 111, 124
Orthogonal families of curves, 73, 281, 282, 295, 305

Parameterization of curve, 21, 111, 114, 119
Parametric equations, 21, 111, 114, 324
Partial fractions, 132, 214
Partial sums, 164
Path, 113, 114
Path of integration, 114
Period of function, 85, 90
Period strip, 86, 256
Point at infinity, 50

Poisson kernal, 156
Poisson's integral formula, 154, 277, 279
Polar coordinates, 10
Polar form of
  Cauchy-Riemann equations, 64
  complex number, 10
  Laplace's equation, 75
Pole, 199, 205, 236
  of order k, 199, 202, 203, 206, 211, 236
  residue at, 211
  simple, 200, 211
  at singular point, 199
Polynomial
  coefficients of, 20
  factorization of, 19, 20, 151
  Legendre, 146
  quadratic formula of, 18
  roots of, 19, 151
  zeros of, 19, 20, 151
Positive orientation, 111, 124
Potential, 73, 281
  complex, 281, 304, 328
  electrostatic, 282, 295, 328
  velocity, 282, 302, 305, 306
Powers, 15, 101
  complex, 101
DeMoivre's Formula for, 15
Power series, 186
  Cauchy product of, 196
  convergence of, 187, 191, 192
  differentiation of, 176, 192
  division of, 198
  integration of, 176
  multiplication of, 196
  radius of convergence of, 188, 191, 192
  uniqueness of, 195, 198
Principal branch of
  log z, 96
  square root, 78
  z^{α}, 102
  $z^{1/n}$, 40, 83
Principal nth root, 83
Principal square root, 39, 78
Principal value of
  arg z, 11
  definite integrals, 220, 224, 227, 232
  log z, 96
Principle
  maximum, for analytic functions, 147, 149
  maximum, for harmonic functions, 147, 149
Product of series, 196
Pure imaginary number, 2

Quadratic formula, 18
Quotient of series, 198

Radius of convergence of power series, 188, 191, 192
Range of function, 27, 77
Rational function, 220, 236
Ratio test, 171, 189
Real axis, 5
Reciprocal transformation, 49
Region, 25
Removable singularity, 199
Reparameterization, 119
Residue, 208, 211
  applications of, 216, 224, 227, 232
  calculation of, 211
  at poles, 211
  at singular points, 208
Residue theorem, 209
Re z, 1, 6
Riemann
  mapping theorem, 264
  sphere, 50
  surface for log z, 83
  surface for $z^{1/2}$, 80
  theorem of, 205
Riemann surface, 80, 83
RLC circuit, 93
Root
  of numbers, 18
  test for series, 189
  of unity, 16
Rotation transformation, 32
Rouché's Theorem, 237

Scale factor, 247
Schwarz-Christoffel
  formula, 314
  theorem, 314
  transformation, 314
Sequence, 164
Series
  binomial, 179
  comparison test, 170
  convergence of, 164, 173, 180, 188, 192
  differentiation of, 176, 192
  divergence of, 164
  Fourier, 156, 186
  geometric, 167
  identity theorem for, 195, 198
  Laurent, 180, 208
  Maclaurin, 173, 176
  power, 186
  product of series, 196
  quotient of series, 198
  ratio test for, 171, 189
  representation of $f(z)$, 173, 186
  Taylor, 173, 176

Series (*Cont.*)
  uniqueness of, 195, 198
Simple closed curve, 21, 111
Simple pole, 200, 211
Simple zero, 200
Simply connected domain, 124
Sine function, 88, 90, 178, 216, 261
  inverse of, 105
Singular point, 67, 199
  essential, 199
  isolated, 199, 208
  pole at, 199
  removable, 199
  residue at, 211
Sink, 328
sin z, 88, 90, 178, 216, 261
Smooth curve, 112
Solenoidal vector field, 67
Source, 328
  image of, 331
Square root function
  branch of, 39, 77, 259
  derivative of, 79
  Riemann surface, 80
Square roots, 17, 37, 39, 78, 259
  branch of, 78
  principal, 39
Steady state temperatures, 282
Stereographic projection, 50
Stream function, 305
Streamlines, 73, 282, 305
Strip, period, 86, 256
Sum, partial, 164

Table of integrals, 317
Tangent function, 89, 260
Tangent vector, 112, 243, 314
tan z, 89, 260
Taylor series, 173, 176
Temperature, steady state, 282
Terminal point, 11
Transformations
  bilinear, 249, 251, 254
  composition, 255, 260
  conformal, 243, 272, 287, 305
  by exp z, 255
  inversion, 49
  linear, 33, 249
  by Log z, 97, 257
  Möbius, 249, 251, 254
  by $1/z$, 49
  reciprocal, 49
  rotation, 32
  Schwarz-Christoffel, 314
  by sin z, 261
  by trigonometric functions 260
  by z^n, 36, 40, 246, 258
  by $z^{1/n}$, 36, 39, 77, 83, 259
Translation, 31
Triangle inequality, 7, 9
  for integrals, 119
Trigonometric functions, 88
  derivatives of, 89
  identities for, 90, 91
  integrals of, 216, 224
  inverse of, 105, 288
  mapping by, 260
  zeros of, 90
Two-dimensional
  electrostatics, 295, 328
  fluid flow, 72, 302, 324, 328
  mathematical models, 280, 295, 302, 328

Unbounded set, 25
Uniform convergence, 166
Uniqueness of power series, 195, 198
$u(x, y)$, 27, 61, 108, 282

Vector field
  irrotational, 67, 304
  solenoidal, 67
Velocity
  of fluid, 72, 302, 305, 306
  potential, 305
$v(x, y)$, 27, 61, 108, 282

Winding number, 239

z^α, 102
z^n, 15, 36
$z^{1/n}$, 18, 40, 83
$1/z$, 3, 49
Zero
  of a function, 151, 199, 200, 236
  isolated, 205, 236
  of order k, 200, 202, 236
  of polynomial, 19, 20, 151
  simple, 200
  of trigonometric function, 90
Z-transform, 186